In the Wilds
of Climate Law

Edited by Rosemary Lyster

www.
AUSTRALIANACADEMIC**PRESS**
.com.au

First published in 2010 from a completed manuscript presented to
Australian Academic Press
32 Jeays Street
Bowen Hills Qld 4006
Australia
www.australianacademicpress.com.au

National Library of Australia cataloguing-in-publication entry:

Title:	In the wilds of climate law / edited by Rosemary Lyster.
Edition:	1st ed.
ISBN:	9781921513640 (pbk.)
Subjects:	Climatic changes--Law and legislation.
	Environmental law, Inter national.
Other Authors/ Contributors:	Lyster, Rosemary.
Dewey Number:	341.762

CONTENTS

(continued over)

CONTENTS
(cont.)

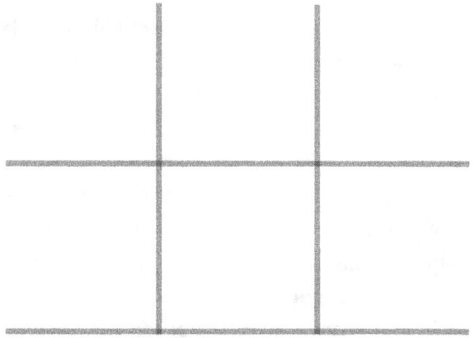

Preface

Rosemary Lyster

Why the title *In the Wilds of Climate Law*? What can one encounter 'in the wilds'? Unchartered territory, new frontiers and experiences, extreme weather with sudden changes, surprises, the need to solve problems quickly as they arise, and perhaps a touch of anxiety about the unknown. So too when entering the relatively new territory of Climate Law. Many of the authors of chapters in this book have not published in the area of Climate Law until now. They have done so graciously at my invitation. In 2008, I invited colleagues at Sydney Law School to collaborate on the establishment of a Climate Law and Policy Group. The intention behind the establishment of this group was deliberate — to bring together a group of legal academics who would investigate the intersections between their legal disciplines and global climate change. The group comprises academics with expertise in the following areas: Environmental Law, International Law, International Human Rights Law, WTO Law, Administrative Law, Corporations Law, Taxation Law, Contracts Law and Torts. The list is not exhaustive.

The Climate Law and Policy Group was launched by John Connor, CEO of the Climate Institute in March 2008. Then in August that year, the Australian Centre for Climate and Environmental Law (ACCEL) at Sydney Law School hosted a conference entitled 'Intersections between global climate change, law and policy'. Members of the group, as well as other invited speakers, presented the papers that now comprise the chapters of this edited collection. As I have already said, for some this was their first ever foray into the relationship between an area of law, in which they had been immersed for years, and global climate change. For others, the conference provided them with an opportunity to expand upon previous publications and ideas to make a truly original contribution to the area of Climate Law.

Our keynote speaker at the conference was Professor Armin Rosencranz from Stanford University. In his address, reproduced and updated here, Rosencranz provided an insight into key climate change issues that need to be resolved in the United States and the ways in which this might happen. He has compared the

Presidential leaderships of Bill Clinton, George Bush and Barack Obama on climate change and has drawn some parallels between the United States and Australia.

In Chapter 2, Paul Babie from the University of Adelaide, provides a convincing argument for the adoption of a new property rights theory in 'Climate Change and the Concept of Private Property'. He argues that climate change is a private property problem in the sense that it permits the self-interested and unequal exercise of power over the use and control of goods and resources. Babie's contribution provides the insight that, although a liberal understanding of private property is dominant, decisions about goods and resources occur in a web of social relationships. A 'relationship' theory of property is proposed to reflect on how the concept of private property may need to adapt in the face of global climate change.

The book then moves to investigate various international legal perspectives on climate change, including WTO Law, the role of international courts in resolving climate change-related disputes, and an international law of distributive justice in the face of climate change and resource scarcity. Professor Gillian Triggs, Dean of Sydney Law School, discusses 'World Trade Organization and Climate Change: A Clash of Civilisations?' in Chapter 3. Professor Triggs argues that although the legal cultures of WTO Law and Environmental Law, such as developed countries' obligations under the *Kyoto Protocol*, might appear to clash there is no necessary contradiction between them. Given this, the challenge is to guard against the rise of 'green protectionism' where states resort to trade measures against imports so as to offset the costs of domestic climate change measures, and to ensure that international trade occurs on equal terms. Continuing the international theme, Tim Stephens writes in Chapter 4 about 'International Courts and Climate Change: "Progression", "Regression" and "Administration"'. The chapter reflects the growing public interest in bringing climate change litigation before international courts and tribunals, given the slow pace of international negotiations to deliver meaningful action on climate change. With 'progressive' international proceedings, litigants seek to highlight the need for a more robust policy response to climate change, while 'regressive' proceedings may be invoked to prevent states from adopting climate change policies that interfere with trade liberalisation. The links here to Chapter 3 are obvious. Finally, 'administrative' litigation might be resorted to under the compliance procedure established under the *Kyoto Protocol*. Stephens predicts that this type of litigation is likely to be the most fruitful arena for climate change litigation in international courts. Finally, in Chapter 5 Associate Professor Ben Saul reflects on how international law remains substantially ineffective at pursuing a program of redistributive justice in the face of increasing climate change-induced resource scarcity. Nevertheless, he looks to specialised areas of international law, such as environmental law, the law of the sea, and elements of international economic law, as starting points for reorientating international law towards a (re-)distributive system capable of addressing what he describes as 'absolute' rather than 'relative' scarcity of key resources.

The next three chapters of the book cover topics which relate to emissions trading schemes. My Chapter 6, entitled 'Reducing Emissions from Deforestation and Degradation plus: Key Legal Issues', provides an overview of the proposals for REDD+

going into Copenhagen as well as in the *American Clean Energy and Security Act 2009* (the Waxman-Markey Bill). It then provides an in-depth analysis of the key legal questions that need to be resolved for the equitable and effective implementation of REDD+. Chapter 7, entitled 'Contractual Perspective of Climate Change Issues' contains an assessment by Professor Elisabeth Peden of the interface between Contract Law and carbon trading, which in the main occurs Over the Counter. The consequences of transfer failure and liquidated damages clauses are highlighted. The limitations of Contract Law to protect local communities, in the context of REDD+, for example, are also exposed. In Chapter 8, Celeste Black considers 'Climate Change and Tax Law: Tax Policy and Emissions Trading'. As Black states, the taxation system 'has the potential to either distort or support the scheme'. After describing the tax regime that is proposed to apply to carbon pollution permits, an evaluation of the proposed Carbon Pollution Reduction Scheme against stated tax policy objectives and the compliance timeline is offered. The analysis extends to: assistance to emissions-intensive trade-exposed industries and strongly affected industries; price caps; foreign ownership of permits; and the development of forward markets.

Chapters 9 and 10 analyse the interface between climate change, Corporations Law and Employment Law. In Chapter 9, Susan Shearing addresses the topic 'Climate Governance and Corporations: Changing the way "Business does Business"'. The obligations of corporate managers in Australia to consider the business risks of climate change are assessed, including with regard to physical, regulatory, reputational and litigation risks. Shearing argues that in some cases, if directors of companies fail to consider these risks they might breach their obligations under Corporations Law. In other situations, the risks of climate change might require a proactive approach to management on the part of directors. In Chapter 10, Victoria Lambropoulos discusses the implications of climate change for the workplace, particularly in the context of Occupational Health and Safety Regulation. As she points out, we are likely to witness a greater coalescing of these issues in the future, given that many workplace arrangements were resolved prior to more recent concerns about climate change.

Finally, in Chapters 11 and 12 the book moves to an analysis of climate change litigation. In Chapter 11, Justice Brian Preston provides an overview of the types of challenges which litigants have brought to the courts including: at the national level, tort law, administrative law or constitutional law; and at the international level applications to the International Court of Justice, the International Tribunal for the Law of the Sea or regional human rights courts. The links between this chapter and Chapters 4 and 5 are clear. Professor Peter Cashman and Ross Abbs in Chapter 12 discuss 'Liability in Tort for Damage Arising From Human-Induced Climate Change'. They question whether a new generation of tort cases will arise out of personal injury, property loss and environmental damage caused by human-induced climate change, while exploring some of the theoretical and practical problems litigants might be required to confront. The conclusion reached is that tort law is ill-adapted to dealing with problems of such magnitude as global climate change and that regulation by international and national governments provides a more appropriate response than piecemeal and ad hoc tort litigation.

Given the lapse of time between the conference in late 2008 and the publication of this edited collection in early 2010, it must be acknowledged that there have been some developments which are not referred to in authors' chapters. In December 2009, world leaders at the Fifteenth Conference of the Parties (COP 15) at Copenhagen failed to reach a legally binding agreement on a post-Kyoto framework. Developed and developing countries have subsequently made voluntary commitments under the accord which represent approximately 78% of global greenhouse gas emissions. The Accord endorses use of market mechanisms to meet commitments. As well, $US30 billion during 2010–2012 and US$100 million a year by 2020 will assist developing countries with mitigation and adaptation. The funds will come from public and private sources and will flow through the Copenhagen Green Climate Fund. Significantly, the Accord recognises developing countries' efforts towards Reducing Emissions from Deforestation and Degradation and requires funding for this to be mobilised immediately.

Meanwhile at the domestic level, both the United States and Australia have seen legislation on emissions trading pass through their respective House of Representatives, only to stall in the Senate. The fate of the Carbon Pollution Reduction Scheme Bill 2010 (Cth) will no doubt depend on the outcome of the Federal elections in late 2010. In the United States, meanwhile, the *American Clean Energy and Security Act 2009* (the Waxman-Markey Bill) which was passed by the House of Representatives has been superseded by a bill proposed by Senators Kerry, Graham and Lieberman (the Kerry-Graham-Lieberman Bill). It is not clear at the time of writing when the Bill will go before the US Senate.

On behalf of all of the authors, I hope that you will immerse yourselves in the chapters of this collection and appreciate the insights which we bring to this rapidly evolving area of the law. The chapters in no way represent the authors' final or definitive contributions to the field. Such is the nature of conference proceedings. I would like to thank Caroline Falshaw from the Sydney Law School's Publishing Unit for her outstanding contribution to the editing and finalising of the collection. Without her keen eye for detail, and her considerable experience in the publishing industry, getting this collection to press would have been far more arduous. Thanks also to Barry Passaris for his earlier editing work. Finally, the authors would like to thank Australian Academic Press and their editor Roberta Blake, for their support and belief in the collection.

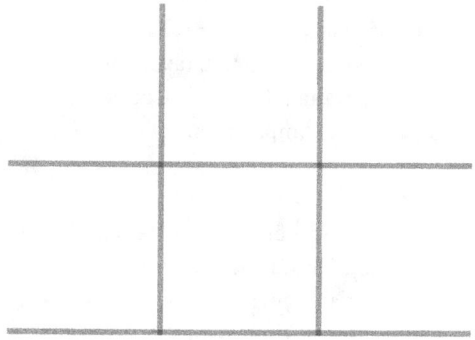

About the Authors

Rosemary Lyster

Rosemary Lyster is a Professor in the Faculty of Law, University of Sydney and Director of the Australian Centre for Climate and Environmental Law (ACCEL). In the area of environmental law, Rosemary specialises in energy and climate law, water law, and genetically modified organisms and environmental law. She has published widely in the areas of energy and climate law. Rosemary is the Energy and Water Special Editor of the *Environmental Planning and Law Journal,* which is the leading environmental law journal in Australia. She is also a member of the IUCN Commission on Environmental Law, and the Commission's Special Working Groups on Water and Wetlands, Energy and Climate Change, and Forests.

Ross Abbs

Ross Abbs is presently reading for the degree of Bachelor of Civil Law at the University of Oxford (Merton College). He was previously an Associate to the Hon Michael Kirby AC CMG, Justice of the High Court of Australia (2007–2008).

Paul Babie

Paul Babie is Associate Dean of Law (Research) and Senior Lecturer at the University of Adelaide Law School. He is also a Founder and Director of the University of Adelaide Research Unit for the Study of Society, Law and Religion. His research focuses on legal theory, especially the nature and concept of property as it relates to globalisation, climate change and community, and he is widely published in this area. Paul is currently writing a book for UBC Press in Canada titled *Private Property, Climate Change and the Children of Abraham.*

Celeste Black

Celeste M. Black is a Senior Lecturer at the Faculty of Law, University of Sydney, where she teaches taxation law. Her current research interests include the taxation implications of market-based mechanisms used to address climate change, and the use and design of environmental taxes, including carbon taxes. She has recently commenced a large Australian Research Council funded project analysing the income and consumption tax implications of emissions trading schemes.

Peter Cashman

Dr Peter Cashman is a barrister and Professor of Law (Social Justice) at the University of Sydney. He was formerly: Commissioner in charge of the civil justice review with the Victorian Law Reform Commission; Commissioner jointly in charge of the reference on class actions with the Australian Law Reform Commission; founding Director of the Public Interest Advocacy Centre; founder and senior partner of the firm Cashman & Partners (now Maurice Blackburn Pty Ltd); Governor of the American Trial Lawyers' Association (now the American Association for Justice) and National President of the Australian Plaintiff Lawyers' Association (now the Australian Lawyers Alliance). He holds a degree in law and a diploma in criminology from the University of Melbourne, and a Master of Laws degree and a PhD from the University of London. He has practised law in the United Kingdom, the United States and Australia and is the author of numerous publications, including Class Action Law and Practice (Federation Press: 2007).

Victoria Lambropoulos

Victoria is a lecturer at the school of law at Deakin University in Melbourne where she teaches workplace and company law. She holds the unit chair of workplace law at the university. Victoria is investigating whether action on climate change can be taken through our industrial and/or employee arrangements. Victoria's most recent publication in this area is 'Greening Australian Workplaces: Workers and the Environment' (2009) Alternative Law Journal 189. Victoria is a regular presenter at academic and practitioner conferences both in Australia and internationally.

Elisabeth Peden

Professor Elisabeth Peden is an international expert on the law of contracts. She has published widely in the area and has written books, contributed to international journals and given many conferences papers. She is interested in the overlap between contract law and climate change, particularly the use of agreements in minimising impact on the world's climate. She has given conference papers on related topics, and has received government funding as a joint researcher looking at deforestation in Indonesia.

Brian Preston

Justice Preston is the Chief Judge of the Land and Environment Court of New South Wales. Prior to his appointment in 2005, he practised as a solicitor and a barrister in environmental, planning, administrative and property law, being appointed senior counsel in 1999. Justice Preston has lectured in postgraduate environmental law for over 17 years, is currently an Adjunct Professor at Sydney Law School, and a member of the Australian Centre for Climate and Environmental Law. Justice Preston has published widely in the areas of environmental law and climate change litigation.

Armin Rosencrantz

Armin Rosencranz is a lawyer and political scientist, and has taught a variety of environmental policy courses at Stanford since 1995. He taught one of India's first courses on environmental law, and his book *Environmental Law and Policy in India* (2nd ed., 2001) is widely used throughout India. Until 1996, Armin headed Pacific Environment, an international NGO that he founded in 1987. He is co-editor of *Climate Change Science and Policy* (2010) and currently teaches a course on energy and climate at Stanford.

Ben Saul

Associate Professor Ben Saul is Co-Director of the Sydney Centre for International Law at the Faculty of Law, the University of Sydney and a barrister. He has published widely on a range of international law issues and taught at UNSW, Oxford, and in Nepal, Hong Kong and Cambodia. Ben is editor-in-chief of the *Australian International Law Journal* and an Associate of the Australian Centre for Climate and Environmental Law.

Susan Shearing

Susan Shearing is a Lecturer at the Faculty of Law, the University of Sydney. She teaches undergraduate and postgraduate environmental law. Her teaching and research interests include biodiversity law, heritage law and protected areas management, water law and sustainable business and environmental regulation. Susan has published in a range of areas, including climate change impacts on natural and cultural heritage, environmental taxation and biodiversity. Susan is a member of the Australian Centre for Climate and Environmental Law, the National Environmental Law Association and the World Commission on Protected Areas.

Tim Stephens

Dr Tim Stephens is a Senior Lecturer at the Faculty of Law, the University of Sydney. He holds a PhD in Law from the University of Sydney, and an M Phil in Geography from the University of Cambridge. He has published widely on the international law of climate change, and international environmental law more generally. He is the author of *International Courts and Environmental Protection* (Cambridge University Press: 2009) and The International Law of the Sea, co-authored with Donald R Rothwell (Hart Publishing: 2010).

Gillian Triggs

Gillian Triggs is the Dean of the Faculty of Law and Professor in Law, University of Sydney. She is a Public International Lawyer and a former Director of the British Institute of International and Comparative Law and General Editor of the *International and Comparative Law Quarterly*. Her most recent books are *International*

Law: Contemporary Principles and Practices (2006) and *International Frontiers and Boundaries* (2008). Gillian was admitted to practice in 1969, has maintained an international commercial legal practice and is a Barrister with Seven Wentworth Chambers, Sydney.

PART I
Keynote

ACCEL Climate Change Law and Policy Conference: Keynote Talk on America's Challenges

Armin Rosencranz[1]

When Adlai Stevenson was running for President in the 1950s, one of his aides approached him and said, 'Governor, all the thinking people of the country are behind you!' Stevenson replied, 'That won't be enough to get me elected.'

The thinking people of America have been a distinct minority, and mostly voiceless in national government, for the past 12 years. Beyond sending his Vice President, Al Gore, to Kyoto in 1997, Bill Clinton did very little to lead his country or the world on climate change. He spent little on research and development and did not reconfigure the energy system.

George W Bush not only continued Clinton's weak leadership but added his own special stamp of climate denial. For several years, under White House direction, no public document could be issued using the words 'climate change', 'global warming', or '*Kyoto Protocol*'. Bush publicly mentioned the words 'climate change' for the very first time in his 2007 State of the Union address.

Ignoring a unanimous internal staff recommendation, Bush's Environmental Protection Agency administrator refused to grant a waiver to California to enable that state to go beyond federal clean air standards and, inter alia, regulate auto emissions.

George W Bush's climate policy was a combination of denying that climate change is a serious problem, relying exclusively on voluntary emissions reductions by industry and the power sector, spreading fear of loss of jobs and reduced gross domestic product, providing financial support primarily to the coal, oil, gas and nuclear industries, and offering virtually no leadership to the international community.

Aside from a 5–4 decision of the *US Supreme Court in Massachusetts v EPA*[2] that declared greenhouse gas (GHG) emissions to be within the purview of the US clean air regime, no litigation in the US federal courts has required any industry or government agency to change behaviour and reduce GHG emissions. Nor has any climate change-driven legal proceeding in any non-US forum, such as the Inuit petition to the Inter-American Commission on Human Rights, resulted in any assumption of responsibility or changed behavior.

California's various laws and regulations, such as the *Global Warming Solutions Act* of 2006, have led the nation, and encouraged other states to follow suit. For example, California has a 'renewable portfolio standard' under which 33% of electricity must be produced from renewable sources by 2020. Moreover, California may not import dirty power from other states. Presumably President Obama will give free rein to these subnational initiatives.

During the US presidential campaign of 2008, John McCain and Barack Obama both advocated 'cap and trade' climate change legislation. Indeed, McCain first proposed cap and trade legislation in the *McCain-Lieberman Climate Stewardship Act of 2003*, which failed to pass the Senate by a 56 to 43 vote. But in the ensuing years, McCain, supposedly a free marketeer, voted against federal support or subsidies for renewable energy at every turn. His favorite sources of energy were coal, oil, gas and nuclear power, and, despite his belief in the free market, he voted to give them tax breaks. In the summer of 2008, when oil prices reached US$147 a barrel, McCain's refrain was 'Drill Here! Drill Now!'.

By contrast, Obama has favored subsidies and tax breaks for solar and wind power, and observed that US land and offshore oil reserves, even if targeted for extraction, would not add measurably to supply; any benefit from domestic drilling wouldn't change America's energy balance for at least 8 years. Mandatory GHG emissions ceilings combined with energy efficiency and US$150 billion in renewable energy infrastructure would, in Obama's view, create new product lines and new jobs. Moreover, Obama's public statements indicate that he understands the need for a strong market price for carbon, and the need also to auction GHG allowances.

In the transport sector, Obama pledged to repeal tax breaks for oil companies and use the savings to invest in fuel-efficient cars, help US automakers retool, and have a million plug-in hybrids on the road by 2010. He noted that hybrid electric vehicles are expected to get 150 miles (250 kilometres) per gallon in the near future.

Both Obama and McCain publicly supported nuclear power. But McCain envisioned several hundred new nuclear plants, without regard to their large capital costs and the possibility that each new plant would be a target for terrorists. Obama's support was much more muted, saying that nuclear power should be among the nation's mix of energy options.

On the international front, the candidates appeared to have different agendas. McCain rarely referred to America's international obligations, whereas Obama said the United States had a responsibility to resume global leadership, renounce unilateralism, revive diplomacy and work toward cooperative international action. Obama called climate change 'one of the great moral challenges of our generation'. He also called on America to provide for the least advantaged people (and countries) among us. Obama's emphasis on justice and equity could be a departure from the utilitarianism that has driven most of our decisions about who gets what, when, where and how for the better part of the last 200 years.

Over the last 18 months, the idea of geo-engineering has entered the public discourse. Climate scientists tend to scorn geo-engineering, but what seemed a short while ago to be outlandish is now talked about in rational terms. One scholar has written about 'solar radiation management', including solar collectors above the troposphere and the deliberate release of aerosols to dim the sun. Decision-makers need to weigh the short and long-term costs and benefits of these proposals, because they seem to offer cheap and quick fixes that have obvious political appeal.

The Obama administration will undoubtedly engage with China and India over the issue of climate change. Rather than India, China and the United States saying, in effect, 'we won't act unless you act first,' all three countries, together with the EU, Canada, Russia, Brazil, South Africa, Japan and Australia need to act in concert to mitigate GHGs in the atmosphere. Richer countries also need to help poorer countries adapt to the effects of climate change — including rising temperatures and sea levels, changing growing belts and lower food production. India will very soon begin to see lower levels and flows in its rivers from Himalayan deglaciation. China, having now surpassed the United States as the world's leading GHG emitter, and recognising the health consequences of increased coal burning, seems determined to pursue low- and non-carbon energy technologies, including nuclear, solar and wind power.

Until November 2007, Australia's Prime Minister and America's President were in tandem on ignoring the effects of climate change. But after a national election in which climate change was a factor in Prime Minister Kevin Rudd's victory, the Australian delegation was greeted warmly at the Bali climate negotiation in December 2007, while the leader of the US delegation was booed off the stage. Kevin Rudd and Barack Obama are a different breed than their predecessors. The United States and Australia are likely to be close allies at the 'post-Kyoto' global climate negotiations in Copenhagen in December 2009.

The United States has done some things right, such as its success in bringing business into partnership with government and environmentalists on the climate issue. Also, the United States has in the past and undoubtedly will continue to use production tax credits to subsidise wind and solar energy, and to use renewable portfolio standards to encourage power companies to buy renewable energy.

Australia has done a terrific job of educating its people about climate change. Ninety-six per cent of Australians favour climate action. By contrast, only a small minority of Americans even understand the effects of climate change, and barely 10%t of Americans favor taking specific action. In the State of Victoria, a coastal development was rejected in August 2008 because of the dangers of coastal inundation from climate change. Nothing like that has happened in the United States.

It's a pleasure for me to have had this opportunity to explore America's challenges in shaping its own and international climate change policy, and to learn how each of you is contributing to climate law and policy.

Endnotes

1   Armin Rosencranz, a lawyer and political scientist, is co-editor of *Climate Change Science and Policy* (2009). He has taught climate policy at Stanford, University of Maryland, University of Virginia and Johns Hopkins-SAIS, and climate law at Vermont Law School and Georgetown Law Center. He is the founder of Pacific Environment, an international environmental NGO.

2   549 US 497 (2007).

Climate Change and the Concept of Private Property

Paul Babie

As agents of planetary change, humans play a significant role in the processes responsible for climate change. Yet, we either know little about or overlook completely the complex interplay between how humans effect such change and its outcomes for others.[1] Some suggest that the processes of anthropogenic (or human-caused) climate change involve a complex interplay of human activities normally the preserve of individual disciplines, including ecology, climatology, economics, anthropology, psychology, political science, philosophy, and neuroscience.[2] Law, of course, has a role to play in explaining and understanding this complexity. And within the discipline of law, the challenge of climate change concerns legal theorists, too. The claim advanced in this essay, then, is legal, theoretical and novel:[3] climate change is a private property problem.[4]

In four parts, this chapter argues that the dominant liberal conception of private property, implemented and operating in legal systems worldwide, permits power — or choice — over the use and control of goods and resources so as to prioritise self-interest over obligation towards the community, both local and global. This, in turn, is one of the components of modern social life making possible the complex processes that produce both anthropogenic climate change and its consequences for humanity.

Parts 1 and 2 explore what are sometimes mistakenly thought to be alternative conceptions of private property but which, in fact, are two complementary and necessary components of the modern concept of private property.[5] Part 1 outlines the core of the dominant liberal understanding of private property, referred to here as 'choice'. Part 2 examines the 'relationship' dimension, providing the crucial insight that decisions permitted by the choice component occur not in a vacuum but in a web of social relationships. This reveals a choice/consequence nexus between those with and those without choice (private property), turning our attention to attempts to regulate private property so as to control the consequences of climate change. But, while those theorists who explore that nexus typically do so as a matter of exploring the domestic law of nation states, as is well known, the effects of climate change do not stop at jurisdictional boundaries. Ultimately, the consequences of choices predicated upon private property affect both those within and those beyond the jurisdictional boundaries of those nations which instantiate the private property that conferred and protected choice in the first place. Part 3, therefore, offers some brief reflections on how the concept of private property may need to adapt to meet the complex challenge of climate change. Part 4 concludes.

Choice

The claim is this: the modern liberal conception of private property confers the choice that underpins anthropogenic climate change. In order to demonstrate this somewhat bold stance, we must examine, briefly, the process of climate change itself. Having done that, the remainder of this section explores, first, the notion that private property is liberalism's means of securing choice in relation to goods and resources and, second, the role this plays in anthropogenic climate change.

Climate Change

The *United Nations Framework Convention on Climate Change* (UNFCC) art 1, defines 'climate change' as 'a change of climate which is attributed directly or indirectly to human [anthropogenic] activity that alters the composition of the global atmosphere and which is in addition to natural climate variability observed over comparable time periods.'[6] 'Climate' is the average weather — usually temperature and precipitation — over a set period of time.[7] While the United Nations definition distinguishes between anthropogenic and natural causes,[8] the two are related through the 'greenhouse effect', which describes how the earth and its atmosphere is warmed through a complex process of energy exchange.[9] This begins with the sun's solar radiation that falls on the earth. Atmospheric molecules and the land and ocean surface return a small percentage of this energy back to space, while the majority heats the surface. To balance incoming energy, the earth must return thermal radiation into space;[10] a perfect balance would result in a surface temperature of –6°C. In fact, the average surface temperature is 15°C. The 'greenhouse effect' explains the temperature difference.

The 'natural greenhouse effect', which warms the earth, both making possible and sustaining organic life, operates as part of the earth's atmosphere, consisting of nitrogen, oxygen, water vapour, carbon dioxide (CO_2), methane, nitrous oxide, chlorofluorocarbons, and ozone. All but nitrogen and oxygen are 'greenhouse gases' and these, which absorb and act as a 'blanket' for some of the thermal radiation, account for the 21°C difference between the actual average surface temperature of 15°C and the –6°C of perfect balancing. It is called 'natural' because the greenhouse gases that cause the effect were there long before human activities began to affect the climate.[11]

Human activities 'enhance' the natural effect, increasing the surface temperature of the earth.[12] To take but one example, since the Industrial Revolution, industry and deforestation have increased atmospheric CO_2 levels by as much as 30%. Such activities add to the thermal blanket, warming the atmosphere and the surface above the 15°C of the natural effect. Projections indicate that in the next 100 years, in the absence of any mitigation, human activities will increase the atmospheric concentration of CO_2 to double that of its pre-Industrial Revolution value.[13]

The human role raises significant questions: 'Can we adapt tools forged to create wealth and use them for sustainable development?', 'Is ever-increasing wealth the only viable measure of progress?' 'Is it possible to have global governance that protects our future?', and 'Can individuals really be rational and modify behaviours accordingly?'.[14] In other words:

> [c]limate change has successfully put anthropogenic change on the global agenda, resulting in demands to cut greenhouse gas emissions. But this is like calling for bans on smoking, or issuing warnings about overeating or having unprotected sex, without taking into account the human or social dimensions.
>
> How can we ask the world to, say, give up flashy gas guzzlers if we don't understand the complex interplay between the psychology/physiology of car ownership and driving or the economic, social and psychological effects of stopping a region from making gas guzzlers?[15]

This chapter argues that human activities that enhance the natural greenhouse effect, and which lie at the heart of questions about wealth, sustainable development, and global governance, are the product of choice, enshrined in and conferred by the concept of private property. The next section explores this proposition.

Private Property as Choice

The dominant modern conception of private property is liberal. Liberalism is concerned with promoting and protecting freedom of choice for the individual — natural or legal — and groups of them.[16] Private property is liberalism's vehicle for achieving that objective — conferring choice — in the allocation and control of goods and resources — natural or manufactured, tangible or intangible — among individuals.[17]

The aim here is not to rehash the standard liberal account of private property as rights in or to things[18] but, rather, to explore the notion of choice as the core of the liberal conception. In other words, it is to show that private property involves the creation, conferral and protection of choice and control over goods and resources. The notion of creation, conferral and protection is traceable to Jeremy Bentham, who first noted the importance of law to the structure of private property arguing that rather than being a natural right, '[p]roperty and law are born together and die together. Before laws were made there was no property take away laws, and property ceases.'[19] Whatever private property is, law creates, confers and protects it.

Distinct from creation, though, is content; for this, one begins with the work of Wesley Newcomb Hohfeld. In two seminal articles, Hohfeld outlined what was, for its time, a novel means of describing legal rights.[20] By identifying four basic types of legal right — rights, privileges, powers, and immunities — Hohfeld's 'jural relations' offered a useful way of understanding and analysing private property. Yet, Hohfeld's was not specifically a theory of private property and, for that reason, left unresolved the identification of the specific rights that constitute the liberal conception. In a seminal article on ownership, however, Anthony Honoré provided the foundations of what has today become the modern liberal conception of private property.[21]

Combining[22] Honoré's 11 standard incidents of ownership[23] conceptually freed Hohfeld's analysis so as to disaggregate the previously undifferentiated Blackstonian model[24] of private property,[25] which posited the consolidation in one person of a cluster of rights and an aggregate of power over a good or resource.[26] This produced the 'sophisticated' or 'legal'[27] 'bundle of rights' picture or metaphor.[28] The bundle metaphor — the liberal conception[29] — dominates contemporary scholarship and judicial decision-making[30] and is largely understood, at least in outline, by most people.[31]

According to the liberal conception, private property comprises individual rights created, conferred and enforced by law, usually including at a minimum the 'liberal triad' of use, exclusivity, and alienability.[32] While the bundle may be held exclusively by one individual or group of them[33] — as would be the case under the ownership model — that is not the only way in which the rights can be held. The bundle may be and frequently is split so as to distribute its constituent rights among many individuals or groups.[34]

Once combined with Honoré's incidents, therefore, Hohfeld's analysis reveals that the rights of private property, whatever they are, however they are bundled, and by whomever they are held, allow the holder to act in certain ways in relation to the rights of other people or groups of them.[35] This is typically expressed in this way:

> [private property is a] claim that other people ought to accede to the will of the owner, which can be a person, a group, or some other entity. A specific property right amounts to the decision-making authority of the holder of that right.[36]

But decision-making authority in terms of what? We tend to think in terms of others and their rights; but it is just as important to remember that the relationship with others still concerns a good or resource — an object of social wealth. This, after all, the object of private property. Lametti, discussing the analytic role of objects in property theory, puts it this way:

> [a]ny act on an object of property [good or resource] — even its consumption — changes the normative status of another person. That is, a person can unilaterally eliminate or change the duties owed by persons to others not to interfere with a particular object.[37]

Private property rights, in other words, confer on and secure to their holder *choice* about how to use, or not to use, a good or resource.[38]

In their recent book, Richard H Thaler and Cass R Sunstein describe what they refer to as 'choice architecture', which is the activity of any institution of either the private or the public sector that attempts 'to steer people's choices in directions that will [result in decisions] that will improve their lives.'[39] This is a helpful way to understand the bundle of rights notion of private property as choice in relation to the control of goods and resources. While Thaler and Sunstein are primarily concerned with the architecture that affects and structures choice, they implicitly make an important point: choice architecture presupposes the existence of choice itself. They say as much: 'people should be "free to choose"', society and its legal system 'strive[s] to design policies that maintain or increase freedom of *choice*', the point of choice architecture is 'to steer people's *choices*'.[40] There are two components to choice architecture: first, the creation, conferral and securing of choice and, second, the identification of those rules that steer, structure and control it. The aggregate of private property rights over a given resource represents a part of the first component.

Before applying this analysis to climate change, one other point must be made in relation to viewing private property as liberalism's means of securing choice over goods and resources. Each of the rights conferred by private property represent a choice permitting its holder to act upon a good or resource *in any way* the holder sees fit. Typically described as 'preference-satisfaction', 'self-interest', 'self-seekingness', or 'self-regarding acts',[41] the ability to exercise rights as one sees fit means that the:

> the rules of [a] property institution are premised on the assumption that, prima facie, [a] person is entirely free to do what he will with his own, whether by way of use, abuse, or transfer.[42]

Moreover:

> [h]e may also, within the terms of the relevant property institution, defend any use or exercise of power by pointing out that, as owner, he was at liberty to suit himself.[43]

Self-seeking, preference-satisfying, self-interested choice is the hallmark of private property, and determining when a choice is truly self-regarding — in the sense that its effect is experienced only by the holder of the right — and when it is not — in the sense that they affect others, or create 'externalities' — is part of its normative content.[44]

All of this ought not surprise us: as we have seen, liberalism secures choice to individuals and puts in place choice architecture to structure that choice,[45] so we would expect that private property is the vehicle through which this process is achieved in the case of allocating and controlling goods and resources. Applied to the enhanced greenhouse effect, this demonstrates that climate change can be seen as the product of choice in relation to goods and resources.

Climate Change as Private Property Choices

This section traces the connection between the greenhouse gases attributable to humans, and the underlying private property choices about goods and resources that produce them. The argument is straightforward: the 'human activity' of the enhanced greenhouse effect is predicated upon choices made in relation to the control and use of resources. The liberal concept of private property — choice — is the principle means by which choice over goods and resources is created, conferred and protected. Private property is, therefore, one of the economic and political drivers of the enhanced greenhouse effect, which is a clear externality. In fact, it is at least arguable that, being a principal foundation of the political distribution of goods and resources and the operation of a functioning market economy, this concept is *the* driver.

The Greenhouse Gases

Of the major atmospheric greenhouse gases — nitrogen, oxygen, water vapour, CO_2, methane, nitrous oxide, chlorofluorocarbons, and ozone[46] — human activities substantially affect only the concentrations of CO_2, methane and nitrous oxide.[47] This section considers each.

The carbon cycle and CO_2 represent the dominant means by which nature transfers carbon between the three natural carbon reservoirs — the atmosphere, oceans, and the soil and land biota. To offset this cycle, the photosynthetic process of plants and trees use light to take in CO_2, retain carbon for growth and return oxygen to the atmosphere. There is much more carbon stored in the land and ocean reserves than in the atmosphere; thus, a small change in the latter two reservoirs can have a large impact on atmospheric concentrations.[48] Humans and animals contribute to this cycle through natural processes: respiration — transferring carbon from food into exhaled CO_2 — fires, rotting wood and the decomposition of organic material in soil. Carbon emitted into the atmosphere in this way is simply redistributed among the three reservoirs.[49] But intentional human activities far exceed natural emissions. The largest increase comes from burning coal, oil and gas, the generation of energy for electricity and heating, transportation,[50] industry, land use change and waste.[51]

While the enhanced greenhouse effect caused by a molecule of methane is about eight times greater than that of CO_2,[52] its lifetime in the atmosphere is much shorter.[53] Apart from natural wetlands, the largest sources of methane are intentional human activities.[54] Agriculture accounts for almost 15% of total anthropogenic greenhouse gas emissions — more than all of the emissions of CO_2 attributable to the world's transportation[55] — and of this almost half is methane from rice paddies, livestock and manure.[56] The remaining direct and indirect human sources include leakage from natural gas pipelines and oil wells, the decay of rubbish in landfill sites, and burning of wood and peat.[57]

Nitrous oxide is an extremely potent greenhouse gas with a very small but increasing atmospheric concentration (currently about 16% greater than in pre-industrial times).[58] While biomass burning and the chemical industry play a small part, the most significant intentional human activity contributing to this increase is agriculture through its use of nitrogen as a fertiliser.[59] While only a small proportion escapes into the atmosphere as nitrous oxide, its long atmospheric lifetime produces a significant greenhouse impact.[60]

The Choices

Rather than focusing on the sorts of human choices responsible for the enhanced greenhouse effect, the available literature typically discusses in general terms the relevant emissions and their human-caused sources.[61] This chapter, however, seeks to demonstrate the link between private property — individual choices about goods and resources — and anthropogenic climate change. Interestingly, the best way to do this is to examine those things that humans are increasingly told *not* to do in order to determine what is presently being done. In other words, what we are told *not* to do tells us something about the choices that we *are* making. Of the available sources of such information, the most accessible are found in the popular news media. This section relies upon a special double issue of *TIME Magazine* entitled '51 Things You Can Do to Make a Difference'.[62]

Of the 51 things that one can do to mitigate climate change, many relate directly or indirectly to CO_2, although some relate to methane and nitrous oxide.[63] Choices about what we eat, for instance, contribute to the production of methane and nitrous oxide. The international meat industry produces 18% of the world's greenhouse gas emissions, most of that from the nitrous oxide in manure and the methane by-product of bovine digestion.[64] The use of money — a private property choice[65] — to purchase beef as opposed to other meats or no meat at all is, therefore, a private property choice.

Of course, CO_2 is the most significant human-generated greenhouse gas, and most of *Time's* admonitions concern where we live, what we do there, and how we travel from place to place. In the case of homes or other dwellings, any choices we make about what happens there depends, ultimately, on the existence of choices created, conferred and protected by the holding of a conventional real property interest, typically a fee simple or leasehold. Within that context, it is a private property choice to change from conventional incandescent to compact fluorescent light bulbs,[66] to check the energy rating of appliances,[67] to alter the way one washes one's clothes (opting for warm rather than hot water, one large rather than many small loads, and clothes lines rather than electric clothes dryers).[68]

An energy audit can provide a long list of small changes resulting in carbon savings,[69] including opening a window instead of running an air-conditioner, adjusting the thermostat to a slightly higher temperature in summer and a slightly lower one in winter, sealing gaps in doors and windows, insulating walls and ceilings, using the dishwasher only when full, installing low flow showerheads, insulating[70] and turning down the thermostat on the hot water heater,[71] turning off all electrical appliances that have a 'standby' function, and turning off lights in unused rooms.[72] In every case, these small choices impact upon the enhanced greenhouse effect through reducing the use of electricity, which depends in many cases on the use of CO_2 producing fossil fuels. Indeed, one can change from electricity to green energy, such as wind or geothermal.[73]

Private property allows choices about land and the structures built on it. We might opt to build a smaller house that requires less energy to cool than a larger one,[74] working from a set of blueprints that maximise energy efficiency,[75] for a 'passive house' (one that implements extra insulation and state-of-the-art ventilation to recycle energy from passive sources such as body heat, the sun and household appliances in order to warm the air).[76] Or one might choose to live in a high-rise building, cutting both transportation and residential energy emissions.[77] In every case, as with the small changes that can be made within the home, the choice of where and how to live is a private property one.

The initial decision about where to live may impact heavily on transportation decisions. If, for instance, one does decide to live in a highly urbanised environment it is very likely that that choice will result in the reduced use of transportation options that produce CO_2.[78] If, however, one chooses to live in a non-urban environment, then public transportation may not be an option;[79] or, if one chooses not to make use of public transportation, or if it is unavailable, other decisions become possible, each having an impact, negative or positive, on the enhanced greenhouse effect. Suggestions about personal transportation — for instance, how to drive (planning routes ahead of time to conserve fuel),[80] the use of green fuel,[81] investing in vehicles that limit carbon emissions,[82] joining a carpool,[83] ensuring proper automobile maintenance,[84] and, when necessary, renting a green car[85] — tell us something about the sorts of decisions that are currently being taken. Longer trips that involve air travel are also relevant to this calculus: commercial air travel contributes substantially to carbon emissions.[86]

Personal lifestyle choices obviously also involve private property. Clothing involves a choice — recycling old fleece jackets[87] or purchasing second-hand clothing,[88] for instance, both result in production and transportation energy savings and so reductions in carbon emissions. The options are endless: paying bills online,[89] purchasing local produce,[90] avoiding plastic bags,[91] 'green' weddings,[92] the beauty products one uses,[93] using recycled paper,[94] or raking leaves[95] can all reduce carbon emissions and all are private property choices.

Private property is also held by corporations, which also, therefore, have choices available to them predicated on private property.[96] Such corporate choices are important, for they structure the range of choices open to individuals, thus conferring on corporations the power to broaden or restrict the meaning of private property in the hands of individuals. Green energy (solar or wind power), for instance, remains unavailable to the individual consumer if no energy producer — a corporation — is willing to produce it.[97] Similarly, the production of ethanol is a corporate property decision. Ethanol was initially viewed as a solution to the use of fossil fuels in transportation. It soon became apparent, however, that the use of corn and other foodstuffs for the production of ethanol was having an adverse effect on the global price of food and that, in fact, its production was more energy intensive, and greenhouse gas producing, than fossil fuels themselves. Both the effect on food prices and the availability of non-fossil fuel sources affect the available choices open to the individual consumer. Some corporations have since increased research into the use of municipal waste, wood pulp and leftover grain and corn husks in the production of cellulosic ethanol.[98] Even the ongoing use of fossil fuels, though, leaves open private property choices for corporations. Greenhouse gas emissions from coal burning power stations, for instance, may be captured through a process known as 'carbon sequestration' (CS).[99] While cost is a factor, CS is already a reality in some Scandinavian countries.[100]

Corporations have available to them choices similar to those open to individuals about their 'homes' and those who 'live' there. In relation to the former, the way in which a building is constructed, for instance, can achieve substantial greenhouse gas savings. Choices here might include the way in which the concrete is produced, the way in which waste water is utilised, and the use and production of electricity.[101] As to the latter, the way in which employees work is a corporate private property choice — an employee sitting in a car in traffic burns fossil fuels and therefore contributes to greenhouse gas emissions. Allowing the same employee to work from home results in a saving; this is the idea behind what is known as 'proximate commuting'.[102]

This non-exhaustive list of admonitions typically issued to individuals or corporations is enough to demonstrate the kinds of choices already being made, each predicated on private property. Yet private property is more than choice alone — choices do not end at the borders, physical or legal, of the object of social wealth in question. They are not made in a vacuum.[103] Rather, as foreshadowed, choices occur in a web of social relationships within which consequences, or externalities, are visited upon others. Relationship, therefore, is central not only to what private property is, but also to understanding its role in anthropogenic climate change.

Relationship

This part has three objectives. First, it explains why the notion of relationship is central to understanding the concept of private property. Second, it outlines how the relationships that give life to private property also allow us to identify externalities — those circumstances where the choices of those who hold private property have consequences for those who do not. The argument advanced here is that understanding private property as a relationship productive of externalities reveals a 'climate change relationship'. Finally, this section will outline how the concept of private property traditionally responds to externalities and how current responses to the challenge of climate change follow this pattern.

Private Property as Relationship

People tend to associate the dominant liberal conception sketched in Part 1 with the notion that private property does or ought to secure unfettered, and therefore absolute, choice in relation to the use of goods and resources.[104] Consider the 'liberal triad' of rights: use, exclusivity and disposition. According to the absolutist view, the right to use secures an absolute choice about any use or non-use; the right to exclusivity the ability to choose absolutely who to exclude and who not to exclude; and the right to disposition the unfettered choice about any alienation or non-alienation. Law establishes the right or rights that constitute the bundle while the notion of self-seekingness or preference-satisfaction ensures that it affords an unfettered and absolute choice.[105] While this is a widespread and popularly held view — Margaret Jane Radin argues that those who subscribe to it consider society to be 'monolithically socialized into [the] conventional Blackstonian [absolutist] view'[106] — it overlooks the reality that one makes choices about a resource. Put another way, decisions predicated upon private property occur within a broader framework or network of relationships between people. Viewed in this way, such decisions therefore hold the potential for positive or negative outcomes for others — in short, choices produce externalities.[107]

In reality, private property as unfettered or unrestrained and so an absolute choice never has existed and does not now exist.[108] Even the 'avatar of absolutism',[109] William Blackstone — whose axiomatic statement that private property is 'sole and despotic dominion' — wrote that private property consists of the '... free use, enjoyment, and disposal of all [a person's] acquisitions, without any control or diminution, save only by the laws of the land.'[110] As Robert Burns concluded in his seminal work on the topic:

> [a]lthough [Blackstone] calls [private] property an 'absolute' right, he does not mean that government — or at least the legislature — is without power to remold the historically conditioned and socially recognised rights of the individual in property.[111]

In this pithy statement, Burns explains why governments may limit what otherwise appears to be the absoluteness of private property choices: the liberal position is nothing more than 'a simple and *non-social*[112] beginning. Private property, while established and protected by law, has its origins not in a gift of the creator,[113] but in social relationships.[114]

The American legal realists gave early shape to this social view. Morris Cohen argued that private property confers a form of 'sovereignty' on rights-holders, creating a relationship between the person holding such sovereignty *and* others,[115] while Felix Cohen added this important modification to the bundle metaphor:

> [p]rivate property is a *relationship among human beings* such that the so-called owner can exclude others from certain activities or permit others to engage in those activities and in either case secure the assistance of the law in carrying out [that] decision. [116]

Building upon the Cohens, realist descendants demonstrated that the source, origin, and constitution of private property lay in *relationships* between people.[117] Private property is, in other words, a dynamic social construct, 'a cultural creation and a legal conclusion.'[118] Rather than being acontextual, it flows from and has meaning according to social context[119] depending for its content on the cultural, political and ideological beliefs of a particular society.[120] 'Understanding what is private property is an inductive and iterative process, one that looks to the chaos of real world relations ....'[121] Private property, then, is the product of relationship,[122] legal as well as social, including both legally recognised private property interests — fees simple, leaseholds, easements, copyright, money, shares, and so forth — as well as the informal reliance and trust which many people regularly place in the positions taken by others[123] to create social and legal entitlements.[124]

Failing to account for the social origin of private property, and the rights that comprise it, causes us to succumb to a dangerous metaphor inherent to the liberal account — that '[r]ights define boundaries others cannot cross and [that] it is those boundaries, enforced by the law, that ensure individual freedom and autonomy'.[125] This metaphorical boundary misconstrues the relationship of liberalism and communitarianism, neither of which, in their early conceptions, fostered incompatibility.[126] This was certainly the case with private property in the early American experience.[127] Yet, with the modern form of individualistic liberalism that prevailed over its earlier, inclusive, form,[128] came also the misconceived basis for individual autonomy that private property purports to protect and foster.[129]

In attempting to recapture the inclusiveness of early liberalism, however, Joseph William Singer argues that not individualism but '*interdependence* is the foundational characteristic of free individuals';[130] people best function within a web of social relationships that allow their own abilities to flourish.[131] Thus:

> [The] human interactions to be governed [by law should not be] seen primarily in terms of the clashing of rights and interests, but in terms of the way patterns of rela-

tionship can develop and sustain both an enriching collective life and the scope for genuine individual autonomy.[132]

This is an entirely 'different concept of individual well-being and autonomy: one that recognizes the individual's need for freedom as well as the need for the development and expression of that freedom in the context of relatedness to others.'[133]

The social origin of private property requires us to reconceptualise rights as establishing *socially contingent* boundaries. Far from being rigid and unyielding, the boundaries between rights actually operate within a social context of relationships involving mutual dependence and obligation.[134] At least five modern relational models — Aristotelian,[135] republican,[136] citizenship,[137] good neighbour or environmental,[138] and democratic[139] — view private property as encompassing choice and obligation. Singer argues that:

> [r]ather than understanding rights and autonomy as 'an effort to carve out a sphere into which the collective cannot intrude,' we understand that because rights conflict, we must define them partially in terms of the relationships they instantiate. Property law can therefore be seen as 'a means of structuring the relations between individuals and the sources of collective power so that autonomy is fostered rather than undermined.'[140]

Thus, private property as choice and private property as relationship are not two conflicting or mutually exclusive models but, rather, two parts of the totality of the concept. Private property rights — choice — have their origins in, exist, operate, and are protected by law within a social context — relationship.[141] Relationship means that private property rights overlap with the rights, property or otherwise, of others, owners and non-owners.[142]

This 'imbricated' nature of private property rights[143] — the overlap between private property rights and the non-private property rights of others — allows for externalities.[144] The next section argues that climate change is such an externality.

Climate Change as Private Property Relationship

The enhanced greenhouse effect is an excellent example of this: private property makes possible the decisions behind the greenhouse gases that drive anthropogenic climate change. The aim of this section is twofold. First, it briefly explores the human externalities of the enhanced greenhouse effect. Second, it argues that establishing a link between individual choice predicated upon private property and the broader human community[145] reveals a 'climate change relationship'.

Externalities

Although the externalities of climate change are both environmental and human,[146] this chapter is concerned only with the latter. The available science demonstrates a clear link between the human conduct — private property choices — that causes climate change and the consequences of that process for other humans. Those consequences fall into four categories: human security, health, water resources and food systems.[147]

Human Security

Most analysts now agree that 'human security' is achieved when individuals and communities have: '[(i)] ... the options necessary to end, mitigate, or adapt to threats to their human, environmental, and social rights; [(ii)] ... the capacity and freedom to exercise these options; and [(iii)] [a]ctively participate in attaining these options.'[148] The term therefore encompasses many of the human dimensions of global change — health, resource availability, vulnerability to hazards, and environmental degradation, including crime, drugs and assorted other problems. Climate change affects each of these factors because it may have implications for resource availability, agricultural productivity and economic output.[149] In addition to these direct consequences, though, climate change also indirectly affects human security, principally through the displacement of people and the resulting erosion of political stability.

Predicted coastal flooding will displace large numbers of 'climate' or 'environmental refugees'. Sixty per cent of the human population lives within 100 kilometres of the ocean, with the majority in small- and medium-sized settlements on land no more than 5 metres above sea level. Even the modest sea level rises predicted for these places will result in a massive displacement of people.[150] In Egypt and Bangladesh, for example, where a large percentage of both population and productive capacity is located less than one metre above sea level, the predicted rise in sea level of between 0.2 and 0.6 metres caused by a doubling of CO_2 levels would cause significant displacement of people.[151]

Political instability will not only follow the displacement of people, but will also be driven by the increasing frequency of extreme weather events and by periodic droughts in arid and semi-arid regions. Extreme weather events are difficult and costly to prepare for and cause significant social disruption. This is particularly so in places such as Southeast Asia, which is especially prone to coastal flooding, and where such events are predicted to increase in frequency and magnitude.[152] In relation to droughts, examples abound. In four years of fighting in the Darfur region of Sudan, for example, more than 200,000 people have been killed and 2.5 million more have been made refugees. Typically characterised by the popular media as genocide waged by the Arab Janjaweed and their backers in the Sudanese Government, evidence is increasingly pointing to the role climate change has played in causing the drought at the root of the conflict. Complicating the tragedy is a fight over a shrinking land base, evaporating water and food supply, and a lack of shelter.[153]

The poor suffer most acutely when human security diminishes. The United Nations Intergovernmental Panel on Climate Change (IPCC) reports that

> [t]he impacts of climate change will fall disproportionately upon developing countries and the poor persons within all countries, and thereby exacerbate inequities in health status and access to adequate food, clean water, and other resources.[154]

A sobering thought when read with the knowledge that over one billion people live in absolute poverty in the developing world, 64% of them in Asia.[155] And poverty is the main cause of malnutrition in the world, presently affecting almost one billion people. Climate change exacerbates this problem, decreasing global food production and water availability, and increasing the potential for longer and more severe droughts, and by compounding land degradation through erosion, flooding, and pollution.[156]

Health

Climate change is predicted to affect human health both directly and indirectly.[157] In the case of the latter, temperature change, floods, storms and sea-level rise are responsible for a host of human health problems, including: (i) death caused by heat waves; (ii) asthma due to gradual climatic changes; and (iii) physical trauma associated with melting ice.[158]

The direct consequences — changes in existing health risks[159] — include:[160]

(i) increased spread of infectious diseases caused by widening social inequalities and changes in biodiversity;[161]

(ii) increases in malaria, dengue fever, yellow fever, and certain types of encephalitis carried by mosquito populations boosted by increasing temperatures;[162]

(iii) prolonged drought followed by early heavy rainfall resulting in birds that would otherwise control hantavirus-carrying rodents being bitten by the Culex pipiens West Nile virus-carrying mosquito in turn increasing human incidences of hantavirus and West Nile virus;

(iv) extremely wet weather causing flooding which produces disease clusters;

(v) prolonged drought, weakening tree defences against infestations, promoting wildfires that can cause injuries, burns, respiratory illnesses, and deaths;[163]

(vi) illnesses affecting wildlife, livestock, crops, forests, and marine organisms, the resulting biologic impoverishment of which may have negative consequences for air, food, and water;[164] and

(vii) major coastal storms which trigger toxic algal blooms or 'red tides', creating hypoxic 'dead zones' in gulfs and bays and harbour pathogens.[165]

The poor and disadvantaged disproportionately bear the brunt of the human consequences of climate change. In 2005, Hurricane Katrina, for example, left hundreds of thousands homeless along the Gulf Coast of the United States with no means to relocate. Following days of utter chaos it became clear that the poor and marginalised had borne the brunt of these consequences.[166]

Water

Approximately two billion people, or one-third of the world's population, currently live in countries considered water stressed, meaning that there are already problems with the security of water quality and quantity. Climate change will exacerbate these stress levels. As the world's population continues to grow, water shortages will increase — from over two dozen countries today exhibiting water stress or water scarcity, to 50 by 2025, and 54 in 2050 with a combined population of over four billion. Projections for Africa, the Mediterranean and Australia show less rainfall. Because irrigation accounts for over 70% of water use, higher demand for water for food production, loss of soil moisture, and the increase in evaporative demand will increase the stress. Flood and drought will further add to these stresses.[167]

Food

Changes in crop yields due to climate change will vary by region, depending on crop type, soil moisture, and other factors. For small temperature increases, the consequences will either be positive or neutral as farmers adapt to new conditions. For changes above 2°C, however, the consequences will likely be negative, including heat stress, decreased yields, increased prices, and increased severity of droughts.

The ability of a region to adapt will play a role in the severity of impact. North America and Europe, for instance, can adjust planting dates, crops, and fertilization rates to minimise impact. In developing countries, however, such options do not exist. Moreover, in the developing world one of the biggest impacts will be decreased prices for what is produced, driven by increased production in developed countries. As with the other human consequences, therefore, the impact of climate change will weigh disproportionately on the food systems of the poor and disadvantaged of the developing world.[168]

A 'Climate Change Relationship'

The process of identifying the choices that humans make — predicated upon the dominant liberal conception of private property — and linking those to specific human externalities reveals a 'climate change relationship' made possible by private property. Joseph William Singer, writing in general terms, notes that:

> property owners and the public are linked to each other through individual actions [choices] and laws affecting the use of property (which can ... be both beneficial and detrimental). From this perspective, we could conceive of property as a type of ecosystem, with every private action and legislative mandate potentially affecting the interests of other organisms.[169]

This succinctly captures the climate change relationship revealed by Parts 1 and 2 in this essay. We have seen, based upon a brief outline of both the dominant conception of private property and the science of climate change that the human choices that enhance the greenhouse effect can be characterised as 'private property choices'.[170] Yet, such choices do not occur in a non-social vacuum; rather, they are socially situated within a network of relationships. And we have seen that the enhanced greenhouse effect, driven by individual and corporate private property choices, produces significant human consequences, or externalities. These human externalities of climate change reveal the social structure within which private property choices occur. Making the link between choice and externality completes the relationship.

The next section turns to the role of law in regulating externalities in order to protect the wider community. Before doing that, however, an example will assist with understanding the climate change relationship and serve to identify the place of law in protecting the community against externalities. Consider a person who takes a holiday, travelling to the chosen destination by plane. This journey contributes to the increase in the concentration of atmospheric CO_2, in turn contributing to the enhanced greenhouse effect, thereby increasing the global temperature. Choice of course lies behind these environmental externalities; and so, too, the human consequences. The increase in global temperature contributes to Arctic sea ice melt, producing a rise in the sea level in Southeast Asia, where significant numbers of people live in overcrowded and poor towns only barely above sea level. Those people become environmental or climate refugees, producing political instability not only in the nation of their origin, but also, possibly, in the place where they resettle. This is just one sequelae of human externality that may follow from the same environmental consequences, produced by the choice to travel by air on a holiday.

Yet, the counter-argument might run, one person surely cannot have this impact. To be sure, alone, one person cannot produce environmental refugees, or any of the other human externalities outlined in this Part. But the aircraft in our example is not otherwise empty; and it is not the only plane. It is trite to say, of course, but the cumulative effect of choices made by individuals is often overlooked. In the past, cheaper oil prices caused a burgeoning of air travel, currently making this form of transportation the single biggest contributor to global CO_2 emissions next to cars, adding 3.9 billion tonnes of CO_2 to the atmosphere in 2004, and projected to be 7.4 billion by 2020.[171] And this is just air travel: each of the human activities considered in Part 1 can be traced, through environmental externalities, to the human consequences outlined in this Part.

Community

Using the conceptual banner of private property, liberalism bundled together the rights that allow humans to choose how to use goods and resources. By allowing choices that produce externalities, private property is, perhaps, the primary means of bringing people into relationship at every level, local and global.[172] Specifically, through the externalities created, it brings those who hold such bundles over goods and resources into relationship with those who do not. The resulting relationship therefore requires a means of preventing harm to the have-nots. And it is law which, we have seen, is so crucial to the very existence of private property, that provides this palliative in the form of regulation to safeguard the community against the excesses of the individual. We find concrete examples of this in the burgeoning body of 'climate law' — regulation aimed at mitigating or ameliorating the externalities of climate change. This section considers regulation as a component of the concept of private property and offers examples drawn from climate law.

Regulation

What is meant here by 'community' is the totality of the relationships instantiated by private property between individuals and others.[173] It is, in other words, context for both the existence and the externalities of property.[174] Seen in this way, the rights of the individual and the community overlap or, as Singer puts it, are imbricated.[175] While tempting to see this as a tension between the individual and the community, however, we must repress that urge for:

> explicitly recognizing th[is] tension … *as a part of the concept of property* … [allows us to] reaffirm [its] … importance … while recognizing the interdependence of the self and others.[176]

Rather than a tension, the overlap between rights in a context of relationship:

> takes for granted that owners have obligations as well as rights and that one purpose of property law is to regulate property use so as to protect the security of neighbouring owners and society as a whole.[177]

Regulation is law's means of mediating the relationships established by private property. To prevent the tyranny of the individual over the community, the underlying social functions and relationships of private property *require* monitoring and regulation of choice by corresponding moral imperatives, duties and obligations.[178] Over time, regulation preserves the social function of a private property right, whatever it might be, within a context of relatedness, thus limiting potentially harmful outcomes for others.[179] Choice, then, because the community is formed through relationships, is not entirely unlimited and unfettered. Rather, because it operates within a network of social relationships that form a community, it is mediated by moral imperatives, duties and obligations imposed by regulation.

Admitting that regulation is an inherent component of private property that seeks to balance the interests of the individual against those of the community,[180] and that it ought to do so in order to promote the social good,[181] is not the product of some outdated socialism, as some would have us believe,[182] but the full consequence of recognising that private property is relational.[183] Every system of private property is inherently limited by moral imperatives, duties and obligations, imposed and enforced by law, so as to allow the holder of private property to choose not only personal preferences but also so as to prevent outcomes inimical to the legitimate interests of others.[184] Thus, while '[private property] … initially appears to abhor obligation, … on reflection we can see that it requires it. Indeed, it is the tension between [unfettered private property rights] and obligation that is the essence of [private] property.'[185] As Singer explains, if private property is inherently social and relational, and serves inherently social purposes, then it is inherently moral:

> Owners have obligations both to share their wealth with the dispossessed and to use their [private] property in a way that is compatible with the interests of non-owners in being able to enter the system to become owners … All this means is that there is no core of [private] property we can define that leaves owners free to ignore entirely the interests of others. Owners have obligations; they have always had obligations. We can argue about what those obligations should be, but no one can seriously argue that they should not exist.[186]

Regulation does not, though, eliminate the core of choice. Comparing the legal protection of choice with the regulation imposed upon it by law always yields a surplus of individual choice and a deficit of regulation. Yet just as regulation is inherent to private property, so too is the privileged position of choice. If that were not the case, there would be no private property. The complete absence of choice is the absence of private property. The holder is always left with choice, and choice has consequences. In the final analysis, the liberal ideal for private property — the conferral and protection of choice, albeit fettered — remains an accurate account. Understanding that private property is also relational, however, adds to this simple picture the difficulty of externalities, and the reality that regulation is not merely necessary, but inherent to the social good, for both the holders of private property and for those who are affected by their actions. With that in mind, the next section offers examples of regulation aimed at the mitigation of anthropogenic climate change drawn from climate law.

Climate Law

'Climate law' is a phrase that has emerged recently 'to describe the regulatory measures which will contain, and perhaps even reverse human-induced global climate change.'[187] The range of techniques used to accomplish this may extend to any existing or proposed legislation that in some way regulates greenhouse gas emissions, such as corporate law or general health law. Any laws that have this incidental outcome do not, however, fall under the banner of climate law. Rather, this term has a narrow compass, being restricted 'to legislation which deliberately seeks to regulate the emission of greenhouse gases into the global atmosphere'[188] by (i) directly controlling the greenhouse gas emissions of the heaviest emitters, including the energy sector and large users of energy; (ii) requiring the adoption of energy efficient measures, or (iii) mandating the achievement of renewable energy targets.[189] Whether directed at individuals or corporations, such measures seek to change behaviour.[190]

Regulation falling within the definition of climate law nudges some individual or corporate decisions in certain directions, limits the range of available choices in others, or altogether removes still others. In sum, the regulatory framework captured by 'climate law' forms the inherent choice architecture of private property,[191] aimed directly at the source of control or power over the goods or resources that contribute to climate change.[192] The primary methods of regulation employed by governments worldwide fall into two broad categories: direct regulation of greenhouse gas emissions[193] and indirect methods of achieving reductions through market-based schemes that allow transfers of greenhouse gas emissions, also known as emissions trading schemes or cap and trade schemes.[194] While many examples of both have been proposed, few operational examples exist.[195] The remainder of this section very briefly outlines two such examples.

Direct regulation is found in the *Change and Greenhouse Emissions Reduction Act 2007* (SA). This legislation sets a state target to reduce the emission of the major greenhouse gases[196] by at least 60% to a level equivalent to or less than 40% of 1990 levels by 31 December 2050.[197] It promotes this commitment by setting specific targets for different sectors of the economy, the development of policies and programs to assist in achieving the stated reductions,[198] and reporting on progress made.[199] Specifically, the legislation encourages the efficient use of energy, promotes research and development of technology that may reduce emissions and or remove such gases from the atmosphere, and supports the commercial development of renewable energy sources.[200]

Emissions trading schemes represent the indirect model of climate law; these tend to face substantial difficulties both in implementation and operation.[201] The European Union has had such a scheme since 2003,[202] the United States is currently attempting to implement one,[203] and in 2008 the Australian Federal Government proposed one.[204] Regardless of jurisdiction, such schemes contain similar essentials; the remainder of this section briefly examines those components of the Australian Carbon Pollution Reduction Scheme (CPRS) relevant to the regulation of private property.

The CPRS highlights the difference between the direct regulation seen in legislation like the *South Australian Climate Change and Greenhouse Emissions Reduction Act*

2007 (SA) and the indirect approach taken by trading schemes. The former impose reductions while the latter rely upon the market to achieve that outcome.[205] The CPRS applies to the approximately 1,000 corporations, or 7.6% of all Australian businesses, which account for 25,000 tonnes of carbon pollution annually[206] and cover the range of industry sectors — stationary energy, transport, fugitive emissions, industrial processes, waste and forestry sectors — responsible for all six greenhouse gases under the *Kyoto Protocol*.[207]

The CPRS requires significant emitters of greenhouse gases to acquire a 'carbon pollution permit' for every tonne of greenhouse gas emitted. In addition to this, the scheme provides for (i) the auditing and monitoring of the quantity of emissions produced by firms; (ii) the annual surrender of a permit for every tonne of emissions from every liable firm (the Government will limit the number of permits issued each year to the total carbon cap for the Australian economy); and (iii) a process whereby firms compete to purchase the number of permits they require. According to the theory behind the CPRS, firms that value permits most highly will be prepared to pay most for them, either at auction or on a secondary trading market, while other firms will find it cheaper to reduce emissions than to buy permits.[208]

The CPRS Green and White Papers repeatedly employ the language of choice and decision. In the case of transport, for instance, the Green Paper notes that over the initial three year period of the scheme 'many people will have the opportunity to make decisions — for example, over the purchase of a new car — informed by the longer term implications of the Carbon Pollution Reduction Scheme, with consequential impacts on their future demand for fuel.'[209] The scheme thus confirms the modification and removal of choices normally conferred by private property by providing assistance to both households and businesses in order to adjust to the increased prices associated with affected goods and services.[210]

Both direct regulation aimed specifically at reducing greenhouse gas emissions and schemes which adopt an indirect market-based approach, such as emissions trading, are part of the choice architecture of private property. Yet, having explored its role within climate change, it may already have struck the reader that the very concept of private property may find it difficult to respond to the challenges presented by such global phenomena. The next Part offers some brief reflections on this difficulty and some possible future directions for property theorists to pursue in order to overcome it.

Future Reflections

What has become apparent from the foregoing discussion is that in its conferral and protection of choice, private property creates and maintains unequal distributions of power — choice — in the control and use of resources.[211] While useful in assessing the role of private property in structuring social interaction within nation states, those critiques of the standard liberal conception that see private property as relational nonetheless struggle to respond to those relationships that have extra-jurisdictional reach — those that extend beyond boundaries founded upon ideas of national sovereignty. Climate change, as we have seen, establishes and is a part of such relationships. This failing is due largely to the fact that the very legal concepts at which these critiques take aim — concepts such as private property — developed largely to account for and explain 'the municipal law of sovereign states, mainly those in advanced industrial societies'.[212] William Twining writes:

> most of the leading Western jurists of the twentieth century have focused very largely on municipal state law, have had strong conceptions of sovereignty, and have assumed that legal systems and societies can be treated as discrete, largely self-contained units. They have either articulated or assumed that jurisprudence and the discipline of law is or should be concerned only with two kinds of law: the domestic municipal law of nation states and of public international law.[213]

Not only do the concepts themselves remain focused on domestic municipal law, but so, too, do the theories that attempt to critique them.[214] In the case of private property, such critiques fail to address the way in which concentrations of power or choice in the control and use of resources not only result in inequalities within the nation state, but also among them, allowing for a small number of people within an even smaller number of selected states negatively to affect vast numbers of others beyond those boundaries. Climate change, seen through the lens of private property, sets this failure squarely before us.

As we have seen, climate change is a truly global phenomena — it involves *global* interaction and interdependence which concerns *all* humankind.[215] The theory of private property, however, rarely explores such extra-jurisdictional choice and relationship — decisions or choices taken by one who holds private property visit their consequences not only on those within the intra-jurisdictional relationships that instantiate private property within a municipal state, but also on those without, forming what we might refer to as inter-jurisdictional relationships. Seen in the global context of climate change, then, private property begins to look rather asymmetrical.[216]

What do we mean by 'asymmetry'? Simply this: the nature of choices as matched against externalities involved in climate change are not limited by national boundaries, as assumed by both the choice and relationship components of the concept of private property. Rather, private property makes possible choices that have consequences far beyond the jurisdictional limits of the systems that created the possibility of such choices. Combined with the fact that the largest holdings of private property are concentrated in western, industrialised, or industrialising nations, and those most vulnerable to and affected by the consequences of the choices made by those who hold private property are those living in the underdeveloped and third world,[217] renders misguided any focus on physical borders and national boundaries as in some way capable of limiting the reach of the consequences of private property choices. Thus, asymmetry refers to that situation which pertains when the consequences of private property choices are visited on those beyond the jurisdictional reach of the legal structures that created, conferred and protected those choices. This is the case when faced with any global phenomena; here we are concerned with anthropogenic climate change.

What follows are four very brief reflections on how property theory may in future seek to address this asymmetry.[218] No attempt is made here fully to explicate these ideas; they are merely intended to be programmatic, laying the foundation for future work. The first reflection is this: the concept of private property could do with some of what Twining refers to as 'belief pluralism'.[219] In a global context, any focus that applies merely to a particular society and its legal culture is unhelpful. Rather, Twining argues, an 'overlapping consensus' is necessary in order to take account of a diversity of belief systems, traditions, and cultures.[220] If private property is relational, it must be possible to substantiate that claim in the only society that really matters, the global society.[221] This may, of course, require shifting the focus of private property away from sovereignty, jurisdiction, and boundaries founded thereon onto the sources and uses of power in a normative and legally plural environment. Using a consciously 'spatial' approach, such a shift is already happening at the jurisdictional level;[222] if private property as a concept is to have meaning in the global context, this may also need to happen at the global level.[223]

Second, combating asymmetry may require a move beyond domestic justice and human rights discourse. Until recently, political liberalism assessed justice in terms of that which could be effected within societies conceived as clearly bounded units. In a global, interdependent world, however, there are no self-contained national societies, and the only closed social system is humanity at large. Thus, a concept of justice and human rights discourse itself must be founded upon background rules or principles formulated at the global level,[224] with private property a key component of achieving that end.[225] Again, the role of space as a concept may play a useful role here, allowing for some account to be taken of and to strive for 'spatial justice.'[226]

Third, a model of private property that takes account of the global consequences of climate change ought at least to account for what Twining calls an ecocentric focus. Anthropocentric actions are those the reasons for which are the provision of a benefit to human beings, while ecocentric ones are those the reason for which is the provision of a benefit to the environment. Twining argues that while most canonical jurists are not indifferent to environmental concerns and do not treat ecocentric reasons as invalid, typically they seem to be anthropocentric in their focus.[227]

Finally, each of these reflections may themselves challenge and so require a reshaping of our very understandings of citizenship and democracy. David Kennedy argues, in the public international law context, that this may involve:

> carry[ing] the revolutionary force of the democratic promise — of individual rights, of economic self-sufficiency, of citizenship, of community empowerment, and participation in the decisions that affect one's life — to the sites of global and transnational authority, however local they may be. To multiply the sites at which decisions could be seen and contested, rather than condensing them in a center, in the hope for a heterogeneity of solutions and approaches and a large degree of experimentation, rather than an improved constitutional process or more stable settlement.[228]

Kennedy offers four examples of what this might mean: experimentation and institutional diversity; mobility (of persons, goods and capital, and national and global citizenship); transnational political will or a sovereignty of open-ended inclusion, both of states and of citizens; and the empowerment and decision-making ability of citizens, polling data serving as the baseline for expert management and international policy juries deciding on issues such as war or peace, poverty or not.[229] I do no more than reiterate these possibilities here. The point is simply that citizenship and democracy itself may need to be reconceived as part of reformulating the concept of private property in an attempt to respond to global challenges like climate change.

Conclusion

This chapter took the position that climate change is a private property problem. It argued that the standard liberal conception of private property — a bundle of rights in or to things — could also be seen as the creation, conferral and protection of individual or group choice in relation to the control and use of goods and resources. Seen this way, most choices made about the use of goods and resources, in one way or another, lie behind climate change.

Such choices, however, are not made in a vacuum; rather, they occur in a web of social relationships. This important legal realist insight reveals that the choices responsible for anthropogenic climate change therefore have negative consequences, or externalities, for the wider community. Generally, private property, as a concept and as a legal institution, attempts to control externalities through the use of regulation. And this technique is the motivating force behind climate law, which defines both direct and indirect attempts to mitigate the effects of climate change.

Seeing climate change through the lens of private property, though, reveals a difficulty: the concentration of choice in a small number of people in an even smaller numbers of sovereign states. Thus, a small number of people in a small number of states make choices that impact a large number of largely poor people the world over. David Lametti refers to this imbalance as the asymmetry of private property, and the essay concluded with four brief thoughts about how future property theory may go about responding to this difficulty. These reflections, set out in no more than an outline for use another day, largely involve shifting our focus away from sovereignty, jurisdiction, and boundaries, towards an overlapping consensus of beliefs, spatial justice, ecocentrism and a reformulation of the very meaning of citizenship and democracy.

Endnotes

1   Niels Röling, 'A Proper Study of Mankind …', *New Scientist* (17 January 2009) 40.

2   Ibid.

3   Although see Endre Stavang, 'Property in Emissions? Analysis of the Norwegian GHG ETS with references also to the UK and EU' (2005) 17 *ELM* 209. See also David E Adelman and Kirsten H Engel, 'Reorienting State Climate Change Policies to Induce Technological Change' (2008) 50 *Arizona Law Review* 835.

4   Clearly, other forms of property — state or public — in both capitalist and non-capitalist systems also pay a role in environmental harm. This chapter, however, limits itself to the role of private property.

5   See Joseph William Singer, *How Property Norms Construct the Externalities of Ownership* (Harvard Public Law Working Paper No 08-06, Harvard Law School, 2008), 2–4.

6   *United Nations Framework Convention on Climate Change*, opened for signature 4 June 1992, 1771 UNTS 107 (entered into force 21 March 1994) ('*UNFCCC*') art 1. The *Kyoto Protocol* was intended to implement reductions based upon the *UNFCCC*: *Kyoto Protocol* to the United Nations Framework Convention on Climate Change, opened for signature 16 March 1998, 2303 UNTS 148 (entered into force 16 February 2005) ('*Kyoto Protocol*') See also Kirstin Dow and Thomas E Downing, *The Atlas of Climate Change: Mapping the world's greatest challenge* (2006) 14.

7   Dow and Downing, above n 6, 14–15.

8   Ibid 15.

9   This is fully canvassed in United Nations Intergovernmental Panel on Climate Change (IPCC), *Climate Change 2007 — The Physical Science Basis: Working Group I Contribution to the Fourth*

Assessment Report (2007). See also IPCC, *Climate Change 2007 — Synthesis Report (2007)*; John Houghton, *Global Warming: The complete briefing* (3rd ed, 2004).

10 Houghton, above n 9, 14–15.

11 Ibid 16.

12 IPCC, *Climate Change 2007 — The Physical Science Basis*, above n 9; IPCC, *Climate Change 2007 Impacts, Adaptation and Vulnerability: Working Group II contribution to the Fourth Assessment Report* (2007); IPCC, *Climate Change 2007 — Mitigation of Climate Change: Working Group III Contribution to the Fourth Assessment Report* (2007); IPCC, *Climate Change 2007 — Synthesis Report*, above n 9; Al Gore, *An Inconvenient Truth: The planetary emergency of global warming and what we can do about it* (2006); Al Gore, *Earth in the Balance: Forging a new common purpose* (1992); Houghton, above n 9, 16. See also Steve Lonergan, 'The Human Challenges of Climate Change' in Harold Coward and Andrew J Weaver (eds), *Hard Choices: Climate change in Canada* (2004) 45, 47–9.

13 Houghton, above n 9, 23–4.

14 Röling, above n 1, 40.

15 Ibid 41.

16 See Andreas Kalyvas and Ira Katznelson, *Liberal Beginnings: Making a Republic for the Moderns* (2008) passim; Paul W Kahn, *Putting Liberalism in its place* (2005) 30; Stephen L Carter, 'Liberal Hegemony and Religious Resistance: An essay on legal theory' in Michael W McConnell, Robert F Cochran and Angela C Carmella (eds), *Christian Perspectives on Legal Thought* (2001) 25, 47–9; Charles Taylor, *A Secular Age* (2007) passim; Richard H Thaler and Cass R Sunstein, *Nudge: Improving decisions about health, wealth, and happiness* (2008) 4–14. On the complexity of the relationship between private property and freedom see Jedediah Purdy, 'A Freedom-Promoting Approach to Private Property: A renewed tradition for new debates' (2005) 72 *University of Chicago Law Review* 1237.

17 Jeremy Waldron, *The Right to Private Property* (1988) 31–40.

18 See Waldron, above n 17; Stephen R Munzer, *A Theory of Property* (1990); Margaret Jane Radin, *Reinterpreting Property* (1993); Singer, above n 5, 7–11; Eric T Freyfogle, 'Property and Liberty' (Illinois Public Law Research Paper No 07–09, University of Illinois College of Law, 2007).

19 Jeremy Bentham, *The Theory of Legislation* (1931) 113.

20 Wesley Newcomb Hohfeld, 'Some Fundamental Legal Conceptions as Applied in Judicial Reasoning' (1913) 23 *Yale Law Journal* 16; (1917) 26 *Yale Law Journal* 710; Wesley Newcomb Hohfeld, *Fundamental Legal Conceptions as Applied in Judicial Reasoning* (1919); Wesley Newcomb Hohfeld, *Fundamental Legal Conceptions as Applied in Judicial Reasoning* II (1923).

21 AM Honoré, 'Ownership' in AG Guest (ed), *Oxford Essays in Jurisprudence* (1961) 107–47.

22 See JE Penner, 'The "Bundle of Rights" Picture of Property' (1996) 43 *UCLA Law Review* 711, 712–13, who argues that the subsequent combination by a line of commentators of Hohfeld's jural relations and Honoré's incidents of ownership produced the bundle metaphor or picture. And see Elizabeth M Glazer, Response, 'Rule of (Out)Law: Property's Contingent Right to Exclude' (2008) 156 *University of Pennsylvania Law Review PENNumbra* 331 <http://www.pennumbra.com/responses/01–2008/Glazer.pdf>.

23 The eleven standard incidents of ownership are: possession, use, manage, income, capital, security, transmissibility and absence of term, duty to prevent harm, liability to execution, and the incident of residuarity: Honoré, above n 21.

24 Private property is 'that sole and despotic dominion which one man claims and exercises over the external things of the world, in total exclusion of the right of any other individual in the universe.': William Blackstone, *Commentaries on the Laws of England: Volume II, Of the Rights of Things* (1766) University of Chicago Press (1979) 2 (emphasis added). See also David B Schorr, 'How Blackstone Became a Blackstonian' (2009) 10 *Theoretical Inquiries in Law* 103; Munzer, above n 18, 17; Stephen R Munzer, 'Property as Social Relations' in Stephen R Munzer (ed), *New Essays in the Legal and Political Theory of Property* (2001) 36, 36.

Blackstone has held a particularly strong hold over the American legal mind. See Daniel J Boorstin, *Mysterious Science of the Law: An Essay on Blackstone's Commentaries* (1941); Dennis R Nolan, 'Sir William Blackstone and the New American Republic: A study of intellectual impact' (1976) 51 *New York University Law Review* 731; Julian S Waterman, 'Thomas Jefferson and Blackstone's Commentaries' (1932) 27 *Illinois Law Review* 629; William Blake Odgers, 'Sir William Blackstone

(Part 1)' (1918) 27 *Yale Law Journal* 599; William Blake Odgers, 'Sir William Blackstone (Part 2)' (1919) 28 *Yale Law Journal* 542.

In relation to property, many have restated and critiqued the 'Blackstonian' view. See Robert P Burns, 'Blackstone's Theory of the "Absolute" Rights of Property' (1985) 54 *University of Cincinnati Law Review* 67; Schorr, above; Robert W Gordon, 'Paradoxical Property' in John Brewer and Susan Staves (eds), *Early Modern Conceptions of Property* (1995) 95; Carol M Rose, 'Canons of Property Talk, or, Blackstone's Anxiety' (1998) 108 *Yale Law Journal* 601, 603; Robert C Ellickson, 'Property in Land' (1993) 102 *Yale Law Journal* 1315, 1362, n 237; Frederick H Whelan, 'Property as Artifice: Hume and Blackstone' in J Roland Pennock and John W Chapman (eds), *Nomos XXII: Property* (1980) 101, 114–25; Michael A Heller, 'Critical Approaches to Propert Institutions' (2000) 79 *Oregon Law Review* 417, 418–19, who cites Frank I Michelman, 'Ethics, Economics, and the Law of Property' in J Roland Pennock and John W Chapman (eds), *Nomos XXIV: Ethics, Economics, and the Law* (1982) 3, 5; Jeremy Waldron, 'Property Law' in Dennis Patterson (ed), *A Companion to Philosophy of Law and Legal Theory* (1996) 6; Thomas W Merrill and Henry E Smith, 'What Happened to Property in Law and Economics?' (2001) 111 *Yale Law Journal* 357, 360–6.

25 Some would say disintegrated. See Thomas C Grey, 'The Disintegration of Property' in *NOMOS XXII: Property,* above n 24, 69.

26 Joseph William Singer, 'Property and Social Relations: From Title to Entitlement' in Charles Geisler and Gail Daneker (eds), *Property and Values: Alternatives to public and private ownership* (2000) 3, 4.

27 On the idea that rights are the background knowledge of modern property, see Robert C Ellickson, 'Unpacking the Household: Informal property rights around the hearth' (2006) 116 *Yale Law Journal* 226, 236–40. And on the extent of informational knowledge about the rules of property, see Henry E Smith, 'The Language of Property: Form, context and audience' (2003) 55 *Stanford Law Review* 1105.

28 See Laura S Underkuffler, *The Idea of Property: Its meaning and power* (2003) 13; Lawrence C Becker, *Property rights: Philosophic foundations* (1977) 11–4, 21–2; Munzer, above n 18, 17–27; Gerald F 'Gaus, Property, Rights, and Freedom', in Ellen Frankel Paul, Fred D Miller Jr and Jeffrey Paul (eds), *Property Rights* (1994) 209, 212–25; Penner, above n 22.

29 Radin, above n 18, 121–3. For a full account of the liberal conception of private property, see Waldron, above n 17; Munzer, above n 18; Radin, above n 18.

30 Munzer, above n 18, 22–36; Munzer, above n 24, 36; Heller, above n 24, 418–19; Michael A Heller, 'The Boundaries of Private Property' (1999) 108 *Yale Law Journal* 1163, 1191–2; Penner, above n 22, 713, n 8; Michelman, above n 24, 5; Waldron, above n 17, 26–61; Becker, above n 28, 11–21; John Christman, *The Myth of Property: Toward an Egalitarian Theory of Ownership* (1994) 3–27; Underkuffler, above n 28, 19.

31 Heller, above n 30, 1191–4. See also Ellickson, above n 24; Smith, above n 27.

32 Radin, above n 18, 121–3. On the issue of essential rights, see Thomas W Merrill, 'Property and the Right to Exclude' (1998) 77 *Nebraska Law Review* 730, 734–5. See also Lior Jacob Strahilevitz, 'Information Asymmetries and the Right to Exclude' (2006) 104 *Michigan Law Review* 1835.

33 Singer, above n 5, 7–11.

34 Singer, above n 26, 8–10.

35 Sally Falk Moore, *Law as Process: An Anthropological Approach* (1978) 70. And see Cass R Sunstein, *The Second Bill of Rights: FDR's unfinished revolution and why we need it more than ever* (2004).

36 C Edwin Baker, 'Property and its Relation to Constitutionally Protected Liberty' (1986) 134 *University of Pennsylvania Law Review* 741, 743. See also Joseph William Singer, 'The Reliance Interest in Property' (1988) 40 *Stanford Law Review* 611, 655; Joseph William Singer, 'Sovereignty and Property' (1991) 86 *Northwestern University Law Review* 1, 2, 15–16; Joseph William Singer, *The Edges of the Field: Lessons on the obligations of ownership* (2000) 18–37; Joseph William Singer, *Entitlement: The paradoxes of property* (2000) 134–9.

37 David Lametti, 'The Concept of Property: Relations Through Objects of Social Wealth' (2003) 53 *University of Toronto Law Journal* 325, 346.

38 This is clear from David Lametti's seminal work, above n 37. See also David Lametti, 'The Morality of James Harris's Theory of Property' in Timothy Endicott, Joshua Getzler and Edwin Peel (eds), *Properties of Law: Essays in Honour of Jim Harris* (2006) 138, 159–62. And see Becker, above n 28, 1.

39 Thaler and Sunstein, above n 16, 5.

40 Ibid (emphasis added).

41 This begins with John Stuart Mill's 'self-regarding act': John Stuart Mill, *On Liberty* (Gertrude Himmelfarb ed, 1974 (1859)). See especially: Singer, above n 5, 7–11 (outlining how property norms assist in determining the difference between a truly self-regarding act and when not); Munzer, above n 18, 3–9; Gregory S Alexander, 'Property as Propriety' (1998) 77 *Nebraska Law Review* 667, 699; JW Harris, *Property and Justice* (1996) 29, 31, 105; Singer, *Entitlement*, above n 36, 30.

42 Harris, above n 41, 29.

43 Ibid 31.

44 See Singer, above n 5.

45 Thaler and Sunstein, above n 16, 4–14; Stephen L Carter, above n 16, 45–9.

46 Other greenhouse gases are of minimal significance, although their greenhouse effect is likely to increase in the coming decades; these include hydrochlorofluorocarbons (HCFC), hydrofluorocarbons (HFC), perflurocarbons, and sulphur hexafluoride: Houghton, above n 9, 46–7.

47 See Peter Aldhous, 'Genes for Greens: A new generation of genetically modified crops could reduce greenhouse gas emissions more than grounding all the aircraft in the world', *NewScientist* (5 January 2008) 28, 31. While other gases attributable to human activities influence the enhanced greenhouse effect through their chemical action on the major greenhouse gases, their overall effect is much less than that of the major direct anthropogenic contributors: Houghton, above n 9, 47–8.

48 Houghton, above n 9, 29–30.

49 Ibid 39–42.

50 On energy and transport, see Houghton, above n 9, 268–321.

51 Aldhous, above n 47, 31.

52 Houghton, above n 9, 42. See also Aldhous, above n 47, 28–31.

53 Houghton, above n 9, 44.

54 Ibid.

55 Aldhous, above n 47, 29.

56 Ibid 31; Houghton, above n 9, 43, Table 3.2.

57 Houghton, above n 9, 43.

58 Ibid 44.

59 Ibid.

60 Aldhous, above n 47, 29.

61 See eg Houghton, above n 9; IPCC, *Climate Change 2007 — The Physical Science Basis*, above n 9; IPCC, *Climate Change 2007 — Impacts, Adaptation and Vulnerability*, above n 12; IPCC, *Climate Change 2007 — Mitigation of Climate Change*, above n 12; IPCC, *Climate Change 2007 — Synthesis Report*, above n 9; Gore, *An Inconvenient Truth*, above n 12; Gore, *Earth in the Balance*, above n 12; Coward and Weaver, above n 12.

62 'The Global Warming Survival Guide: 51 Things You Can Do to Make a Difference' *TIME Magazine* (9 April 2007), 47–68. See also *TIME Magazine: Global Warming* (2007); Peter Miller, 'Saving Energy: It Starts at Home' *National Geographic* (March 2009) 60; Dave Reay, 'Your Top 10 Ways to Take on Global Warming', *New Scientist* (10 September 2005) 36.

63 'The Global Warming Survival Guide', above n 62.

64 Ibid 58 '22 Skip the Steak', 59 '25 Support Your Local Farmer'.

65 Harris, above n 41, 47–50.

66 'The Global Warming Survival Guide', above n 62, 48 '3 Change Your Lightbulbs'.

67 Ibid 57 '20 Check the Label'.

68 Ibid 50 '7 Hang Up a Clothesline'.

69 Ibid 56 '18 Ask the Experts for an Energy Audit of your Home'.

70 Ibid 57 '21 Cosy Up to Your Water Heater'.

71 Ibid 56 '17 Open a Window'.

72 Ibid 61 '30 Shut Off Your Computer', '32 Kill the Lights at Quitting Time'.

73 Ibid 57 '19 Buy Green Power, at Home or Away'.

74 Ibid 50 '6 Ditch the McMansion'.

75 Ibid 48 '2 Get Blueprints for a Green House'.

76 Ibid 68 '50 Be Aggressive About Passive'.

77 Ibid 54 '15 Move to a High-Rise'.

78 Ibid.

79 Ibid 54 '14 Ride the Bus'.

80 Ibid 65 '45 Make One Right Turn After Another'.

81 Ibid 48 '1 Turn Food into Fuel'. While there has recently been some controversy over the use of food for fuel, increasingly it is becoming possible to use waste products to produce green fuels.

82 Ibid 58 '23 Copy California'.

83 Ibid 64 '41 Fill 'er up with Passengers'.

84 Ibid 65 '44 Check Your Tires'.

85 Ibid 66 '48 Drive Green on the Scenic Route'.

86 Ibid 59 '27 Straighten Up and Fly Right'.

87 Ibid 51 '8 Give New Life to Your Old Fleece'.

88 Ibid 52 '11 Take Another Look at Vintage Clothes'.

89 Ibid 56 '16 Pay Your Bills Online'.

90 Ibid 59 '25 Support Your Local Farmer'.

91 Ibid 58 '24 Just Say No to Plastic Bags'.

92 Ibid 60 '28 Have a Green Wedding'.

93 Ibid 61 '31 Wear Green Eye Shadow'.

94 Ibid 62 '34 Rake in the Fall Colors'.

95 Ibid 62 '35 End the Paper Chase'.

96 See Harris, above n 41, 100–2, outlining the differences between private property held by individuals and that held by corporations. For present purposes, however, the issue of choice as the core of private property, held by individuals or corporations, is unaffected.

97 'The Global Warming Survival Guide', above n 62, 57 '19 Buy Green Power, At Home Or Away'.

98 Ibid 48 '1 Turn Food into Fuel'.

99 Ibid 53 '12 Capture the Carbon'.

10 Ibid 53 '12 Capture the Carbon'.

101 Ibid 52 '9 Build a Skyscraper'.

102 Ibid 54 '13 Let Employees Work Close to Home'.

103 Singer, above n 5, 3.

104 See, eg, Richard Epstein, *Takings, Private Property and the Power of Eminent Domain* (1985) and Radin's analysis, above n 18, 121–3. For a history of this approach, see Singer, above n 5.

105 This is based on Radin, above n 18, 121–3, and her analysis of Epstein, above n 104, 22–9, 58–73, 112–21, 195–215, 230–1, 252–3, 304, 324–9. See also Richard Epstein, 'Why Restrain Alienation?' (1985) 85 *Columbia Law Review* 970. For a vigorous critique, see Thomas C Grey, 'The Malthusian Constitution' (1986) 41 *University of Miami Law Review* 21.

106 Radin, above n 18, 123. See also Singer, above n 5.

107 Singer, above n 5; Lametti, above n 37, 342–7; David Lametti, 'Property and (Perhaps) Justice. A Review Article of James W Harris, *Property and Justice*, and James E Penner, *The Idea of Property in Law*' (1998) 43 *McGill Law Journal* 663, 670–2.

108 There has yet to be any example in the history of human society where William Blackstone's 'sole and despotic dominion' described the on-the-ground distribution of resources or social wealth; see Gordon, above n 24, 95; Rose, above n 24, 603; Ellickson, above n 24, 1362, n 237; Whelan, above n 24, 114–25; Heller, above n 24, 419. Even the Romans — to whom the notion of absolute dominium in things is often attributed — did not in practice recognise such a possibility: Joshua Getzler, 'Roman Ideas of Landownership' in Susan Bright and John Dewar (eds), *Land Law: Themes and perspectives* (1998) 81–106. The limitations can be found in different sources — see for instance the 'Aristotelian' understanding of the ownership as being for the common good: Ronald J Colombo, 'Ownership, Limited: Reconciling traditional and progressive corporate law via an Aristotelian understanding of ownership' (2008) 34 *Journal of Corporation Law* 247.

109 Schorr, above n 24, at Abstract, who calls Blackstone the 'avatar of absolut[ism]'.

110 William Blackstone, *Commentaries on the Laws of England: Volume I, Of the Rights of Persons* (1765) (ed, 1979) 138 (emphasis added). Blackstone went on to explain the many ways in which fetters

were placed upon private property in 18th century England: William Blackstone, *Commentaries on the Laws of England: Volume III, Of Private Wrongs* (1768) (University of Chicago Press, 1979) 212–15, 217–18. But Book II itself was an account of English real property law that 'took pains to point out both that this right of "property in its highest degree [the fee simple]" was always "held of some superior, on condition of rendering him service; in which superior or ultimate property of the land resides," and that lesser interests were frequently vested in some other person or persons ... Not only did "absolute" ownership not exist in England, it was hardly discussed even as a mythological ideal type.': Schorr, above n 24, 107.

111 Burns, above n 24, 69.

112 Gregory S Alexander, *Commodity & Propriety: Competing visions of property in American legal thought, 1776–1970* (1997) 321 (emphasis added).

113 As was the case for Blackstone: see Burns, above n 24.

114 This view can be traced to the father of the bundle concept, Wesley Newcomb Hohfeld (1913), above n 20; (1917), above n 20; (1919), above n 20; (1923), above n 20.
Hohfeld's thinking was subsequently taken up by Robert L Hale, 'Coercion and Distribution in a Supposedly Non-Coercive State' (1923) 38 *Political Science Quarterly* 470; Morris R Cohen, 'Property and Sovereignty' (1927) 13 *Cornell Law Quarterly* 8; Robert L Hale, 'Bargaining, Duress, and Economic Liberty' (1943) 43 *Columbia Law Review* 603; Felix S Cohen, 'Dialogue on Private Property' (1954) 9 *Rutgers Law Review* 357.
Contemporary scholars, especially those of the Critical Legal Studies movement, have extensively developed and expanded the early realist work on property: see CB Macpherson, 'The Meaning of Property' and 'Liberal-Democracy and Property' in CB Macpherson (ed), *Property: Mainstream and critical positions* (1978) 1, 199; CB Macpherson, 'Capitalism and the Changing Concept of Property' in Eugene Kamenka and RS Neale (eds), *Feudalism, Capitalism and Beyond* (1975) 104; Jennifer Nedelsky, 'Law, Boundaries, and the Bounded Self' (1990) 30 *Representations* 162; Jennifer Nedelsky, 'Reconceiving Autonomy: Sources, thoughts and possibilities' (1989) 1 Yale *Journal of Law and Feminism* 7; Jennifer Nedelsky, 'Reconceiving Rights as Relationship' (1993) 16 *Review of Constitutional Studies/Revue d'études constitutionelles* 1; Duncan Kennedy, 'The Stakes of Law, or Hale and Foucault!' (1991) 15 *Legal Studies Forum* 327; Singer, *Edges of the Field*, above n 36; Singer, *Entitlement*, above n 36; Joseph William Singer, *Introduction to Property* (2nd ed, 2005); Joseph William Singer, 'The Legal Rights Debate in Analytical Jurisprudence from Bentham to Hohfeld' [1982] *Wisconsin Law Review* 975; Singer, 'Reliance Interest', above n 36; Joseph William Singer, 'Re-Reading Property' (1992) 26 *New England Law Review* 711; Joseph William Singer and Jack M Beermann, 'The Social Origins of Property' (1993) 6 *Canadian Journal of Law and Jurisprudence* 217; Singer, 'Sovereignty and Property', above n 36; Carol M Rose, *Property & Persuasion: Essays on the history, theory, and rhetoric of ownership* (1994); Baker, above n 36; Underkuffler, above n 28; Laura S Underkuffler, 'On Property: An essay' (1990) 100 *Yale Law Journal* 127. This view is not, of course, without its critics; see Munzer, above n 18.

115 Morris R Cohen, above n 114, 8.

116 Felix S Cohen, above n 114, 373 (emphasis added).

117 Joseph William Singer, the modern exemplar of social relations, offers this succinct summary: '[p]roperty concerns legal relations among people regarding control and disposition of valued resources. Note well: Property concerns relations *among* people, not relations between people and things.': Singer, *Introduction to Property,* above n 114, 2 (footnote omitted and emphasis in the original). See also Macpherson, 'The Meaning of Property' and 'Liberal-Democracy and Property', above n 114; Macpherson, 'Capitalism and the Changing Concept of Property', above n 114, 104–24; Nedelsky, 'Law, Boundaries, and the Bounded Self', above n 114; Nedelsky, 'Reconceiving Autonomy', above n 114; Nedelsky, 'Reconceiving Rights as Relationship', above n 114; Kennedy, above n 114; Singer, *The Edges of the Field*, above n 36; Singer, *Entitlement*, above n 36; Singer, 'The Legal Rights Debate', above n 114; Singer, 'Reliance Interest', above n 36; Singer, 'Re-Reading Property', above n 114; Singer and Beerman, above n 114; Singer, 'Sovereignty and Property', above n 36; Rose, above n 114; Baker, above n 36; Underkuffler, above n 28; Underkuffler, above n 114.

118 Baker, above n 36, 744.

119 Underkuffler, above n 114 128; Harris, above n 41; Munzer, above n 18; Radin, above n 18, 9 and *passim*; JE Penner, *The Idea of Property in Law* (1997); Waldron, above n 17. For more recent acceptance of the relevance of social context, see Alexandra George, 'The Difficulty of Defining "Property"' (2005) 25 *Oxford Journal of Legal Studies* 793; Daniel Fitzpatrick, 'Evolution and Chaos

in Property Right Systems: The Third World tragedy of contested access' (2006) 115 *Yale Law Journal* 996. On the fact that most of these scholars, in one way or another, adopt a bundle of rights approach, whether they say so or not, see Lametti, above n 107.

120 Merrill, above n 32, 737–9. This can be traced to the seminal work of Grey, above n 25. See also Christman, above n 30, 20; Bruce A Ackerman, *Private Property and the Constitution* (1977) 26–9; Gordon, above n 24, 95–110. But see Waldron, above n 17, 47–53.

121 Heller, above n 24, 432.

122 Nedelsky, 'Reconceiving Rights as Relationship', above n 114, 8.

123 Singer, 'Reliance Interest', above n 36, 618–21 and 751.

124 Singer, *Entitlement*, above n 36, 56–139.

125 Nedelsky, 'Reconceiving Rights as Relationship', above n 114, 7–8.

126 Kalyvas and Katznelson, above n 16, *passim*; Kahn, above n 16, 37–65.

127 See Jennifer Nedelsky, *Private Property and the Limits of American Constitutionalism: The Madisonian framework and its legacy* (1990); Alan Freeman and Elizabeth Mensch, 'Property' in Jack P Greene and JR Pole (eds), *The Blackwell Encyclopedia of the American Revolution* (1991) 620, 620–8; Underkuffler, above n 114, 133–42; Gordon, above n 24, 95–110.

128 Kalyvas and Katznelson, above n 16, 1–17.

129 See Singer, *Entitlement*, above n 36, 130–1; Baker, above n 36.

130 Singer, *Entitlement*, above n 36, 131.

131 Ibid.

132 Ibid, citing Nedelsky, 'Reconceiving Rights as Relationship', above n 114, 8.

133 Underkuffler, above n 114, 129.

134 Singer, *Entitlement*, above n 36, 131 (footnotes omitted).

135 See Colombo, above n 108.

136 See Margaret Jane Radin, 'The Liberal Conception of Property: Crosscurrents in the jurisprudence of takings' (1988) 88 *Columbia Law Review* 1667, republished as 'The Liberal Conception of Property: Crosscurrents in the Jurisprudence of Takings' in Radin, above n 18, 120–45. On this model, see also William H Simon, 'Social-Republican Property' (1991) 38 *UCLA Law Review* 1335; Underkuffler, above n 114; Frank I Michelman, 'Possession vs Distribution in the Constitutional Idea of Property' (1987) 72 *Iowa Law Review* 1319.

137 Joseph William Singer, 'The Ownership Society and Takings of Property: Castles, investments, and just obligations' (2006) 30 *Harvard Environmental Law Review* 309.

138 Singer, above n 5.

139 Singer, above n 137.

140 Singer, *Entitlement*, above n 36, 131, citing Nedelsky, 'Reconceiving Rights as Relationship', above n 114, 8.

141 Singer, above n 5, 3.

142 Singer, *Entitlement*, above n 36, 6; Singer, above n 139, 139–40.

143 Singer, above n 5, 3.

144 Ibid.

145 See Amnon Lehavi, 'How Property Can Create, Maintain, or Destroy Community' (2009) 10(1) *Theoretical Inquiries in Law* <http://www.bepress.com/til/>.

146 These consequences are well-documented; see IPCC, *Climate Change 2007 — The physical science basis*, above n 9; IPCC, *Climate Change 2007 — Impacts, adaptation and vulnerability*, above n 12; IPCC, *Climate Change 2007 — Mitigation of climate change*, above n 12; IPCC, *Climate Change 2007 — Synthesis report*, above n 9; Gore, *An Inconvenient Truth*, above n 9; Sir Nicholas Stern, *Stern Review on the Economics of Climate Change* (2006); Ross Garnaut, *The Garnaut Climate Change Review — Final report* (2008); Andrew J Weaver, 'The Science of Climate Change' in Coward and Weaver, above n 12, 13–43, 25, Figure 2.8.

147 Lonergan, above n 12, 51. See also IPCC, *Climate Change 2007 — Impacts, adaptation and vulnerability*, above n 12; Garnaut, above n 146, 120–52.

148 Lonergan, above n 12, 51. See also Garnaut, above n 146, 145–6.

149 Lonergan, above n 12, 51. See also IPCC, *Climate Change 2007 — Impacts, Adaptation and Vulnerability*, above n 12; Garnaut, above n 146, 120–52.

150 Michael S Northcott, *A Moral Climate: The Ethics of Global Warming* (2007) 29–31; Garnaut, above n 146, 147–50.

151 Lonergan, above n 12, 51–3; Garnaut, above n 146, 147–50.

152 Lonergan, above n 12, 51–3; Garnaut, above n 146, 147–50.

153 See Stephan Faris, Forecast: *The Consequences of Climate Change, from the Amazon to the Arctic, from Darfur to Napa Valley* (2008).

154 IPCC, *Climate Change 2001 — Impacts, Adaptation and Vulnerability: Working Group II Contribution to the Third Assessment Report* (2001) 7.

155 Lonergan, above n 12, 52.

156 Ibid 52–3.

157 Ibid 53; Paul R Epstein, 'Climate Change and Human Health' (2005) 353 (14) *New England Journal of Medicine* 1433.

158 This list has been compiled from Epstein, above n 157, 1433–5.

159 Lonergan, above n 12 at 53; Epstein, above n 157.

160 This list has been compiled from Epstein, above n 157, 1435–6. See also Garnaut, above n 146, 147.

161 For a list of specific factors and the diseases that might be affected see Lonergan, above n 12, 54.

162 IPCC, *Climate Change 2001 — Impacts, adaptation and vulnerability,* above n 154, 9–11; Lonergan, above n 12, 53

163 IPCC, *Climate Change 2001 — Impacts, adaptation and vulnerability,* above n 154, 9–11.

164 Ibid.

165 This list has been compiled from Epstein, above n 157, 1435–6.

166 For a general account, see Douglas Brinkley, *The Great Deluge: Hurricane Katrina, New Orleans, and the Mississippi Gulf Coast* (2006); Mike Tidwell, *The Ravaging Tide: Strange Weather, Future Katrina, and the Coming Death of America's Coastal Cities* (2006). See also Joseph William Singer, 'After the Flood: Property and Equality in Property Regimes' (2006) 52 *Loyola Law Review* 101.

167 IPCC, *Climate Change 2001 — Impacts, Adaptation and Vulnerability,* above n 154, 9–11; Lonergan, above n 12, 54–5.

168 IPCC, *Climate Change 2001 — Impacts, Adaptation and Vulnerability,* above n 154, 9–11; Lonergan, above n 12, 55–6. See also Garnaut, above n 146, 146.

169 Singer, above n 137, 334, n 82. See also Craig Anthony Arnold, 'The Reconstitution of Property: Property as a Web of Interests' (2002) 26 *Harvard Environmental Law Review* 281.

170 This list is summarised from Reay, above n 62.

171 Dow and Downing, above n 6, 46.

172 Even China, a legally communist country, both politically and economically, is turning increasingly to capitalism and its foundation, private property, in order to fuel an ever-expanding economy: Property Code of the People's Republic of China, 2007. See Lei Chen, 'The New Chinese Property Code: A Giant Step Forward?' 11.2 *Electronic Journal of Comparative Law* (September 2007) <http://www.ejcl.org/112/art112–2.pdf>. Thus, while even 20 years ago it could be said that state or public property made a major contribution to the enhanced greenhouse effect and to climate change — in places such as China, East Germany and the Soviet Union — its dominance is decreasing.

173 See Lehavi, above n 145, 8.

174 See Community and Property, Special Issue (2009) 10(1) *Theoretical Inquiries in Law.*

175 Singer, *Entitlement,* above n 36, 203.

176 Underkuffler, above n 114, 147–8 (emphasis added).

177 Singer, above n 5, 3 (emphasis in original).

178 Lametti, above n 37, 346–8. See also Lametti, above n 107.

179 Singer and Beermann, above n 114, 228.

180 Singer, *Entitlement,* above n 36, 208; Lametti, above n 38, 164; Underkuffler, above n 114, 143–4.

181 See Macpherson, 'Capitalism and the Changing Concept of Property', above n 114, 121.

182 See, eg, Jon Meacham and Evan Thomas, 'We Are All Socialists Now', *Newsweek* (16 February 2009).

183 Gregory S Alexander in Singer, *Entitlement,* above n 36.

184 Singer, *Entitlement*, above n 36, 204. And see Honoré, above n 21, especially regarding the duty to prevent harm and the liability to execution. Singer, *Entitlement*, above n 36, 78–9, makes this point in relation to the United States' system of private property, although it can easily be extended to the legal system of any Western, capitalist, market economy.

185 Singer, *Entitlement*, above n 36, 204 (emphasis added).

186 Ibid 18.

187 Rosemary Lyster, 'Chasing Down the Climate Change Footprint of the Public and Private Sectors: Forces converge — Part I' (Sydney Law School, Legal Studies Research Papers No 08/39, 2008) 4.

188 Ibid 5.

189 Ibid (footnotes omitted).

190 See Julia Black, 'Critical Reflections on Regulation' (2002) 27 *Australian Journal of Legal Philosophy* 1, 12, citing Better Regulation Taskforce, *Principles of Better Regulation* (Cabinet Office, London, undated) 1. See also Thaler and Sunstein, above n 16.

191 Thaler and Sunstein, above n 16.

192 See Grey, above n 25, 69–85.

193 Lyster, above n 187, 6.

194 Ibid. See also Rosemary Lyster, 'Chasing Down the Climate Change Footprint of the Public and Private Sectors: Forces converge — Part II' (Sydney Law School, Legal Studies Research Papers Research Paper No 08/40, 2008); Reuven S Avi-Yonah and David M Uhlmann, 'Combating Global Climate Change: Why a Carbon Tax is a Better Response to Global Warming than Cap and Trade' (University of Michigan Law School Public Law and Legal Theory Working Paper Series, Working Paper No 117, 2008).

195 See Lyster, above n 187, 6–22.

196 *Climate Change and Greenhouse Emissions Reduction Act 2007* (SA) s 4.

197 *Climate Change and Greenhouse Emissions Reduction Act 2007* (SA) s 3(1)(a)(i) and 3(5)(1).

198 *Climate Change and Greenhouse Emissions Reduction Act 2007* (SA) s 3(1)(b).

199 *Climate Change and Greenhouse Emissions Reduction Act 2007* (SA) s 3(1)(i) and (k).

200 *Climate Change and Greenhouse Emissions Reduction Act 2007* (SA) s 3(1)(c), (d) and (e).

201 See Stavang, above n 3.

202 See Lyster, above n 187, 10–11.

203 See Avi-Yonah and Uhlmann, above n 194.

204 Commonwealth Government, *Carbon Pollution Reduction Scheme Green Paper, Full Report and Summary* (July 2008); Commonwealth Government, *Carbon Pollution Reduction Scheme: Australia's Low Pollution Future White Paper* (15 December 2008); Draft Carbon Pollution Reduction Scheme Bill 2009 (Cth).

205 *Carbon Pollution Reduction Scheme Green Paper, Summary,* above n 204, 9.

206 Ibid IV, 19.

207 Ibid 15–9.

208 Ibid 12, Box 1.

209 Ibid, 16–7; *Carbon Pollution Reduction Scheme: Australia's Low Pollution Future White Paper,* above n 204, vol 2, [17.6].

210 *Carbon Pollution Reduction Scheme Green Paper, Summary,* above n 204, 24–31.

211 See Hale, 'Coercion and Distribution', above n 114; Hale, 'Bargaining, Duress, and Economic Liberty', above n 114; Kennedy, above n 114, Sunstein, above n 35, 1–5 and 17–34.

212 William Twining, *Law, Justice and Rights: Some implications of a global perspective* (Working Paper Draft 1/07, UCL Faculty of Law) 4: <http://www.ucl.ac.uk/laws/academics/profiles/index.shtml?twining>

213 Ibid 7–8.

214 For an amusing example of this blindness, see William Twining, *Globalisation and Legal Theory* (2001) 59, n 9, where he discusses his attendance at the 1982 Harvard Critical Legal Studies Conference, where he noted that one of the failings of CLS was its inward-lookingness. While some modern theorists are beginning to reveal the flaws in such thinking in the case of both domestic municipal law — see Twining, above n 212; Twining, above — and public international law — David

Kennedy, 'The Mystery of Global Governance' (Kormendy Lecture, Ohio Northern University, Pettit College of Law, 25 January 2008) (copy on file with the author) — it remains a problem.

215 Twining, above n 212, 2–4.

216 This phrase is adapted from the seminal work of Lametti, above n 37. See also Lametti, above n 38.

217 See United Nations Development Programme, *Human Development Reports*, retrieved 27 March 2009 from <http://hdr.undp.org/en/statistics/>; James B Davies, Susanna Sandström, Anthony Shorrocks and Edward N Wolff, *The World Distribution of Household Wealth* (Discussion Paper No. 2008/03, UNU-WIDER, 2008).

218 The work of William Twining, who has written extensively about private law and general jurisprudence in a global world, forms the foundation of these thoughts: Twining, above n 212; Twining, above n 214; to a lesser degree, so, too, does that of David Kennedy, whose concern lies with public international law: Kennedy, above n 214.

219 Twining, above n 212, 12–13.

220 Ibid 12.

221 Ibid 12–13.

222 See the Stanford Law Review Symposium: Surveying Law and Borders (1996) 48 *Stanford Law Review* No 5, Introduction and 1037–429; Nicholas K Blomley, *Law, Space, and the Geographies of Power* (1994); Nicholas Blomley, David Delaney, and Richard T Ford (eds), *The Legal Geographies Reader* (2001); Andreas Philippopoulos-Mihalopoulos (ed), *Law and the City* (2007).

223 See David Harvey, *Spaces of Global Capitalism: Towards a Theory of Uneven Geographical Development* (2006) 119–48.

224 Twining, above n 212, 13–14.

225 See Harris, above n 41; Singer, *Introduction to Property,* above n 114, 19.

226 On this see Henri Lefebvre, *Writings on Cities/Henri Lefebvre* (1996); Henri Lefebvre, *The Production of Space* (1974, 1991); Henri Lefebvre, *Critique of Everyday Life* — Volume I: Introduction (1991); Henri Lefebvre, *The Survival of Capitalism* (1976); Harvey, above n 223; Edward W Soja, *Postmodern Geographies: The reassertion of space in critical social theory* (1989); Edward W Soja, *Thirdspace: Journeys to Los Angeles and other real-and-imagined places* (1996); Edward W Soja, *Postmetropolis: Critical studies of cities and regions* (2000); Stanford Law Review Symposium: Surveying Law and Borders, above n 222.

227 Twining, above n 212, 19, citing Bebhinn Donnelly and Patrick Bishop, 'Natural Law and Ecocentrism' (2007) 19 *Journal of Environmental Law* 89.

228 Kennedy, above n 214, 26.

229 Ibid 26–7.

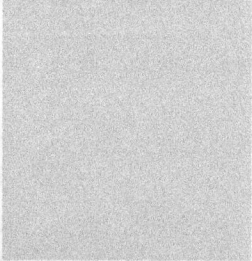

PART II
International Perspectives

World Trade Organization and Climate Change: A Clash of Civilisations?

Gillian Triggs

The 2006 *Stern Review on the Economics of Climate Change* found that greenhouse gases (GHG) reflect 'the greatest and widest-ranging market failure ever seen' as the social costs they cause are not calculated into the price of goods.[1] While this remains true, states have been stimulated to adopt trade measures to neutralise the loss of their competitive edge resulting from compliance with *Kyoto Protocol*[2] targets to reduce greenhouse gas emissions.[3] Carbon taxes, subsidies, border tax adjustments and energy efficiency standards, adopted to 'level the playing field' in international trade may, however, prove to be inconsistent with the traditional rules of the World Trade Organization (WTO).[4] It has thus become necessary to consider how the *GATT*/WTO rules 'fit' with climate change obligations. The title to this chapter suggests that the WTO and *Kyoto* obligations might risk a 'clash of civilisations'; a clash of the legal cultures of trade and environmental laws. My suggestion is that, while the *WTO Agreement* and *Kyoto Protocol* have different objectives, there is no necessary contradiction between them. The practice of the Appellate Body of the WTO indicates that it will seek to harmonise the rules and to defer to national governmental policies to achieve desirable environmental outcomes. But, and here is the practical if undesirable point, almost all WTO disputes relating to health or the environment that are submitted to the dispute resolution process fail, and fail ignominiously. This is not because of any inconsistency between the relevant rules, but because the trade measure adopted is judged to be a disguised restriction on international trade. The challenge is thus to prevent the rise of 'green protectionism'.[5]

The threats posed by global warming and climate change have stimulated a remarkable effort within the international community, led by the United Nations, to create an entirely new regime to reduce the emission of GHGs. The *United Nations*

Framework Convention on Climate Change[6] was adopted at the Rio Conference in 1992 and attended by almost all the world's premier ministers. The *UNFCCC* was, as its name suggests, the skeleton upon which evolving laws were to be built. In 1997 the *Kyoto Protocol* was negotiated under which the Annex I states (26 developed states and 13 states in transition to market economies) accepted a commitment to reduce their GHG emissions by 5% below 1990 levels and to meet defined targets by 2012. Since those optimistic days, states have grappled with the practical task of achieving their targets through the flexibility mechanisms such as emissions trading, joint implementation and the clean development mechanism. Of central concern to the Annex I states has been the realisation that to impose obligations on their national businesses to reduce GHGs is likely also to reduce their international competitiveness; this, at a time of unparalleled instability of the global economy. It is not surprising therefore that states have resorted to trade measures against imports to ameliorate the costs of national climate change measures and to ensure international trade is conducted on equal terms.

While recognising the distortion of the real price of goods, the *Stern Review* also observed that the practical consequence of factoring in the carbon cost is to risk losing a competitive edge in the global market. It was reported, for example, that the EU's mandatory emissions reductions led to significant increases in the cost of electricity. The *Stern Review* concluded that a shift of carbon-intensive production to developing states is 'already happening' and that there has been a loss of competitive edge for energy-intensive goods such as iron, steel, aluminium, cement, glass, chemicals, pulp and paper.[7] Despite these troubling aspects of the *Stern Review*, the Director-General of the WTO, Pascal Lamy, argued in May 2008 that the WTO trade rules are not a substitute for environmental regulation.[8] Rather, he stressed that a consensus on climate change issues, including all major emitters, should be sought through the *UNFCCC*, not the *WTO Agreement*. Lamy conceded that:

> the relationship between international trade, the WTO, and climate change, would be best defined by a consensual international accord on climate change. One that successfully embraces all major emitters. In other words, until a truly global consensus emerges on how best to tackle the issue of climate change, WTO Members will continue to hold different views on what the multilateral trading system can and must do.[9]

While a global consensus on GHG emission reduction seems illusory at present, the WTO's Appellate Body is applying WTO rules with sensitivity to the need to balance environmental priorities with established trade law. Many trade measures adopted to 'level the playing field', where there has been a loss of competitiveness as a result of differing commitments to GHG emission reduction, will, on their face, be technically inconsistent with WTO rules. A crucial issue has become whether these rules permit any justification if a trade measure is adopted in good faith to achieve an environmental objective. Article XX of the *GATT*, and similar provisions under the *Agreement on Technical Barriers to Trade*[10] provide some flexibility in the application of core WTO principles in respect of environmental and health measures. A trade measure might, for example, be justified if it relates to the conservation of exhaustible natural resources, including clean air. The trade disputes determined thus far by the WTO panels and Appellate Body indicate that they are striving to harmonise international trade rules with environmental measures within the wider norms of public international law. This chapter will explore the ways in which such harmonisation might be achieved.

Tension Between WTO Rules and the *Kyoto Protocol*

There are many practical examples of the tensions between the WTO trade principles and the *Kyoto Protocol*. States will typically adopt trade measures to offset the costs of internalising the obligation to reduce GHG emissions by imposing similar costs on imports. The following are examples of such trade measures:

- In 2007, *The Times* of London called on its readers to buy French wines instead of New Zealand wines as part of a 'low carbon diet master plan' under which products with a low carbon footprint would be promoted above those with high GHG emitting transport costs.[11]

- Also in 2007, the United States company Walmart introduced a 'Global Sustainable Sourcing Initiative' with respect to food miles and packaging. Where, for example, flowers are flown to London from Kenya, they will be taxed at a higher rate than flowers brought by ship from Holland. It is arguable that the concept of 'food miles' does not withstand scrutiny because 90% of internationally traded goods are carried by sea, the most carbon efficient means of transport. For example, a Kenyan flower flown to Europe emits one third the CO_2 of the same flower grown in Holland.[12]

- The US has threatened to impose a border tax adjustment on steel produced by China, as China does not impose emission reduction requirements on its domestic manufacturers.

- Process and production methods (PPM) including eco-labelling, technical regulations, energy efficiency product standards, percentage of wind, solar tidal and solar energy.

- The EU building directives, renewable energy requirements, voluntary agreements (as imposed by the OECD and Japan) domestic emissions trading, public procurement and subsidies.

- Carbon taxes adopted by Denmark, Finland and Sweden.

Put more starkly, are the following taxes consistent with WTO rules?

- State A applies lower taxes on the import of a hybrid car than on a standard fuel car as the hybrid car has better fuel efficiency.

- State B decreases tariffs on imported furniture that has been manufactured using wind and solar energy, favouring it above other imported furniture that uses fossil fuels.

The legal questions posed by these examples are whether a trade measure is consistent with WTO rules or whether it is, in reality, a disguised restriction on trade inconsistent with established trade rules. More broadly, the question is whether there is a necessary tension between state sovereignty and protectionism on the one hand and WTO governance and free trade on the other. If that tension is not a necessary one, a question for the future is whether the WTO rules are capable of harmonisation with international environmental standards.

Framework Convention and WTO Rules

The risk of tension between trade rules and emission reduction efforts was recognised during negotiations in 1992 for the *UNFCCC* which provides that:

> [m]easures taken to combat climate change, including unilateral ones, should not constitute a means of arbitrary or unjustifiable discrimination or a disguised restriction on international trade.[13]

In addition, the *Kyoto Protocol* requires Annex 1 states to adopt policies:

> to minimize adverse effects, including ... effects on international trade ... and economic impacts on other parties, especially developing country Parties.[14]

By contrast, the WTO 'rule book' does not deal directly with climate change despite the reality that many aspects of the multilateral trading system are vulnerable to conflict with climate change measures. The *Marrakesh Agreement*, for example, recognises the link between sustainable development and disciplined trade liberalisation and requires that free markets should not jeopardise environmental and social objectives:

> *Recognizing* that their relations in the field of trade and economic endeavour should be conducted with a view to raising standards of living, ensuring full employment and a large and steadily growing volume of real income and effective demand, and expanding the production of and trade in goods and services, while allowing for the optimal use of the world's resources in accordance with the objective of sustainable development, seeking both to protect and preserve the environment and to enhance the means for doing so in a manner consistent with their respective needs and concerns at different levels of economic development.[15]

The more recent *Doha Communiqué* also provides that:

> the aims of upholding and safeguarding an open and non-discriminatory multilateral trading system, and acting for the protection of the environment and the promotion of sustainable development can and must be mutually supportive.[16]

Potential Conflicts with WTO Rules

The provisions of *GATT 1994* and the *TBT Agreement* that are most likely to conflict with state regulatory policies on climate change are:

- *GATT* art I: The most favoured nation principle imposes an obligation not to discriminate among 'like products' of different member states (punitive tariffs).
- *GATT* art II: Obliges members to agree to 'bind' tariffs and to ban all higher tariffs.
- *GATT* art III.4: National treatment requires that imported goods 'shall be accorded treatment no less favourable than that accorded to like products of national origin in respect of all laws, regulations and requirements affecting their internal sale, offering for sale, purchase, transportation, distribution or use'.

This provision is similar to art 2.1 of the *TBT Agreement*, the aim being to ensure that members do not apply 'internal taxes and regulations in a manner which affects the competitive relationship … so as to afford protection to domestic production' and constrain the ability of states to adopt protectionist measures.

- Article XI: Prohibition on quotas, including bans on products from member states (for example, the threatened US import ban on Chinese steel products produced with coal powered energy).

Other relevant provisions include art 2 of the *TBT Agreement*, non-actionable subsidies under the *Agreement on Subsidies and Countervailing Measures*[17] and the *General Agreement on Trade in Services*[18] where there is a conflict with emission reduction units traded as 'securities' allowed only for Annex I states. Of course, there will be no conflict if states are parties to both treaties and one of them agrees to forego their entitlements under the *WTO Agreement*.[19]

In the interests of simplicity, emphasis in this chapter will lie on the application of art III.4 in the context of trade measures to achieve climate change objectives. The core legal issues are:

- Is the challenged trade measure covered, for example, by *GATT*?
- Are the imported and domestic products 'like products' for the purposes of art III.4?
- Is the treatment of the 'like product' no less favourable than in respect of other like products?

Is a Trade Measure to Reduce GHG Emissions Caught by *GATT* or *TBT*?

Some trade measures that, for example, set mandatory emissions, voluntary agreements and product characteristics, appear to be a 'law, regulation and requirement' for the purposes of the national treatment obligation under art III.4. This provision may have a broader scope than art 2 of the *TBT Agreement* as not all technical regulations will be covered by art III.4. Fuel efficiency and mandatory eco-labelling provisions, for example, will be technical regulations, so that trade measures relating to purchase, distribution and transport may fall only within the *GATT* rules.

It is, however, probable that PPM regulations with respect to energy efficiency, usually imposed through eco-labelling programmes, will be inconsistent with *GATT*, unless they are considered to be a legitimate differentiation of 'like products'. So too, it is likely that non-discriminatory border adjustments will be consistent trade measures under art III.2 of *GATT*. Carbon taxes or emission credits, designed to offset advantages to competitors, which impose the same costs on imports as domestic climate change legislation imposes on local production, might be valid measures if they apply equally to all imports and exports.

If the trade measure to reduce GHG emissions is covered by *GATT* and *TBT*, the next question is whether the standard of 'like products' has been met.

'Like Products'

The question of whether the traded goods are 'like products' and thus within the ambit of art III.4 is vital to a determination of whether the contested trade measure is inconsistent with the WTO rules. The test of 'like products' has been considered in several disputes by the Appellate Body, which has developed a considerable body of jurisprudence on the question. Its findings in *EC — Measures Affecting Asbestos and Products Containing Asbestos* in 2001 provided a 'breakthrough' in thinking about trade measures that have been adopted to equalise the effects of climate change regulation.[20] France had imposed a ban on asbestos products on the grounds that they can cause cancer, prompting a complaint from Canada that the trade measure was inconsistent with the national treatment obligation under art III.4 and the prohibition on quantitative restrictions under art XI. A preliminary legal issue was whether the Canadian asbestos is a 'like product' compared with the substitute glass fibres that were permitted imports under French law. The Appellate Body found that the two products were not 'like products'; put brutally, this was because one product causes cancer and the other does not. As the French ban on asbestos was not inconsistent with the obligation under art III.4, as the products were not 'like products', there was no need to rely on any justification under art XX.

The test of 'like products' has been interpreted by the panels and Appellate Body as meaning a 'competitive relationship between the products'. Four factors will be taken into account in determining whether this relationship exists: substitutability, end uses, consumer tastes and habits, and tariff classification.[21] In short, the Appellate Body has adopted a market-based test. In the context of the *EC — Asbestos case*, the Appellate Body found that the asbestos fibres in issue are not 'like' the substitute polyvinal alcohol, cellulose and glass fibres. Despite their similar end uses, the products are therefore different.

The jurisprudential value of the findings in the *EC — Asbestos case* in the area of health policy is that it leaves open the possibility, even probability, that environmental aspects might also qualify as a 'difference' between products. An environmentally-related trade measure in respect of a product may possibly render it unlike another product and thus not in a direct competitive relationship. Any such trade measure will consequently not violate the national treatment obligation under art III.4 of *GATT*.

This said, it is also true that the findings in the *EC — Asbestos* case are obvious as the scientific evidence demonstrates that asbestos is seriously harmful to health. A more difficult question for the future is whether two products that differ only as to their energy use or GHG emissions are 'like products' for the purposes of art III.4. In *US — Autos*, a panel (in a report that was not adopted) found that a 'gas guzzler tax' imposed by the US was not inconsistent with the national treatment obligation as cars that used more than one gallon for 22.5 miles were not 'like' cars that used significantly less. The precedent value of this case for future trade measures on climate change remains limited however.

Scientific evidence of the effect of the product on the environment is likely to have an important persuasive influence on the Appellate Body. In the *EC — Asbestos* case, for example, the strength of the evidence that asbestos causes cancer was sufficient to make the products different in the eyes of the informed consumer. This was a clear case of difference in the market place. Mandatory emissions, voluntary agreements, eco-labelling schemes in respect of the means by which a good is produced (PPMs) probably do not amount to a distinguishing feature so that the products remain 'like' products for the purposes of *GATT* and *TBT*. If so, the products must be treated equally. It remains to be seen whether any differentiation will be recognised with respect to fuel efficiency. One of the few instances in which such a differentiation has been challenged under the trade rules concerns Japan's proposed taxes on categories of car based on engine size and fuel efficiency. The EU claimed that the effect of the measure would be discriminatory against European cars as distinct from cars made in Japan. The complaint prompted Japan to decide not to impose the tax and a dispute was avoided. Recent petrol price rises and a public commitment to climate change policies suggest that trade measures that distinguish between vehicle fuel efficiency might, in today's political climate, be valid on the grounds that the products are not 'like'.

In support of the analysis that goods with a small carbon footprint are not 'like' those with large foot prints is recognition by the Appellate Body in *EC — Asbestos* of the sovereign autonomy and right of states to set their own levels of protection. It seems that a wide measure of discretion will be granted to the regulating state. Also vital to the outcome in the *EC — Asbestos case* was the Appellate Body finding that there is no need to quantify the risk. Rather, a qualitative assessment will be sufficient, so long as the relevant evidence is clear.

In addition to the requirement that the products must be 'like' each other to attract the obligation of national treatment, conformity with art III.4 of *GATT* also requires that the treatment should be 'no less favourable' than that afforded to like products.

What is 'Less Favourable' Treatment?

The requirement that a 'like product' does not receive 'less favourable treatment' means that domestic products should not be 'protected' as against the imported product. It remains, nonetheless, possible to treat 'like products' differently on the grounds of a 'legitimate regulatory category' so long as this does not modify the conditions of competition in the relevant market to the detriment of the imported product. The Appellate Body will thus examine whether any less favourable treatment is justified by an appropriate regulatory goal. In *EC — Asbestos* the Appellate Body understood art III to mean that the measure should not be applied so as to 'afford protection' to the domestic product. It is not clear whether an environmental objective, including PPMs, will save a trade measure that would otherwise be a violation of the national treatment obligation under art III.4.

Justifications for Inconsistency with a WTO Rule

If a measure designed to support a reduction in GHG emissions to equalise uncompetitive trade is found to be inconsistent with the WTO rules, do these rules provide any form of justification? The primary defence lies in art XX of *GATT* which provides that:

> Subject to the requirement that such measures are not applied in a manner which would constitute a means of arbitrary or unjustifiable discrimination between countries where the same conditions prevail, or a disguised restriction on international trade, nothing in this Agreement shall be construed to prevent the adoption or enforcement by an contracting party of measures …

    (b) necessary to protect human, animal or plant life or health ...

    (g) relating to the conservation of exhaustible natural resources if such measures are made effective in conjunction with restrictions on domestic production or consumption.[22]

Similarly, art 2.2 of the *TBT* provides:

> technical regulations shall not be more trade restrictive than necessary to fulfill a legitimate objective ... [including] national security requirements; the prevention of deceptive practices; protection of human health or safety; animal or plant life or health, or the environment. In assessing such risks, relevant elements of consideration are ... available scientific and technical information, related to processing technology or intended end-uses of products.[23]

The first question that arises under art XX is whether the disputed trade measure on GHG emissions falls within one of the listed exceptions. If so, it becomes necessary to determine whether the measure is 'necessary' or 'relating to' the exception. The panels and Appellate Body have had several opportunities to interpret and apply the exceptions under art XX in the areas of health and the environment. A typical example is the WTO dispute settlement, *United States — Standards for Reformulated and Conventional Gasoline*.[24] Here, the panel found that the United States had discriminated against imported gasoline by establishing differential baselines for olefin reductions (an impurity causing pollution) by foreign producers. The panel found that the measure could not be justified under art XX and that, while WTO Members could set environmental standards, these had to be implemented through measures that are WTO consistent. On appeal, the Appellate Body concluded that the policy to reduce the depletion of clean air was within the meaning of art XX(g) and was 'primarily aimed at' this objective. In an expansive analysis of the chapeau, the Appellate Body concluded that the regulations were an arbitrary and unjustifiable discrimination between members and a disguised restriction on international trade. One reason was that other WTO-consistent courses of action were available. The contested trade measure was thus found to be consistent with art III.4 and, as the regulations did not comply with the *chapeau* to art XX, it could not be excused on the grounds of one of the listed exceptions. It was therefore recommended that the United States should bring its regulations into consistency with the rules. Indeed, the United States then altered its regulations to permit individual baselines that could then be used by foreign importers. This case has proved to be typical in the sense that while the trade measure might satisfy the art XX(g) as relating to the conservation of an exhaustible natural resource — clean air — the measure will be inconsistent with the WTO rules if it is applied in an arbitrary or unjustifiable way. It is notable that, as in the *US — Reformulated Gasoline* case many of the environmental and health cases have proved to be little more than a disguised restriction on trade.

    Another example of the application of art XX that is relevant to the validity of climate change trade measures is the *US — Shrimp* case where a 1987 US prohibition on the import of shrimp under the *Endangered Species Act of 1973* was found not to be justified under art XX.25. The United States had justified its prohibition on the grounds that harvesting led to the drowning of a turtle species that is protected by the *United*

Nations Convention on International Trade in Endangered Species of Wild Flora and Fauna.[26] While the United States had negotiated the *Inter-American Convention for the Protection and Conservation of Sea Turtles*[27] no similar attempt was made outside its region to negotiate an agreement with any other state. India, Thailand, Pakistan and Malaysia complained to the Dispute Settlement Body that the ban was inconsistent with *GATT* and was not justified under art XX. These original complainants were then joined by other WTO members, including the EU and Australia.

The Appellate Body found, on appeal from the panel, that the US ban was arguably necessary to protect animal life under art XX(b) or that was related to the conservation of an exhaustible natural resource under art XX(g). The measure was, nonetheless, inconsistent under the WTO rules because it failed to meet the conditions of the *chapeau* and was, yet again, arbitrary and discriminatory in its application. The emphasis of the Appellate Body in the *US — Shrimp* case lay with the need for a state to engage in serious negotiations with other states to achieve regional or multi-lateral agreements in relation to such environmental measures. Ironically, the *US — Shrimp* case was seen as a victory for environmental groups. Developing nations of Asia, by contrast, saw the findings as an attempt to move the trade 'goal posts' and to impose environmental standards on developing, low income nations (depriving them of their natural comparative advantage). While there are many aspects of this decision that warrant critical comment, the vital point is that, in applying art XX, the Appellate Body has upheld the view that a trade measure can be adopted to meet environmental or natural resource protection objectives under art XX. The only limitation upon this power is that the measure must be applied in a non-discriminatory manner.

Many other disputes before the WTO disputes process follow the pattern of analysis in the *US — Reformulated Gasoline case*. In *Australia — Measures Affecting Importation of Salmon*, for example, Australia had imposed a ban on uncooked salmon to prevent disease entering the country.[28] The panel found that the ban violated the *Agreement on the Application of Sanitary and Phytosanitary Measures*[29] because the required risk assessment failed to show a 'probability' of entry of the disease. The ban also failed the *chapeau* test and was a disguised restriction on trade to protect the Tasmanian salmon industry. Similarly, in the *Japan — Measures Affecting the Importation of Apples*, Japan failed to meet the high standard of scientific proof needed to justify its stringent condition on the import of apples from the United States.[30] The Appellate Body found that the ban could not be justified under art XX or the *SPS Agreement* as the scientific assessment was inadequate and such science as was available overwhelmingly showed there was no risk. The findings in this case will doubtless be vital to the current complaint by New Zealand against Australia's longstanding ban on New Zealand apple imports to prevent the spread of fire blight.

Of recent jurisprudential value are the findings of the Appellate Body in the environmentally related complaint before the WTO in *Brazil — Measures Affecting Imports of Retreaded Tyres*.[31] Here, the European Communities complained that Brazil's ban on the import of retreaded tyres, and exemption of the MERCOSUR Member States, were inconsistent with arts XI:I and III:4. Retreaded tyres have a shorter life span than completely new tyres and become 'waste' sooner. Accordingly, the ban was intended to reduce future waste. Brazil argued that the trade ban was justified under art XX(b) on the grounds that it was necessary to protect human, animal and plant life and health from the risks posed by the accumulation of waste. The panel agreed that the trade measure was inconsistent with *GATT* but justified under art XX(b). Importantly, the Panel also agreed with Brazil that the test of 'necessary' had been met. But here again was the fatal flaw. The exemption for state parties to MERCOSUR meant that the ban was a disguised restriction on international trade, as it was clearly discriminatory in its application contrary to the *chapeau*.

On appeal, the Appellate Body upheld panel findings and confirmed that it is within the sovereign mandate of each WTO Member to set its public health or environmental objectives. The Appellate Body considered that the 'necessity' test requires weighing and balancing factors including the contribution made by the ban to enforcement of the objective, the importance of common interests or values, the impact of the ban on imports or exports, and the relative importance of the ban. While the Appellate Body identified the elements of 'necessity', it also recognised the problem of demonstrating that the trade measure has been effective in achieving the desired climate change outcome.

> In the short term, it may prove difficult to isolate the contribution to public health or environmental objectives of one specific measure from those attributable to the other measures ... the results obtained from certain actions — for instance, measures adopted in order to attenuate global warming and climate change, or certain preventive actions to reduce the incidence of diseases that may manifest themselves only after a certain period of time — can only be evaluated with the benefit of time. In order to justify an import ban under Article XX(b), a panel must be satisfied that it brings about a material contribution to the achievement of its objective ... a panel might conclude that an import ban is necessary on the basis of a demonstration that the import ban at issue is apt to produce a material contribution to the achievement of its objective. This demonstration could consist of quantitative projections in the future, or qualitative reasoning based on a set of hypotheses that are tested and supported by sufficient evidence.[32]

It seems the Appellate Body will accept that a rational, qualitative analysis by a state when deciding to impose a trade measure can meet the test of necessity. The Appellate Body also confirmed that it is not necessary to show there are no reasonable alternatives to the ban adopted. Rather, the member has a 'right to achieve its desired level of protection with respect to the objective pursued'. With these standards in mind, the Appellate Body concluded that, on balance, the import ban was necessary to achieve the allowable objective but that the discrimination between MERCOSUR countries and other WTO Members in the application of the import ban failed to satisfy the terms of the chapeau.

How Will a Climate Change Trade Measure Meet the Test of Necessity?

The obligation to meet the test of 'necessity' is the potentially fatal weakness in the argument that art XX can justify a trade measure to achieve a climate change objective. It is arguable that the scientific evidence has been uncertain, or at least not sufficiently clear that a trade measure to reduce GHG emissions is 'necessary'. There remain many uncertainties regarding the impact of GHG emissions on agriculture, habitat, forests, oceans and the economy and there is little evidence of the effectiveness of voluntary approaches or eco-labelling. By contrast, the evidence in respect of the need to protect a listed species of turtle, the cancer-causing properties of asbestos and pollution from gasoline has been strong. Despite these uncertainties, the Appellate Body has taken a flexible approach in deferring to governments acting in good faith.[33] *The EC — Hormones Measures Concerning Meat and Meat Products (Hormones)* case in 1998 provides an example of this flexibility.[34] The EU had imposed an import ban on meat from cattle treated with growth hormones. While the Appellate Body found the ban was too general to meet the scientific risk assessment standard under art 5 of the *SPS Agreement*, it recognised the need to recognise a good faith judgment by government:

> responsible and representative governments may act in good faith on the basis of what, at a given time, may be a divergent opinion coming from qualified and respected sources.[35]

A similar approach was adopted in the *US — Asbestos* case. Further:

> In justifying a measure under Article XX(b) of the *GATT* 1994, a Member may also rely, in good faith, on scientific sources which, at that time, may represent a divergent, but qualified and respected, opinion. A Member is not obliged, in setting health policy, automatically to follow what, at a given time, may constitute a majority scientific opinion.[36]

In summary, it is probable that a trade measure adopted to respond to global warming and to meet *Kyoto* targets is likely to be covered by art XX(b). The Appellate Body has confirmed that states may adopt in good faith a precautionary approach to reduce GHG emissions. There is, moreover, no need to rely on majority scientific opinion, so long as the trade measure is based on creditable evidence and is reasonably related to the exception listed in art XX. Finally, the WTO rules will presume validity of a trade measure if it is based on a relevant international standard.[37] In light of the Appellate Body's concern to ensure consultation and regional and multilateral negotiation, the *Kyoto Protocol* provides an international benchmark justifying adoption of a trade measure that is otherwise inconsistent with the WTO rules.

WTO Initiatives on Climate Change

The WTO has adopted several initiatives that have a direct bearing on state efforts to reduce GHG emissions. The Doha Development Round includes, for the first time, explicit negotiations on multilateral trade and the environment to achieve a 'harmonious coexistence'.[38] While these negotiations have stalled, it is hoped that the round, including agriculture and fisheries subsidies, will ultimately contribute to efforts to adapt to climate change. Other initiatives include the establishment in 1994 of the WTO's Committee on Trade and Environment, which is charged with advancing the trade and environmental agenda and promoting sustainable development. This committee has concluded that *GATT 1994* is 'not to be interpreted in clinical isolation from international law'. The recognition of the wider international legal context within which the WTO rules apply will help decision-makers to integrate trade rules with environmental objectives.

WTO Members have also sought the adoption of a uniform approach to taxes on energy and GHG emissions, such as the equalisation of carbon taxes, to eliminate the leakage arising from uneven treatment by members.

A World Bank study has found that if tariffs and non-tariff barriers to clean technologies, such as wind and hydropower turbines, solar water heaters and tanks for biogas production were to be eliminated, world trade would increase by 14%.[39] Accordingly, the EU and UK have proposed that barriers to trade on clean goods, technology and services that benefit the environment be eliminated. In December 2007, they made formal proposals giving import priority to environmental goods and services, a change that could improve market access for some developing states.

Mechanisms for information exchanges and improved coordination have been made between the respective secretariats of multilateral environmental agreements and the WTO, for example, the *UNFCCC* now takes part in the deliberations of the WTO Committee on Trade and the Environment. The work of the *TBT* Committee on Labelling and the International Organisation for Standardisation also provide options for promoting environmental standards.

Conclusion

There is a significant risk of trade disputes over measures adopted to reduce GHG emissions where the measure is inconsistent with WTO rules. In the interests of harmonisation with environmental obligations, the Appellate Body has been willing to apply art XX flexibly to justify an otherwise inconsistent trade measure. In practice, however, attempts to rely on this provision have failed because evidence indicated the measure was discriminatory or a disguised restriction on trade. The absence of any example of a genuine legal dispute over an environmental trade measure indicates that the fear of tension between the WTO rules and the *Kyoto Protocol* is overstated.

While a multilateral solution to potential conflicts between trade and environmental law is the ideal strategy, the prospect of achieving an international consensus is slight so long as the United States remains out of the *Kyoto Protocol,* and so long as China, among others, does not accept binding environmental obligations. The imbalance of obligations to reduce GHG emissions under the *Kyoto Protocol* is thus a powerful stimulus for states to adopt WTO inconsistent trade measures to redress the balance. It is this imbalance that prompts breaches of international trade rules, not any failure on the part of international law to integrate WTO trade rules with *Kyoto Protocol* obligations. As there is no necessary tension between trade rules and environmental law, or between state autonomy and free trade, harmonisation is achievable. The WTO Appellate Body has led the way in developing standards to ensure a practical accommodation of these differing legal regimes.

Endnotes

1   Sir Nicholas Stern, *Stern Review on the Economics of Climate Change* (2006) 'Executive Summary', in *Stern Review.*

2   *Kyoto Protocol to the United Nations Framework Convention on Climate Change*, opened for signature 16 March 1998, 2303 UNTS 148 (entered into force 16 February 2005) ('*Kyoto Protocol*').

3   There is significant body of academic analysis of the relationship between WTO trade rules and climate change initiatives, see, eg: Joost Pauwelyn, 'US Federal Climate Policy and Competitiveness Concerns: The limits and options of international trade law' (NI WP 07-02, Nicholas Institute for Environmental Policy Solutions, 2007); Javier de Cendra, 'Can Emissions Trading Schemes be Coupled with Border Tax Adjustments?' (2006) 15 *Review of European Community & International Environmental Law* 131; Jeffrey Frankel, 'Climate Change and Trade: Links between the Kyoto Protocol and WTO' (2005) 47(7) *Environment* 8; Paul Demaret and Raoul Stewardson, 'Border Tax Adjustments under *GATT* and EC Law' (1995) 28(4) *Journal of World Trade* 5; Christian Pitschas, '*GATT*/WTO Rules for Border Tax Adjustment and the Proposed European Directive Introducing a Tax on Carbon Dioxide Emissions and Energy' (1995) 24 *Georgia Journal of International and Comparative Law* 479; Steve Charnovitz, 'Trade and Climate: Potential Conflicts and Synergies', in Joseph E Aldy, John Ashton, Richard Baron et al, *Beyond Kyoto: Advancing the international effort against climate change* (2003) 141 Pew Center on Global Climate Change <http://www.pewclimate.org/global-warming-in-depth/all_reports/beyond_kyoto>; Cinnamon Carlarne, 'The Kyoto Protocol and the WTO: Reconciling Tensions Between Free Trade and Environmental Objectives' (2006) 17 *Colorado Journal of International Environmental Law & Policy* 45; Catherine Redgwell, 'Trade Measures and Environmental Protection', in Joseph A McMahon (ed), *Trade and Agriculture: Negotiating a New Agreement* (2001) 314; Nita Ghei, 'Evaluating the WTO's Two Step Test for Environmental Measures Under Article XX' (2007) 18 *Colorado Journal of International Environmental Law and Policy* 117; Gregory Shaffer, 'The World Trade Organization Under Challenge: Democracy and the Law and Politics of the WTO's Treatment of Trade and Environment Matters' (2001) 25 *Harvard Environmental Law Review* 1.

4   *Marrakesh Agreement Establishing the World Trade Organization*, 1867 UNTS 3 (entered into force 1 January 1995) ('*Marrakesh Agreement*' or '*WTO Agreement*').

5   WTO, *Environmental Requirements and Market Access: Preventing green protectionism* <http://www.wto.org/english/tratop_e/envir_e/envir_req_e.htm>.

6   *United Nations Framework Convention on Climate Change*, opened for signature 4 June 1992, 1771 UNTS 107 (entered into force 21 March 1994) ('*UNFCCC*').

7   In a simulation exercise on the impact of proposed EU Kyoto Protocol border tariff on US exports that are not subject to greenhouse gas standards 'could significantly affect the US trade balance'. It was found that an EU 30% tariff will lead to a 30.5% loss in US energy-intensive exports in steel and cement.

8   Pascal Lamy, A Consensual International Accord on Climate Change is Needed (WTO News, 29 May 2008) < http://www.wto.org/english/news_e/sppl_e/sppl91_e.htm>.

9   Ibid.

10 *Agreement on Technical Barriers to Trade*, 1186 UNTS 276 (entered into force 1 January 1980) ('*TBT Agreement*').

11 Vicki Waye, 'Carbon Footprints, Food Miles and the Australian Wine Industry' (2008) 9 *Melbourne Journal of International Law* 271; Anna Shepard, Low Carbon Diet Masterplan, *The Times* (London), 21 April 2007, Body and Soul, 12.

12 Lamy, above n 8.

13 *UNFCCC* art 3.5.

14 *Kyoto Protocol* art 2.3.

15 *Marrakesh Agreement* Preamble (emphasis in original).

16 Ministerial Declaration, Fourth WTO Ministerial Conference, Doha, 9–14 November 2001, adopted on 14 November 2001, [6] <http://www.wto.org/english/thewto_e/minist_e/min01_e/mindecl_e.htm> (*Doha Communiqué*).

17 WTO, *Agreement on Subsidies and Countervailing Measures*, LT/UR/A-1A/9, 15 April 1994 ('*Subsidies Agreement*') A failure to internalise the carbon cost is not the same thing as a subsidy. Therefore anti-subsidy duties will not be permissible.

18 *Marrakesh Agreement Establishing the World Trade Organization*, 1867 UNTS 3, Annex 1B (entered into force 1 January 1995) ('*General Agreement on Trade in Services*' or '*GATS*').

19 There are limits to employing the technique of imposing anti-dumping duties as they are to be assessed only by reference to the price of the goods in the state of origin, not by reference to the price that has been elevated by carbon emission restrictions in the state of import.

20 WT/DS135/AB/R, 12 March 2001 (adopted 5 April 2001) (*EC — Asbestos*).

21 *Report of the Working Party on Border Tax Adjustments*, GATT, BISD 18S/97, [18]; G Triggs, *International Law: Contemporary Principles and Practices* (2006) [12.34].

22 GATT art XX.

23 *TBT Agreement* art 2.2.

24 WT/DS2/AB/R, 29 April 1996 (*US — Reformulated Gasoline*).

25 WTO Doc WT/DS58/AB/R adopted 6 November 1998.

26 *United Nations Convention on International Trade in Endangered Species of Wild Flora and Fauna*, opened for signature 3 March 1973, 993 UNTS 243 (entered into force 1 July 1975).

27 *Inter-American Convention for the Protection and Conservation of Sea Turtles* (Caracas, 1 December 1996).

28 WT/DS18/AB/R, 12 October 1998 (*Australia — Salmon*).

29 WTO, *Agreement on the Application of Sanitary and Phytosanitary Measures*, LT/UR/A-1A/12, 15 April 1994 ('*SPS Agreement*').

30 WT/DS245/AB/R, 26 November 2003.

31 WT/DS332/AB/R, 3 December 2007.

32 WT/DS332/AB/R, 3 December 2007, [151].

33 The Appellate Body found that 'a panel must consider the relevant factors, particularly the importance of the interests or values at stake, the extent of the contribution to the achievement of the measure's objective and its trade restrictiveness': [178].

34 WT/DS26/AB/R — WT/DS48/AB/R (*EC — Hormones*).

35 WT/DS26/AB/R — WT/DS48/AB/R.

36 WT/DS135/AB/R, [178]

37 *TBT Agreement* art 2.5.

38 WTO, *An Introduction to Trade and Environment in the WTO* <http://www.wto.org/english/tratop_e/envir_e/envt_intro_e.htm>.

39 WTO, *Activities of the WTO and the Challenge of Climate Change* <http://www.wto.org/english/tratop_e/envir_e/climate_challenge_e.htm>.

International Courts and Climate Change: 'Progression', 'Regression' and 'Administration'

Tim Stephens

In late 2008 the Climate Justice Programme, Oxfam International, and Advocates for International Development launched a competition, open to lawyers and law students, to find a compelling legal case that could be brought against developed countries in an international forum for failing to take decisive action to address climate change.[1] The suggestion has also recently been made by the eminent climate scientist James Hansen that international criminal law should be invoked to prosecute directors of energy companies for 'crimes against humanity and nature.'[2] These developments are a clear manifestation of the growing public interest in bringing the most pressing environmental issue of our time before an international court or tribunal, given the slow pace of international negotiations to deliver a meaningful strategy to avert catastrophic human-induced climate change.

This chapter assesses the extent to which this interest in international judicial and quasi-judicial bodies is well-placed, and to this end considers the prospects and pitfalls for international climate change litigation. It is seen that proceedings may be initiated within a multiplicity of courts, tribunals and quasi-judicial bodies that operate on the international plane. This is a consequence of the marked proliferation in international courts and tribunals, including in the environmental field, that took place in the latter part of the 20th century.[3] Critically, the proliferation phenomenon means that climate cases could potentially be heard not only by bodies with a recognised environmental specialisation but also in forums that are focussed on other issue-areas of importance to the international community, such as human rights, trade or investment.

So far as climate change litigation is concerned we can identify three types of international proceedings. First, there is what we might term 'progressive' proceedings — those designed to bring positive outcomes in terms of mitigating, or adapting to, climate change, or compensating its effects. As in domestic settings,[4] the chief value of such litigation is in highlighting the need for a more robust policy response to climate change, rather than in delivering concrete outcomes in terms of reducing greenhouse gas emissions in relation to which international negotiations under the framework of the *United Nations Framework Convention on Climate Change*[5] and the *Kyoto Protocol*[6] remain critical.

A second category of international climate litigation is what can be called 'regressive' proceedings, which may be invoked to prevent states, or groups of states, from adopting national or international climate policies that could interfere with other norms such as those relating to trade liberalisation. The most obvious forum for such regressive proceedings is the World Trade Organization ('WTO'). Despite the considerable 'greening' of the WTO, principally as a result of decisions of the Appellate Body,[7] it seems inevitable that WTO members seeking to adopt unilateral measures to reduce emissions that have trade impacts will be the subject of proceedings in the WTO, given the significant ambiguity that surrounds the implementation of provisions of the WTO and the scope of applicable environmental exceptions.[8]

Most likely to assume prominence in international law is a third type of proceedings, which can be called 'administrative' litigation, within and outside of the climate change regime established by the UNFCCC and the *Kyoto Protocol*, and any successor, which may be utilised to ensure that commitments under the regime are kept. The compliance procedure established under the *Kyoto Protocol* to perform these tasks has already begun to consider certain infractions of technical reporting requirements under the international climate regime. However, the benefits that this procedure can bring are necessarily limited by the extent of the emissions reduction or limitation commitments under the international climate regime.

'Progressive' Proceedings

From the Domestic ...

In understanding the prospects for progressive international climate change litigation it is helpful to examine first the nature and extent of similar domestic climate change litigation where issues that would need to be addressed in any international proceedings have received some attention.

There has been a wave of climate change litigation in domestic courts in several states. The bulk of such cases have been brought in jurisdictions in the United States which, for reasons including that costs do not follow the event, are more friendly to public interest litigation than other forums.[9] However, climate litigation has also been pursued, with some success, in other jurisdictions including in Australia where several cases have resulted in greenhouse gas emissions being considered in planning processes.[10]

The justification given for litigation in this context is that the courts must be turned to by public interest litigants as the other arms of government are not responding promptly and effectively to the grave threat of climate change. Executives are refusing to take on international commitments to reduce emissions, or are assuming only the most modest and ineffective emissions targets. Many legislatures are similarly moving slowly, if at all, to enact regulatory regimes to reduce emissions. In view of this, it is understandable that judicial mechanisms are being utilised, even if they are not necessarily able to produce optimal outcomes in regulating, in a coordinated way, a global public good.[11] As one of the most trenchant critics of domestic climate litigation has noted, 'bad judicial regulation might be better than no regulation at all.'[12] This argument has particular trenchancy given recent assessments that show that climate change poses a more immediate and significant risk[13] than suggested in the Fourth Assessment Report of the Intergovernmental Panel on Climate Change released in 2007.[14]

The domestic litigation that has been brought to date falls into two broad categories: (1) private law actions and (2) public law proceedings. Private, or civil law, actions include proceedings for the torts of negligence, nuisance, conspiracy and misrepresentation and also on the basis of trade practices legislation designed to prevent deceptive and misleading conduct.[15] The primary purpose of such litigation is to obtain compensation or another remedy for injury that is suffered by a plaintiff. The litigation is therefore mostly responsive in character, looking to past or continuing injury, although it may have effects into the future to the extent that it forces corporate or government behaviour to change.

Tort-based climate litigation has been largely unsuccessful to date. This is unsurprising given the difficult hurdles that need to be overcome in establishing the requisite elements of a cause of action, including reluctance of courts in some jurisdictions to step into a policy arena that some argue is better left to the political arm of government. In the United States, courts have invoked the doctrine of non-justiciability in deference to the role of the executive arm of government in addressing global environmental issues such as climate change which involve questions of foreign policy.[16] If and when such litigation is brought in Australia, such concerns are unlikely to be an impediment to litigation between private parties.[17] The situation would, however, be different if tort-based litigation targeted an Australian government at federal, state or local level for failing to adopt and implement policies to reduce greenhouse gas emissions. As seen in *Graham Barclay Oysters Pty Ltd v Ryan*,[18] a case in negligence alleging the failure by the New South Wales government to prevent water pollution contaminating an oyster farm, the High Court emphasised that public policy considerations militate against a court passing judgment on the reasonableness or otherwise of governmental action or inaction.[19]

In any event, other obstacles lie in the road of tort-based climate change litigation in Australia. A successful case in negligence, for instance, requires the establishment of the four main elements: a *duty of care*, and *breach* of that duty that is *causative* of a relevant *injury* to the plaintiff.[20] Satisfying all but the injury criterion in relation to climate change is extremely challenging. Most problematic appears to be the element of causation, given that it is difficult, if not impossible, to assign specific climate change impacts to even the largest emitters, such as electricity utilities that rely wholly or predominantly on coal-fired power plants. Establishing a breach of a duty of care (assuming the identification of the duty in the first place) could be further complicated where a defendant polluter's emissions are countenanced by government through a regulatory regime such as an emissions trading scheme that seeks to achieve a government's emissions reduction targets set pursuant to an international agreement. These issues emphasise the fundamental problem of scale in domestic climate change litigation in appropriately matching plaintiffs and defendants in the context of what is inescapably a global problem.[21]

Nonetheless, it is too early to pronounce negligence, or other tort-based litigation, an impossibility given the capacity for courts to develop principles and rules of civil liability to address new legal challenges. Tort law in the 21st century is virtually unrecognisable to that of the late nineteenth century, despite continued use of basic terminology derived from early cases such as *Donoghue v Stevenson*.[22] It is likely to develop further and faster as climate change accelerates and causes additional impacts, which in turn will prompt fresh litigation. Moreover, scientific understanding of causation in the context of climate change is maturing rapidly, such that increasingly precise estimations can be made of the probability of the occurrence of certain weather events (such as flooding, heatwaves and the like) as a result of given greenhouse gas concentrations.[23]

Turning to public law actions, those intended to prompt or prevent certain government action, there is evidence of substantially more success in a number of different jurisdictions than is the case for private law actions. For instance, in several cases in Australian jurisdictions it has been held that greenhouse implications of development proposals must be addressed under environmental planning legislation.[24] This is despite the fact that such legislation often makes no reference to climate change and requires only that broad consideration be given to environmental impacts and values including the objective of sustainable development.

Notwithstanding these important developments, it is important to recognise their limitations. These cases have concerned the scope of assessment processes, not the merits or otherwise of projects from a climate change perspective. Moreover, even where there is a success there is nothing preventing government from legislating to allow projects to proceed despite a defective assessment of climate change impacts.[25] This litigation has to date, and will continue in the future, to enhance the visibility of climate change in planning decisions. However, as in many other areas of environmental management, in addressing climate change there appears no substitute for clear goals set within a statutory scheme,[26] which is backed by judicial enforcement.

The Howard Government refused to countenance such an approach, preferring instead to rely on voluntary action by firms and consumers to reduce their own contributions to Australia's emissions signature.[27] However, the Rudd Government has committed to a more interventionist legislative agenda on climate change through the proposed Carbon Pollution Reduction Scheme (CPRS).[28] Although legitimate questions surround the appropriateness of the low emissions reduction limit that will underpin the scheme (between 5% and 15% by 2020), and the compensation provided to many trade-exposed high emissions firms, there is no doubt that the legislative move itself signals a critical shift in thinking in terms of addressing climate change in Australia.

The result of legislative changes in the offing at the federal level may leave relatively little room for the courts to be the vanguard of private law actions brought in relation to climate change — given that adherence to the requirements of the CPRS may be a good defence to various private law actions, and may also sideline existing regulatory law under state planning approval systems.[29] As a consequence, the more common litigation that may result may be cases brought in relation to the CPRS in connection with the new form of property it credits in establishing emissions permits, and in the system established for trading in this new commodity. Litigation will also be available to seek judicial and merits review of decisions of the Australian Climate Change Regulatory Authority which is to be established to administer the trading scheme (under the Draft Carbon Pollution Reduction Scheme Bill 2009), the reporting regime (the *National Greenhouse and Energy Reporting Act 2007*) and the renewable energy regime (*Renewable Energy (Electricity) Act 2000*).

Parts 24 and 26 of the Draft Carbon Pollution Reduction Scheme Bill 2009 indicate that merits review (by the Administrative Appeals Tribunal) and judicial review (under the *Administrative Decisions (Judicial Review) Act 1977*) will be available in relation to many decisions taken by the Authority and by the Minister under the suite of legislation establishing the CPRS. However, as the legislation stands it does not go so far as to provide for judicial enforcement of emissions reduction targets adopted by the Australian government, in contrast to the United Kingdom's *Climate Change Act 2008* (UK) which imposes a statutory duty on the Secretary of State to ensure that the United Kingdom's near- and long-term carbon pollution reduction targets are met.[30]

... To the International

On the plane of international law, there are a number of norms of both customary and conventional character that could potentially provide redress for states impacted by climate change. The situation in this context appears on its face more straightforward than that within domestic legal systems, for the reason that international litigation offers a better fit with the scale of the climate change problem, as Adler explains:

> A government-versus-government structure for global warming liability seems particularly promising as a matter of [corrective justice]. For example, Australia might seek redress from the United States for damage to Australia's coral reefs that emissions from the United States, as a whole, caused and that the United States government allowed ... the causal issues, however thorny, are less so than in the case of compensation for private property losses or individual injuries or deaths; and a government's inaction in the face of [greenhouse gas] emissions may well be wrongful, by actually making a difference to global temperatures, even if individual or firm emissions are not.[31]

Longstanding rules of international law relating to liability for transboundary damage have translated the basic principle of governmental liability for harm caused by one state to another to include environmental damage. In the seminal *Trail Smelter Case*[32] an arbitral tribunal concluded that no state has the right to use, or permit the use, of its territory in such a way as to cause serious injury by pollution in the territory of another state:

> [U]nder the principles of international law, as well as of the law of the United States, no state has the right to use or permit the use of its territory in such a manner as to cause injury by fumes in or to the territory of another or the properties or persons therein when the case is of serious consequence and the injury is established by clear and convincing evidence.[33]

There is no more elemental stipulation of customary international law relating to environmental questions,[34] and the Tribunal's statement ultimately found expression in Principle 21 of the *1972 Declaration of the United Nations Conference on the Human Environment*[35] (which was substantially repeated in Principle 2 of the *1992 United Nations Declaration on Environment and Development*).[36] Principles of state responsibility for cross-border environmental harm have also been further developed by the International Law Commission through its *Draft Articles on Prevention of Transboundary Harm from Hazardous Activities*[37] and *Draft Principles on the Allocation of Loss in the Case of Transboundary Harm Arising out of Hazardous Activities*.[38]

This body of international law has to date only been invoked for far simpler and clearer cases of transboundary pollution, such as the situation confronted in the *Trail Smelter Case* where the emission of sulphur dioxide from the lead and zinc smelter at Trail in British Columbia drifted southwards across the border with the United States and caused damage to crops and other agricultural interests. However, with greater understanding of the causes and consequences of climate change, Verheyen has argued that there is no reason in principle why states may not be held to account under this principle of transboundary harm for failing to implement emissions cuts, if this failure can be proven to have had discernable transborder impacts.[39]

Verheyen invokes several case studies in this respect, including a potential claim by Nepal or Bhutan against the United States in respect of flood damage from glacial melting. If as a consequence of global warming, glaciers in the Himalayas are rapidly retreating, and the threat of glacial lake outburst floods increases, then Verheyen argues that the United States, as the developed country with the largest carbon footprint, would assume state responsibility for such damage in these Himalayan nations.[40]

While the causation problems are easier to address in the context of inter-state responsibility claims, they are not entirely eliminated, and it must be recognised that public international law has not grappled with causation issues in the same comprehensive way as in many domestic legal systems. More fundamentally, even if a vulnerable state such as Bhutan did have a good legal case against a major emitter such as the United States, there would still remain jurisdictional and political impediments to proceedings. Unless submitted by joint agreement, which seems unlikely, a case before the International Court of Justice (ICJ) could only proceed where the parties had both accepted the compulsory jurisdiction of the ICJ pursuant to art 36(2) of the court's statute.

Even if the ICJ possessed jurisdiction there would remain major political hurdles to proceedings. Few states have sought to resolve transboundary pollution disputes via the use of inter-state dispute settlement mechanisms, and the *Trail Smelter Case* in this regard is an exceptional scenario rather than the norm. Even the most serious cases, such as the Chernobyl nuclear power plant disaster in 1986 have not prompted inter-state litigation. Nonetheless, it remains a possibility *in extremis*, and several govern-ments have sought to preserve their rights to utilise this body of law if necessary. For instance, Tuvalu, a microstate in the Pacific already experiencing the impacts of rising sea levels from climate change, declared on signing and ratifying the *UNFCCC* and *Kyoto Protocol* that joining the climate regime 'shall in no way constitute a renuncia-tion of any rights under international law concerning state responsibility for the adverse effects of climate change.'[41] Were the ICJ seized of a case concerning climate change it is be hoped that the court would address the dispute directly, rather than seek to avoid the substantive issues as it has done in the context of disputes over another existential threat — the health and environmental impacts of the use[42] and testing[43] of nuclear weapons.

Beyond interstate litigation seeking to invoke state responsibility for climate change damage, there are other international fora that may be used in aid of the pro-gressive form of climate change litigation. These include human rights courts and complaint procedures, such as the Human Rights Committee established by the *First Optional Protocol*[44] to the *International Covenant on Civil and Political Rights*.[45] Under human rights complaints procedures individuals who have exhausted local remedies may initiate proceedings before an international body for breaches of human rights by a state party to the relevant human rights treaty. The body may issue non-binding views calling upon the state to desist from, or provide redress for, a human rights violation.

There has been only one such climate case to date, namely a petition from a group of Inuit people to the Inter-American Commission on Human Rights claiming that the United States' failure to restrain its emissions 'violate[d] the Inuit's fundamental human rights protected by the *American Declaration of the Rights and Duties of Man* and other international instruments.'[46] The rights cited in the petition included 'rights to the benefits of culture, to property, to the preservation of health, life, physical integrity, security, and a means of subsistence, and to residence, movement, and invi-olability of the home.'[47] The issues of human rights law raised by the petition were not addressed by the Commission, and instead the Commission issued a letter[48] to the petitioners in 2007 which stated that the complaint did not provide sufficient infor-mation to enable the Commission to determine that there had been a violation of the *American Declaration of the Rights and Duties of Man*.[49]

Climate-focused human-rights litigation is contingent not only on establishing that a state, through its acts or omissions, is responsible for specified climate change impacts — it must also be proven that these impacts infringe nominated human rights. However, as Cordes-Holland has noted, '[d]emonstrating that [international human rights] can extend to an environmental problem such as global warming is not an easy task.'[50] In the absence of an established right to a clean and healthy environ-ment (or to a 'stable climate'),[51] it is necessary to point to recognised human rights such as the right to life, the right to health, the right to respect for private life and home, and the right to peaceful enjoyment of possessions.

It may be possible to rely on one or more of these rights to found a case given that the existence of a stable climate is necessary in order to enjoy them. In order to do so, the rights violation must be causally connected with climate change and with the action or inaction of the government that is the respondent to the proceedings. With the exception of major emitters such as the United States this may be difficult. However Cordes-Holland has argued that the causation issue in this context turns not on identifying a respondent state as having sole or primary responsibility, but rather whether the state has made a not insubstantial contribution to climate change.[52]

Some cautionary notes should be sounded as to the value of international human rights litigation in the climate change context. Conceptually, human rights protection is inherently anthropocentric, and concerned only with environmental issues to the extent to which they have human impacts.[53] This means that human rights litigation can only be used to address climatic changes affecting individuals and communities, not for climate change damage generally (as would theoretically be possible under principles of state responsibility for transboundary harm). In terms of practical result, human rights litigation may also deliver only very modest benefits. For instance, were a state found to have violated a human right by failing to restrain emissions of greenhouse gases that resulted in sea level rise or other changes that impacted on the life, property and well-being of a complainant, the violation could be cured by assisting the affected citizen adapt to changes already underway, such as by providing compensation, rather than by addressing the cause of the violation, such as by restraining greenhouse gas emissions.

These difficulties highlight the problems of using the language and machinery of human rights protection to deal with new challenges not foreseen in the original establishment of these legal institutions. In this respect Bertram has argued that climate litigation is not only 'unduly stressing existing treaties that were not created to address global warming' but also involves an unjustifiable request of international judges and other decision-makers 'to expend political capital that should be reserved for the core purposes of those treaties.'[54]

'Regressive' Proceedings

A signal feature of contemporary international environmental litigation is that it may be pursued in a range of fora. These include courts or tribunals with specialised environmental jurisdiction (such as the International Tribunal for the Law of the Sea with respect to marine pollution), courts with general subject-matter competence (such as the ICJ and the Permanent Court of Arbitration) and also bodies that have a non-environmental focus (such as those operating under the dispute settlement system of the WTO).

This jurisdictional patchwork is characterised by substantial gaps, but also overlaps which give rise to practical challenges of jurisdictional competition and coordination such as forum shopping.[55] An implication of this heterogeneity is that regimes that do not have an environmental focus may be used strategically to stymie progressive climate change policies adopted by governments. Nowhere are the prospects of this type of litigation more evident than in the context of the WTO; as inevitably there is a high degree of overlap between climate change policies and international trade law that may potentially lead to conflict.[56]

This conflict is of two types: (a) between provisions of the climate change regime and provisions in WTO agreements; and (b) between national climate change laws and policies and provisions in WTO agreements. Substantial efforts have been expended by the international community to ensure that the former type of conflict is avoided, as seen in the holding of Informal Trade Ministers Dialogue on Climate Change at the 13th COP of the UNFCCC in Bali in December 2007, where it was emphasised that the climate and trade regimes *can* and *should* be developed in a mutually supportive manner. And the likelihood for conflict is reduced by the substantial overlap in the membership of the WTO and climate regimes.[57]

Nonetheless it remains a possibility under the *Kyoto Protocol's* flexibility mechanisms that the carbon credits established could potentially be considered 'trade goods' or 'services' for the purposes of WTO rules.[58] As in the interaction between the WTO and the *United Nations Convention on the Law of the Sea*,[59] where there has been litigation under both regimes in relation to the same fundamental dispute,[60] members of the WTO may seek to litigate matters within the purview of the climate regime under the WTO dispute settlement system, in the hope of securing a more favourable outcome.

By contrast there remains a significant likelihood of the second type of conflict, given that governments may begin to lose patience with the slow pace of international negotiations on emissions reductions, and instead seek to forge ahead with unilateral climate policies in an effort to kick start greater international action. Such measures may conflict with WTO disciplines, and could be made subject to the powerful compulsory dispute settlement system established within the WTO.

The WTO was established in 1995,[61] providing an institutional superstructure for the international trade rules that had developed out of the *General Agreement on Tariffs and Trade* (GATT)[62] agreed at the end of World War II.[63] Integral to the WTO is a compulsory and binding system of dispute settlement established by the *Understanding on Rules and Procedures Governing the Settlement of Disputes* (DSU) which affords broad rights of standing to WTO members to challenge the legality of actions by other members.[64] Under this procedure members may initiate consultations and, should they fail, request the establishment of a Panel.[65] Appeals against Panel reports may thereafter be brought before a standing Appellate Body.[66] Unlike the previous *ad hoc* GATT panel system, the decisions of WTO Panels and the Appellate Body are adopted automatically and are therefore binding, unless the WTO Dispute Settlement Body decides against adoption by consensus.[67]

Since it commenced operation in 1995, over 356 complaints have been notified to the WTO, with almost 200 Panel and Appellate Body reports adopted.[68] By any measure the WTO dispute settlement system is highly active, and is vital to the development, consolidation and constitutionalisation of the WTO regime.[69] The system has also encountered environmental issues, with a number of disputes involving essentially environmental questions being brought before GATT panels and the WTO. This has in turn implicated environmental rules and standards.

The key organising principles of the WTO system are the most-favoured nation principle,[70] and the national treatment principle.[71] Together they limit the circumstances in which states may impose trade restrictive measures. Various national climate change policies may conflict with WTO rules applying to tariffs (border measures), the imposition of quotas, the granting of subsidies and labelling of products. However they are subject to Article XX of the GATT, which sets out certain limited exceptions for health and environmental measures:

> Subject to the requirement that such measures are not applied in a manner which would constitute a means of arbitrary or unjustifiable discrimination between countries where the same conditions prevail, or a disguised restriction on international trade, nothing in this Agreement shall be construed to prevent the adoption or enforcement by any contracting party of measures:
>
> ...
>
> (b) necessary to protect human, animal or plant life or health;
>
> ...
>
> (g) relating to the conservation of exhaustible natural resources if such measures are made effective in conjunction with restrictions on domestic production or consumption.[72]

The case law to date in the WTO has been promising in so far as it has recognised the capacity of governments to introduce measures having a restrictive effect on trade so long as these are justified on the basis of the protection of the environment.[73] In this regard the WTO Appellate Body has been responsible for a substantial greening of WTO law, certainly in comparison with decisions of ad hoc panels established under the previous GATT regime. There is no reason to think that disputes concerning climate change measures would be approached any differently,[74] however the outcome of a case will depend largely on the nature of the national measure that is implemented.

A measure that is likely to be introduced will be carbon taxes combined with border tax arrangements to ensure that domestic producers are not placed at a competitive disadvantage. A border tax arrangement would impose a carbon tax on imports equal to that which would have been applied had the import been produced domestically. Theoretically, it is possible to tailor such arrangements in such a way as not to infringe the WTO principles of national treatment and most favoured nation status. However, there remains considerable room for debate over the legality of particular measures on the basis of matters such as the method of calculating the greenhouse gas emissions resulting from the manufacture of a given product.[75] To avoid uncertainty and friction, Ross Garnaut has recommended that there should be a new WTO code on the subject.[76] Garnaut notes that it would be difficult to achieve this as support from WTO members would need to be unanimous. However 'given combined EU-US leadership, the credible threat of unilateral responses if no agreement were reached, the other incentives for cooperation on climate change, and the strong United Nations role ... an agreement may be possible.'[77]

'Administrative' Proceedings

The prospects for progressive and regressive climate change litigation remain somewhat speculative at this early stage. By contrast, we can expect a substantial number of disputes to come before the compliance mechanism established under the *Kyoto Protocol*. For example, where parties to the treaty fail to meet their greenhouse gas reporting obligations and commitments and, where applicable, to limit or reduce greenhouse gas emissions over the 2008–2012 first commitment period. This type of international litigation attracts the label as administrative in character because it is concerned fundamentally with upholding the implementation of the international climate regime, in much the same way as domestic proceedings may be possible to ensure the observance of national climate change laws and policies.

The *Kyoto Protocol* was the one of the earliest multilateral environmental agreements to adopt a non-compliance procedure (NCP).[78] Regardless of the legal form that will be taken by any successor to the *Kyoto Protocol*, it is likely that it will incorporate the same or similar compliance system. NCPs are designed to address disputes under multilateral environmental agreements in a more responsive and non-confrontational manner than traditional methods of inter-state dispute settlement such as international arbitration or judicial settlement.[79] The *Kyoto Protocol* NCP is more detailed and comprehensive than the first NCP to be established, that under the *Montreal Protocol on Substances that Deplete the Ozone Layer*.[80] Compliance is monitored and acted upon by the Compliance Committee which functions through a Plenary session, a Bureau and two branches: the Facilitative Branch and the Enforcement Branch.[81] The Committee is constituted by independent experts, operates according to highly structured procedures, and can make essentially self-executing decisions. As a consequence it is quasi-judicial in character.

The Facilitative Branch provides 'advice and facilitation to Parties' in implementing the *Kyoto Protocol*, promoting compliance having regard to the principle of common but differentiated responsibilities.[82] The Enforcement Branch is responsible for determining whether a party is complying with provisions of the *Kyoto Protocol*, including quantified limitation or reduction commitments,[83] and for applying the consequences set out in the NCP in cases of non-compliance.[84] In applying such measures, the Enforcement Branch is required to aim 'at the restoration of compliance to ensure environmental integrity, and shall provide for an incentive to comply.'[85]

Situations of non-compliance, described somewhat obliquely in the *Kyoto Protocol* NCP as 'questions of implementation', may be raised by expert review teams, any party with respect to itself, or any party with respect to another party.[86] The Bureau is responsible for allocating questions of implementation to the appropriate branch, which then undertakes a preliminary examination of the situation in order to determine whether the matter should proceed.[87] If the matter does proceed, then the party concerned may designate persons to represent it during the consideration of the question of implementation by the relevant branch.[88] Deliberations by the relevant branch are to be based, inter alia, on reports of any expert review team, the party concerned, or the party that raised the question of implementation.[89] Importantly, competent inter-governmental and non-governmental organisations may submit relevant factual and technical information to the relevant branch,[90] and the branch itself may seek expert advice.[91]

Many aspects of the procedure for the Enforcement Branch are judicial in character. Decisions must include conclusions and reasons, the party concerned may request the holding of a hearing (which will ordinarily be public) by the Enforcement Branch at which it will have an opportunity to present its views,[92] and final decisions by the Enforcement Branch as to non-compliance are to be made available to the other parties and to the public.[93] Decisions by each branch within their respective competencies are also self-executing. However, there does exist a system of review, by which a party may appeal to the Conference of the Parties serving as the Meeting of the Parties against a decision of the Enforcement Branch if the party believes that it has been denied due process.[94]

The Enforcement Branch has at its disposal a host of possible measures that can be applied in the event of non-compliance. When the requirements for developing systems for estimating and reporting emission levels have not been met, the Enforcement Branch is to make a declaration of non-compliance and develop a plan in collaboration with the party concerned in order to bring the party back into compliance.[95] The Enforcement Branch may suspend the eligibility of a party to transfer emission reduction units in relation to sinks of greenhouse gases, to participate in the clean development mechanism, or to engage in emissions trading when it determines that the party does not meet the eligibility requirements for these mechanisms.[96]

When the Enforcement Branch determines that a party has failed to meet its emission target, then it is required to declare that the party is not in compliance with its commitments. The defaulting party must then make up for any excess emissions during the second commitment period, and the new target will include a further 30% reduction by way of penalty for non-compliance. The Enforcement Branch must also suspend the eligibility of the party to participate in emissions trading, and develop a compliance action plan to ensure that the party meets its quantified emission limitation or reduction commitment in the subsequent commitment period.[97]

The *Kyoto Protocol* NCP has now been in operation for three years, and in this time has examined several situations of non-compliance. The Third Annual Report of the Compliance Committee to the Fourth Meeting of Parties to the *Kyoto Protocol* in Poznan in December 2008 reveals the types of compliance issues that are occupying the attention of the NCP.[98] The Committee received reports from expert review teams of initial and in-depth reports of fourth national communications as required by Articles 7 and 8 of the *Kyoto Protocol*. Overall there has been a high level of compliance with the reporting and review obligations under the *Kyoto Protocol*, however in the case of two parties compliance issues have arisen.

In December 2007, the Compliance Committee received a question of implementation indicated in the initial review report of Greece. This led ultimately to the Enforcement Branch finding in April 2008 that Greece was in non-compliance with a *Kyoto Protocol* obligation, the very first such finding in relation to any party. Greece was found in non-compliance with national system requirements for Annex B parties, that is those with emissions reduction or limitation targets. A national system is required by the *Kyoto Protocol* to ensure that such parties account for their emissions and establish compliance with their emissions targets. As a consequence of the decision of the Compliance Committee, Greece was required to submit a plan to address its non-compliance and, pending reinstatement, is prevented from participating in any of the *Kyoto Protocol* market mechanisms (Emissions Trading, the Clean Development Mechanism and Joint Implementation). At its sixth meeting, held in October 2008, the enforcement branch requested Greece to submit a revised plan.[99]

In April 2008, the Compliance Committee received a question of implementation indicated in the initial review report of Canada. The compliance issue turned on whether Canada's national registry met the requirements of the *Kyoto Protocol*. Parties are required to maintain a computerised registry in order to document holdings and transfers of carbon credits under the *Kyoto Protocol* market mechanisms, and parties such as Canada listed in Annex B with quantified emissions targets must have registries that meet specified technical standards. After hearing the views of Canada which argued that its registry was in full compliance with Article 7 of the *Kyoto Protocol*, the enforcement branch decided in June 2008 not to proceed further with the 'question of implementation' raised by the initial expert review.[100] At the time of this decision Canada was in compliance with the *Kyoto Protocol*, but the enforcement branch's decision makes clear that at the time of the initial review the status of Canada's national registry resulted in non-compliance with the applicable guidelines and modalities. Canada subsequently made a further written submission to the enforcement branch contending that it was beyond the competence of the enforcement branch to make a determination as to past situations of non-compliance.[101]

The *Kyoto Protocol* NCP is the main mechanism for ensuring compliance with the *Kyoto Protocol*, however it is not the only way in which disputes over obligations under the climate regime may come before an international forum. Under art 14 of the *UNFCCC* states may refer disputes to the ICJ by agreement, or if they have declared that they consent to the jurisdiction of the ICJ for such disputes. Alternatively, a party may request the establishment of a conciliation commission which would have the power to make recommendations which the parties are to consider in good faith.

A further avenue for disputation concerning the *Kyoto Protocol* is between governments and private parties, or between private parties themselves. As Brown has noted, disputes between states and private parties may arise in a variety of situations such as where a private entity enters into a contract with a state agency to carry out a Joint Implementation or Clean Development Mechanism project.[102] A dispute between private parties could arise in circumstances such as the non-delivery of emissions allowances.[103] In these types of mixed and private disputes there is no role for the *Kyoto Protocol* NCP or other intergovernmental dispute settlement system to ensure that the rules under the protocol operate smoothly. Instead the parties will need to turn to domestic courts or, as Brown argues, to international commercial arbitration which has advantages over domestic litigation.[104] These include generally greater efficiency, the availability of a forum perceived to be more 'neutral' and a familiarity by arbitrators with the issues of international law inevitably implicated in such disputes.

Conclusion

Human-induced climate change is taking place at a relentless and increasing pace, and there are few signs that sufficiently stringent international agreement will be reached in time to halt a dangerous rise in global average temperatures. As Lovelock has recently written, 'before long, we may face planet-wide devastation worse even than unrestricted nuclear war between superpowers'[105] which 'could kill nearly all of us and leave the few survivors living a Stone Age existence.'[106] In this context it is understandable and inevitable that all legal avenues will be explored in an attempt to provide a safe and stable climate for future generations. Social, cultural, economic and environmental change has always proven to be a stimulus for litigation to promote positive changes in the law and climate change is no exception.

This chapter has examined some of the opportunities for international courts, tribunals and quasi-judicial bodies to be invoked to address disputes addressing, in one form or another, issues of climate change. What were labelled 'progressive proceedings' are potentially possible under a plethora of regimes, from customary norms concerning transboundary harm, through to the law of the sea dispute settlement system established by the *LOS Convention*, and human rights complaints procedures. However, experience in international environmental litigation suggests that only those mechanisms open to non-state actors, such as the human rights procedures, will be extensively used given the reluctance of states to incur the political costs associated with asserting an interstate claim. Nonetheless, this situation may quickly change as additional effects of global warming are felt. Much as the threat of nuclear weapons promoted international litigation in the late 20th century, climate change may prompt similarly creative proceedings in the early 21st century.

However, before such proceedings are pursued it is possible that states will move unilaterally or in regional groupings to implement climate policies that could have implications for international norms, such as those relating to the protection of foreign investment and the liberalisation of international trade. This raises the spectre of 'regressive' proceedings which would act as a roadblock to innovation in climate change law and policy-making. Such proceedings could only be avoided if investment agreements and WTO rules are amended so as to recognise the necessary priority of climate mitigation policies. While agreement on this will be difficult, it is an indispensable and achievable objective if international consensus is reached on the seriousness of climate change.

Administrative proceedings, those commenced under the rubric of the international climate regime, will be an inevitable part of the implementation of the regime as it becomes progressively more stringent over time, and the market and other mechanisms it establishes become operational. The value of these proceedings, whether in the *Kyoto* NCP or its successor, or in the arbitration of disputes between private parties over the new carbon currencies created by the climate regime, is ensuring that the rule of law is brought to bear on the international regulation of greenhouse gas emissions. Over time this promises to be the most fruitful arena for climate change litigation in international courts.

Endnotes

1 Climate Justice Program, *Writing Competition* <http://www.climatelaw.org/competition> at 15 Apr 2009.

2 Ed Pilkington, 'Put Oil Firm Chiefs on Trial, Says Leading Climate Change Scientist', *The Guardian* (23 June 2008). Note that the criminal legal system is already engaging with climate change issues as climate change activists taking direct action are brought before the courts and raise environmental motives by way of defence or mitigation in sentencing: John Vidal, 'Kingsnorth trial: Will this open the floodgates to protest' *The Guardian* (12 September 2008).

3 See generally T Stephens, *International Courts and Environmental Protection* (2009) 21–62.

4 See generally J Peel, 'The Role of Climate Change Litigation in Australia's Response to Global Warming' (2007) 24 *Environmental and Planning Law Journal* 90; M Averill, 'Climate Litigation: Ethical Implications and Societal Impacts' (2007–08) 85 *Denver University Law Review* 899.

5 *United Nations Framework Convention on Climate Change*, opened for signature 4 June 1992, 1771 UNTS 107 (entered into force 21 March 1994) ('*UNFCCC*').

6 *Kyoto Protocol to the United Nations Framework Convention on Climate Change*, opened for signature 16 March 1998, 2303 UNTS 148 (entered into force 16 February 2005) ('*Kyoto Protocol*').

7 See Stephens, above n 3, 321–44.

8 R Garnaut, *The Garnaut Climate Change Review: Final Report* (2008) 232–4.

9 See generally BC Mank, Civil Remedies in MB Gerrard (ed), *Global Climate Change and US Law* (2008) 183.

10 Leading cases include *Gray v Minister for Planning, Director-General of the Department of Planning and Centennial Hunter Pty Ltd* [2006] NSWLEC 720 (the '*Anvil Hill Case*', which found that downstream greenhouse gas emissions from a proposed coal mine had to be considered during environmental impact assessment). See Anna Rose, '*Gray v Minister for Planning:* The rising tide of climate change litigation in Australia' (2007) 29 *Sydney Law Review* 725.

11 EA Posner, 'Climate Change and International Human Rights Litigation: A critical appraisal' (2007) 155 *University of Pennsylvania Law Review* 1925.

12 Ibid 1938.

13 J Hansen et al, 'Climate Change and Trace Gases' (2007) 365 *Philosophical Transactions of the Royal Society A* 1925.

14 Intergovernmental Panel on Climate Change (IPCC), *Climate Change 2007: The Physical Science Basis, Summary for Policymakers, Contribution of Working Group I to the Fourth Assessment Report* (2007).

15 Such as the *Trade Practices Act 1974* (Cth) s 52.

16 See, eg, *State of Connecticut v American Electric Power Corporation*, 206 F Supp 2d 265 (2004). See generally Averill, above n 4, 908.

17 See for instance *Humane Society International v Kyodo Senpaku Kaisha* [2006] FCAFC 116, a case brought in relation to breaches by a Japanese company of the *Environment Protection and Biodiversity Conservation Act 1999* (Cth) in taking whales in the Australian Whale Sanctuary offshore the Australian Antarctic Territory. See further Tim Stephens and Donald R Rothwell, 'Japanese Whaling in Antarctica: *Humane Society International Inc v Kyodo Senpaku Kaisha Ltd*' (2007) 16 *Review of European Community and International Environmental Law* 243–6.

18 (2002) 211 CLR 540.

19 (2002) 211 CLR 540, 544 (Gleeson CJ).

20 D Hunter and J Salzman, 'Negligence in the Air: The duty of care in climate change litigation' (2007) 155 *University of Pennsylvania Law Review* 1741, 1742.

21 Averill, above n 4, 911.

22 [1932] AC 562.

23 See MR Allen and R Lord, 'The Blame Game: Who will pay for the damaging consequences of climate change?' (2004) 432 *Nature* 551 (arguing that will improved monitoring and better mathematic analysis 'we could one day see Californian farmers suing member states of the European Union for authorizing emissions that threatened the security of their water supplies.')

24 See *Greenpeace Australia v Redbank Power Company* (1994) 86 LGERA 143; *Australian Conservation Foundation v Minister for Planning* [2004] VCAT 2029; *Gray v Minister for Planning* [2006]

NSWLEC 720; *Taralga Landscape Guardians v Minister for Planning* [2007] NSWLEC 59. For discussion of these and other domestic cases see the chapter by Justice Brian Preston in this volume.

25 See, e.g., *Queensland Conservation Council Inc v Xstrata Coal Queensland Pty Ltd* [2007] QCA 328 which overturned a decision of the Queensland Land and Resources Tribunal which refused to impose conditions on the expansion of a coal mine so as to address the mine's greenhouse gas emissions. Shortly after the decision was handed down amendments were passed to the *Mining Legislation and Other Legislation Amendment Act 2007* (Qld) to nullify its effects. See C McGrath, 'The Xstrata Case: Phyrric Victory or Harbinger?' In T Bonyhady and P Christoff (eds), *Climate Law in Australia* (2007) 214.

26 T Bonyhady, The New Australian Climate Law. In T Bonyhady and P Christoff (eds), *Climate Law in Australia* (2007) 8, 26. See generally Rory Sullivan, *Rethinking Voluntary Approaches in Environmental Policy* (2005).

27 T Stephens, 'A Slow Burn: The emergence of climate change law in Australia' in Gavin Birch (ed), *Water, Wind, Art and Debate: How environmental concerns impact on disciplinary research* (2007) 1; Clive Hamilton, *Scorcher: The dirty politics of climate change* (2007); Guy Pearse, *High and Dry: John Howard, climate change and the selling of Australia's future* (2007).

28 See the Draft Carbon Pollution Reduction Scheme Bill 2009 (Cth); Australian Government, *Carbon Pollution Reduction Scheme: Australia's Low Pollution Future* (2008).

29 Bonyhady, above n 26, 24.

30 T Stephens, 'The United Kingdom's Carbon Emissions Reduction Legislation' (2007) 24 *Environmental and Planning Law Journal* 249.

31 MD Adler, 'Corrective Justice and Liability for Global Warming' (2007) 155 *University of Pennsylvania Law Review* 1859.

32 *Trail Smelter Case (Canada/United States of America)* (1938 and 1941) 3 RIAA 1911.

33 Ibid 1965.

34 *Legality of the Threat or Use of Nuclear Weapons* [1996] ICJ Rep 226 [26] ('*Nuclear Weapons Advisory Opinion*').

35 UN Doc A/CONF.48/14/Rev.1 (1973) (adopted on 16 June 1972).

36 UN Doc A/CONF.151/5/Rev.1 (1992).

37 Report of the International Law Commission, 53rd Session, 366, UN Doc A/56/10 (2001).

38 Report of the International Law Commission, 56th Session, 153-6, UN Doc A/59/10 (2004).

39 R Verheyen, *Climate Change Damage and International Law* (2006).

40 Ibid 287.

41 *UNFCCC* Declaration 1.

42 *Nuclear Weapons Advisory Opinion* [1996] ICJ Rep 226.

43 *Nuclear Tests Cases (Australia v France) (Interim Measures)* [1973] ICJ Rep 99, *(Merits)* [1974] ICJ Rep 253; *Nuclear Tests Cases (New Zealand v France) (Interim Measures)* [1973] ICJ Rep 135, *(Merits)* [1974] ICJ Rep 457.

44 Opened for signature 16 December 1966, 999 UNTS 302 (entered into force 23 March 1976).

45 Opened for signature 16 December 1966, 999 UNTS 171 (entered into force 23 March 1976).

46 Petition to the Inter-American Commission on Human Rights Seeking relief from Violations Resulting from Global Warming Caused by Acts and omissions of the United States. Submitted by Sheila Watt-Cloutier on behalf of the Inuit of the Arctic regions of the United States and Canada <http://www.ciel.org/Publications/ICC_Petition_7Dec05.pdf> at 10 March 2008.

47 Ibid.

48 O Cordes-Holland, 'The Sinking of the Strait: The implications of climate change for Torres Strait Islanders' human rights protected by the *ICCPR*' 405 (2008) 9 *Melbourne Journal of International Law* 405, 421. See also Timo Koivurova, 'International Legal Avenues to Address the Plight of Victims of Climate Change: Problems and Prospects' (2007) 22 *Journal of Environmental Law and Litigation* 267.

49 OAS Res XXX (1948).

50 Cordes-Holland, above n 48, 417.

51 See generally Meinhard Doelle, 'Climate Change and Human Rights: The role of the international human rights in motivating states to take climate change seriously' (2004) *Macquarie Journal of International and Comparative Environmental Law* 179.

52 Cordes-Holland, above n 48, 435.

53 See for instance the way in which the European Court of Human Rights addressed a human rights complaint relating to industrial pollution in *Fadeyeva v Russia* [2005] ECHR 376. The court at [68] cited *Kyrtatos v Greece* [2003] ECHR 242 in concluding that '[t]he Court cannot accept that the interference with the conditions of animal life in the swamp constitutes an attack on the private or family life of the applicants.'

54 DR Bartram, 'International Litigation over Global Climate Change: A skeptic's view' (2007) 101 *Proceedings of the American Society of International Law* 61, 65.

55 See generally Stephens, above n 5, 271–303.

56 H Van Asselt, F Sindico and MA Mehling, 'Global Climate Change and the Fragmentation of International Law' (2008) 30 *Law and Policy* 423, 433.

57 A Petsonk, 'The Kyoto Protocol and the WTO: Integrating greenhouse gas allowance trading into the global marketplace' (1999) *Duke Environmental Law and Policy Forum* 185, 194.

58 C Carlane, 'The Kyoto Protocol and the WTO: Reconciling tensions between free trade and environmental objectives' (2005–06) 17 *Colorado Journal of International Environmental Law and Policy* 45, 58.

59 Opened for signature 10 December 1982, 1833 UNTS 397 (entered into force 16 November 1994).

60 *Conservation and Sustainable Exploitation of Swordfish Stocks in the South-Eastern Pacific Ocean (Chile/European Community)* (International Tribunal for the Law of the Sea, proceedings suspended) <http://www.itlos.org> at 16 April 2009; *Chile — Measures Affecting the Transit and Importation of Swordfish*, WTO Doc WT/DS193/2 (2000) (request for the establishment of a Panel).

61 *Marrakesh Agreement Establishing the World Trade Organization* opened for signature 15 April 1994, 1867 UNTS 3 (entered into force 1 January 1995) ('*Marrakesh Agreement*').

62 *1947 General Agreement on Tariffs and Trade.* See now the *1994 Marrakesh Agreement,* annex 1A ('*GATT 1994*').

63 See generally G Goh and T Witbreuk, 'An Introduction to the WTO Dispute Settlement System' (2001) 30 *University of Western Australia Law Review* 51, 52.

64 *Marrekesh Agreement* annex 2 (DSU).

65 Ibid art 4(7) and 5(4).

66 Ibid art 17.

67 Ibid arts 16(4) and 17(14).

68 K Leitner and S Lester, 'WTO Dispute Settlement from 1995 to 2006 — A statistical analysis' (2007) 10 *JIEL* 165.

69 See DH Cass, The 'Constitutionalization' of International Trade Law: Judicial Norm-Generation as the Engine of Constitutional Development in International Trade' (2001) 12 *European Journal of International Law* 39.

70 *GATT 1994* art I requires equality of treatment for like products, and therefore members of the WTO may not discriminate between like products of other members.

71 *GATT 1994* art III stipulates that domestic and imported products should be treated equally in terms of the application of internal regulations and policies.

72 Given the relatively recent emergence of environmental concerns in international law it is unsurprising that the only reference of the *GATT* is to 'natural resources' rather than the environment more broadly: Mark Harris, 'Beyond Doha: Clarifying the Role of the WTO in Determining Trade-Environment Disputes' (2003) 21 *Law in Context* 307, 309.

73 See especially *United States — Import Prohibition of Certain Shrimp and Shrimp Products,* WTO Doc WT/DS58/AB/R.

74 See generally G Triggs, '*Kyoto Protocol*: Compliance and the World Trade Organization' [2002] *AMPLA Yearbook* 49.

75 See M Genasci, 'Border Tax Adjustments and Emissions Trading: The Implications of International Trade Law for Policy Design' (2007) 1 *Carbon and Climate Law Review* 33.

76 Garnaut, above n 8, 233.

77 Ibid 233–4.

78 Decision 24/CP.1, Report of the Conference of the Parties on its Seventh Session, UN Doc FCCC/CP/2001/13/Add.3 (2002) ('*Kyoto Protocol NCP*').

79 G Ulfstein, 'Dispute Resolution, Compliance Control and Enforcement in International Environmental Law' in Geir Ulfstein (ed), *Making Treaties Work: Human rights, environment and arms control* (2007) 115, 132.

80 Opened for signature 16 September 1987, 1522 UNTS 29 (entered into force 1 January 1989).

81 *Kyoto Protocol NCP*, above n 78, s II.

82 Ibid s IV [4].

83 Ibid s V [4], [5].

84 Ibid s V [5].

85 Ibid s V [6].

86 Ibid s VI [1].

87 Ibid s VII.

88 Ibid s VIII [2].

98 Ibid s VIII [3].

90 Ibid s VIII [4].

91 Ibid s VIII [5].

92 Ibid s IX [2].

93 Ibid s IX [10].

94 Ibid s XI.

95 Ibid s XV [1], [2].

96 Ibid s XV [4].

97 Ibid s XV [5].

98 UN Doc FCCC/KP/CMP/2008/5.

99 UN Doc CC-2007-1-10/Greece/EB.

100 UN Doc CC-2008-1-6/Canada/EB.

101 Annual report of the Compliance Committee to the Conference of the Parties serving as the meeting of the Parties to the Kyoto Protocol, UN Doc FCCC/KP/CMP/2008/5 (2008) 45.

102 C Brown, 'The Settlement of Disputes Arising in Flexibility Mechanism Transactions under the *Kyoto Protocol*' (2005) 21 *Arbitration International* 361, 372.

103 Ibid 373.

104 Ibid 384.

105 J Lovelock, *The Vanishing Face of Gaia: A final warning* (2009) 21.

106 Ibid 22.

Climate Change and Resource Scarcity: Towards An International Law of Distributive Justice

Ben Saul

Scarcity, real and imagined, has profoundly influenced the creation of public international law — and propelled its absence. From the outset, the concept of scarcity begs definition and there are various ways of thinking about its properties. It is often thought of as an economic concept, when economists speak of scarcity of labour, capital, resources, commodities and so on.[1] It can be negatively defined by its opposite, abundance. In a literal sense, scarcity simply refers to an object's rarity. In this article, 'scarcity' is discussed contextually as it relates to the material goods (or resources) necessary to sustain human dignity — in particular, food, water and energy, and derivatively, the spatial areas (land, sea and air) on which they depend.

The central questions asked in this chapter are the following: how does international law respond — or fail to respond — and how ought international law to respond, to the special global problems posed by increasing resource scarcity? The chapter first traces some historical patterns in international responses to scarcity, particularly during the pre-1945 era of colonialism when weak norms on the use of military force tolerated territorial expansion and conquest as a hegemonic national economic response to resource scarcity. Resource scarcity was a driver of conflict, in the absence of peaceful distributive mechanisms to adequately respond to scarcity to prevent conflict.

Despite the post-war aspirations towards economic and social development embodied in the Charter of the United Nations, the modern international legal order remains substantially ineffective at pursuing or realising any coherent and effective program of distributive justice in respect of essential resources. If the international legal order is already failing to cope with the lesser historical challenge of relative scarcity of resources, the present era of human-induced climate change is likely to deeply strain that legal order because it is bringing about a shift from relative to absolute scarcity of key resources.

There are, however, some tentative and promising movements towards a distributive ethos in some specialised branches of international law, particularly environmental law, the law of the sea, and elements of international economic law. Such developments provide useful starting points for reorienting international law as a whole towards a (re-)distributive system that is capable of grappling with the resource scarcities that climate change will induce or aggravate.

This chapter is premised on the value of infusing international law with an ethic of global distributive justice. To some, its value is self-evident; for others, it requires much justification. To begin with a rough domestic analogy, in most liberal democratic orders, and often in other legal systems, some notion of distributive justice is at the heart of the political and legal order[2] and is essential to perceptions of its fairness, egalitarianism and ultimately its legitimacy. The social compact depends on maintaining some degree of social inclusion — taking care of those in greatest need — even if it does not necessitate any notion of absolute equality or radical socialist redistribution. Regulatory intervention to moderate the extreme effects of the market, and social welfare programs, are part of the mainstream political consensus in many liberal democracies, even more so in the aftermath of the global financial crisis of 2008–2009. In such nations, the battle is not about whether regulation and welfare are desirable, but over the nature and extent of regulation or welfare in any given context. Distributive justice is both a condition of political legitimacy but also a distinct ethical principle: it is right and just to redistribute some wealth for social purposes. That is not to claim that economic redistribution is sufficient to secure development, but it is part of the developmental trajectory, along with other elements that expand human freedom.[3]

In this context, this chapter considers the extent to which international law is, and has been, animated by any notion of distributive justice in responding to resource scarcity among nations. The domestic analogy is instantly imprecise: one might object that the structural preconditions for an ethic of distributive justice are absent on the international plane. Whereas (some) national political communities are premised on a relatively thick set of shared values, ethical impulses and feelings of solidarity — sufficient to activate a sense of social obligation towards others — in contrast, social relations are far thinner between states in the more rudimentary international community. Sovereign competition has tended to overshadow cooperation, at least when it comes to hip-pocket questions of distributive justice in the economic field.

Yet, the growing complexity and sophistication of the international legal order over time has led to a concomitant thickening of ethical and social relations, including the proliferation of constraints on sovereign freedom of action: human rights law, international criminal law, international environmental law, the quasi-constitutional law of the United Nations, and even an emerging global administrative law. The social conditions for an international ethic of distributive justice are no longer so tenuous: the deepening of social relations through networking by non-governmental organisations and technology that links disparate national populations is generating political demand for distributive justice. The evidence is everywhere, from G20 protests to fair trade campaigns to third world debt relief strategies. Humans are no longer stuck within a pre-modern world of small, face-to-face communities, unable to imagine and empathise with the plight of those outside our village.

There is a growing consensus that the operation of the market alone is not sufficient to achieve distributive aspirations: growth there is, but it is uneven, many are left behind, and some go backwards. Yet, as Alston observes, distributive justice is 'curiously anathema to the vast majority of international law scholars and practitioners' even though '[s]uch an aversion is neither defensible nor sustainable given the central importance of questions of distribution and patent injustice which both generates and flows from existing patterns of distribution'.[4] One need not be a Marxist to agree with Chimni that:

> MILS [Mainstream International Law Scholarship] does not seriously engage with the idea of global justice. The creation of a just world is not seen as the task of international law. It is believed that in the absence of some form of global sovereignty the Rawlsian principles of justice or any other theory of distributive justice, since these anticipate some form of a Global State, is not applicable.[5]

It is understandable that many international law practitioner-scholars are driven foremost by realism rather than a reformist moral sensibility: international commercial arbitration, world trade law and international investment law are lucrative areas of legal practice and there is little incentive to tilt the law towards the global poor; government international lawyers in developed countries realise that their governments would have something to lose in a redistributivist world; and the failed push for a New International Economic Order or stalled progress on liberalisation of world trade in agriculture would give even optimistic reformers cause to throw in the towel.

Even so, this chapter proceeds on the basis that distributive justice is an important global ethical principle that ought to animate and underpin the international legal order. As Pogge observes, 'moral convictions can have real effects — even in international relations'[6] notwithstanding the reductionist claims of realists. In any case, as a pragmatic or realist matter, distributive justice can assist in addressing poverty and defusing resource conflicts. In consequence, embedding a principle of distributive justice is ultimately likely to enhance the legitimacy of the international legal order which, in the long run, will help to establish the conditions for human dignity, shared prosperity and global peace.

Scarcity, Resources and Conflict in an Era of Climate Change

Resource scarcity has long been linked to conflict, in the absence of effective distributive mechanisms for preventing or diffusing recourse to violence to secure either essential needs or economic advantage. One major study which analysed 8000 conflicts over 500 years, between 1400 and 1900, suggested that long-term climate change correlated with an increase in conflict and population decline all around the world.[7] In particular, periods of climatic cooling directly affected agricultural production (by shortening the growing season, restricting the availability of arable land and leading to crop failure), thus restricting food supply (and increasing food prices) and increasing the likelihood of armed conflicts, famines and epidemics, ultimately reducing population size. The many other possible causes of war could not explain the temporal and spatial patterns of war shown by this research, which found that relative food scarcity is the key explanatory variable.[8]

While the historical patterns in that study related to periods of climate cooling, today's higher temperatures (that is, climate warming) will also affect agricultural production in comparably harmful ways. While warming may initially increase total bio-production, the negative effects of warming are likely to outweigh the positive effects overall,[9] particularly when combined with the other impacts documented above (sea level rises, ocean decline, disease, extreme weather, glacial retreat, energy price increases due to fossil fuel scarcity, resource disputes and social tensions). As the study puts it, continuing temperature acceleration 'might break the balance of human ecosystem that has been long established at a lower temperature'.[10]

Until our recent era of climate change, *relative* scarcity has been the norm. There is, for instance, presently enough food, energy and water to sustain the world's 6.7 billion people at a reasonable standard of living. The classical problem has been a distributive one: those who need a share of the world's abundant food stores don't receive it (including the 800 million people worldwide who face food insecurity, or the 50% of Indian children who are malnourished,[11] or the 50% of Indian women who are anaemic); some countries arrogate a disproportionate share of the world's energy to themselves while others go without or face unpredictable electricity brown-outs; and freshwater is diverted to industry and intensive agricultural commodity production while safe drinking water is absent in many parts of the developing world.

That bleak picture is set to darken given the likely shift from relative to absolute scarcity brought about by climate change. As noted earlier, global population growth, combined with the ecological effects of unabated climate change, may lead to absolute global scarcity of water, food, energy and arable land. Alarming predictions have been made about the potential for climate change to fuel war and other forms of violent conflict,[12] substantially due to likely competition over increasingly scarce key resources. At the extreme end, it has been suggested that the uncontrolled effects of climate change could generate conflicts on the scale of the world wars but lasting for centuries.[13] Another study suggests that 46 countries with 2.7 billion people are at high risk of violent conflict due to climate change interacting with existing underlying causes of conflict, while a further 56 countries with 1.2 billion people are at high risk of political instability possibly leading to violent conflict in the longer term.[14] As the UN Secretary-General stated:

> Things are easier in times of plenty, when all can share in the abundance, even if to different degrees. But when resources — whether energy, water or arable land — are scarce, our fragile ecosystems become strained, as do the coping mechanisms of groups and individuals. This can lead to a breakdown of established codes of conduct, even to outright conflict.
>
> Scarcity of resources, especially water and food, could help transform peaceful competition into violence. Limited or threatened access to energy is already known to be a powerful driver of conflict; our changing planet risks making it more so.[15]

Over time, the direct physical effects of climate change (some of which are already manifesting) are likely to include more extreme weather events (such as storms, floods, droughts); the melting of glaciers; desertification; rising sea levels and salination of land; and higher temperatures.[16] In the past two decades alone, there has been an increase from 200 to 400 major disasters per year.[17] These immediate impacts of climate change will be felt differently in different places. Some developing states will be particularly affected by climate change, because of both geographic vulnerability and their relative inability to adapt easily to the negative effects.

In various places, the direct physical effects of climate change are likely to detrimentally impact on the conditions essential for maintaining basic human dignity. First, increasing food insecurity is likely to result from factors such as higher temperatures, changing weather patterns and seasonal monsoons, water scarcity, shorter agricultural growing seasons, and the acidification of the oceans affecting the sustainability of (already stressed) global fish stocks. Already there are 850 million people malnourished worldwide. Yet the price of staple foods has increased by 80% over the past three years[18] and it is estimated that 100 million people have been pushed into poverty in the past two years alone due to rising food (and energy) prices.[19] In India, for example, rice and wheat production has declined due to temperature increases, affecting food security in an agriculturally-dependent and underdeveloped country. The World Bank estimates that 33 countries are at risk of political destabilisation and internal conflict due to food price inflation,[20] while 36 countries with 1.4 billion people are at risk of food or water scarcity by 2025.[21] The problem is aggravated by a turn towards protectionism in some countries, with recent bans on the export of staple foods in some producer countries, aggravating shortages and causing price rises in net import countries.

Second, climate change is making fresh water increasingly scarce in places where it is presently needed by human populations. The Intergovernmental Panel on Climate Change (IPCC) estimates that between 1.1 and 3.2 billion people could experience water scarcity by 2080[22] (unless, for instance, there is mass migration to those places where water may become relatively more abundant due to shifting rainfall patterns following from climate change). In Asia, the large number of people currently facing water scarcity will sharply rise as climate change affects surface water levels and curtails Himalayan glacial water levels in the long term.[23] Around 655 million people in Asia lack safe drinking water, while 1.9 billion people lack access to basic sanitation.[24] Climate change will aggravate the underlying causes of water scarcity in Asia, which hosts 60% of world's population but only 36% of the world's water resources.[25] As the UN Secretary-General has noted, high population growth, rising consumption, pollution and poor water management are contributing to a water crisis which is exacerbated by climate change:

> The consequences for humanity are grave. Water scarcity threatens economic and social gains … And it is a potent fuel for wars and conflict.[26]

Third, according to the World Health Organization ('WHO'), 'climate change poses substantial risks to human health, particularly among the poorest populations'.[27] The WHO estimates that the effects of climate change from the mid-1970s to 2000 caused at least 150 000 deaths.[28] Such risks arise because climate change affects the conditions necessary for good public health, such as the availability of water, food, shelter and social stability. Specific threats to public health include increasing heat waves, rainfall changes affecting water supplies, rising temperatures and the impact on agriculture, rising sea levels and coastal flooding, and population displacement.[29] Certain diseases (such as cholera, typhoid, hepatitis, giardia, malaria, dengue fever, plague and encephalitis) are likely to spread geographically to new areas and persist for longer, and affect populations which do not have immunity or lack health resources.[30] Climate induced migration may shift disease patterns; crop failure and famine may reduce resistance to infection; water scarcity and contamination may increase disease transmission; and public health infrastructure will come under strain.[31] Already most developing countries are unlikely to reach the health targets of the Millennium Development Goals (MDG) by 2015 and the effects of climate change are likely to aggravate inequities in health between developed and developing countries.[32]

The direct impacts of climate change described above will aggravate existing structural weaknesses in some states. For example, economic weakness in some societies narrows income possibilities and deprives the state of resources with which to meet basic needs,[33] and may generate internal conflict over resource distribution. Further, political instability and poor or weak governance make it harder to adapt to the effects of climate change and to resolve conflicts without violence.[34] As a result, conflict over essential but increasingly scarce resources — water, food, land and energy — may increasingly arise.

Already some conflicts have been driven by disputes about land and resources that are at least partly due to desertification (aggravated by climate change) in northern Nigeria, Sudan, Kenya and Ghana. The potential for resource conflict is aggravated by competitive policy responses to increasing resource scarcity and consumer demand, such as the expanding 'resource diplomacy' of China and India as they court resource-rich countries in Africa and the Asia-Pacific.

In addition (and sometimes connected) to conflicts animated by resource scarcity, further (and more conventional) conflicts over territorial and maritime boundary disputes may arise as a result of climate change. In the Arctic region, for example, the rapid melting of polar ice — with a possibly ice-free Arctic by 2060[35] — potentially opens up the region to newly accessible, commercially exploitable maritime trade routes as well as possible resource exploitation.[36] This may aggravate longstanding disagreements about the status of certain Arctic waters as international straights or internal waters (particularly as between Canada and the United States). Russia has asserted continental shelf claims under the Arctic, exemplified by its infamous planting of a Russian flag on the seabed by submarine in 2007. Denmark has also staked Arctic claims, while the area is becoming increasingly militarised, with Canada announcing a $5 billion Arctic defence plan in 2008. Likewise, in Antarctica, increasing activity by major powers (including China) may put pressure on the Antarctic treaty regime, including by undermining the suspension of sovereign claims to territory and the non-exploitation of resources, potentially giving rise to new tensions. The possibility of energy shortages due to a growing oil crisis over time are likely to accentuate competition over scarce oil and gas resources, at least if alternative energy sources are not developed quickly enough.

In the Pacific Ocean region and South Asia, the likely loss of land territory due to rising sea levels may lead to the disappearance of islands or other maritime formations, or result in certain areas becoming uninhabitable. Drought, salination, storms and sea level rises are seriously affecting island states such as Micronesia, Tuvalu and Kiribati, while in Maldives, a two-metre sea rise would inundate 1200 islands.[37] At the same time, the coping capacity of some of these states is being decimated by other economic impacts of climate change; coral bleaching in the Pacific, for instance, may result in up to a 50% loss in GDP in Pacific islands by 2020.[38] Sea-level rises (or changes, driven by climate change, in shared boundary river courses, lakes or wetlands) place pressure on the maintenance and stability of existing maritime and territorial boundaries and claims. On one estimate, only 160 of 365 potential maritime boundaries are already determined,[39] and the shifting of boundaries through rising sea levels may well complicate the settlement of disputes.

Overarching all of the above physical effects of climate change is the prospect of large-scale displacement and migration prompted by the combination of effects. The IPCC suggests 50 million will be displaced by 2010, while the *Stern Review on the Economics of Climate Change* suggests up to 200 million may be displaced by 2050.[40] In Asia, for instance, 40% of the population lives within 60 km of the coast and a 45 cm sea level rise in Bangladesh will displace 5.5 million and inundate 11% of Bangladesh's territory.[41] Mass displacement carries risks of internal and inter-state conflict, including due to political sensitivities about migration control, the inflammation of ethnocentric political agendas, and increasing isolationism.

There is a twofold risk of radicalisation: first, within communities faced with receiving large numbers of 'climate refugees' and second, within displaced communities frustrated by the unwillingness of the international community to adequately respond to their plight. Dissatisfaction may be aggravated by concerns that those who bear a disproportionately large burden of the impact of climate change — developing countries — are not the major historical source of carbon emissions. Any failure by developed (or relatively developed) countries to responsibly respond to climate-induced displacement may generate further tensions.

Climate Change as a 'Threat Multiplier'

Given the potential alarmism surrounding the conflict threats posed by climate change, it is important to acknowledge the risk of positing any kind of automatic, causal connection between climate change and conflict, and not only because the empirical data is still in its infancy.[42] Conflicts are ordinarily the product of a complex array of causes, including broader patterns of environmental change, and it is often difficult to isolate any single cause, let alone climate change, as the key explanatory factor.[43] Even if the various interconnected structural causes of conflict can be identified, it must still be recognised that human agents ultimately choose to wage war and there is never an inevitable chain of causal factors that leads to conflict.

To use one example, the conflict in Darfur, Sudan, is often pointed to as the first example of a 'climate war', including by Britain in the Security Council. However, it is an oversimplification to assert that the conflict there is caused by climatic or broader environmental change. Rather, a complex array of causes are relevant: historical grievances, racial perceptions, unfair distributions of economic and political power, land scarcity and population stress, small arms proliferation, militarisation of young people, weak state institutions, a lack of respect for human rights norms and so on.

Importantly, coping, adaptation and adjustment capacity can often defuse conflict from materialising. The longitudinal study of long-term climate change and conflict mentioned above observes that the impact of climate change depends on the variety of human social mechanisms (including measures of adaptation and mitigation) deployed in response to it, which may include warfare, economic change, innovation, trade, migration, resource redistribution and the capacity of social and political institutions.[44] During the historical periods of climate change noted earlier, such responses often did not keep pace with ecological change, resulting in demographic decline and conflict.

The same study acknowledges that economic changes after 1900 due to industrialisation mean that climate-related changes in global agricultural production are less likely, than in the past, to drive population shock and conflict. Further, today there is greater adaptive capacity than in the pre-modern world due to more sophisticated and robust social institutions for managing conflict (on the national and international planes), more advanced societies and better technology. At the same time, however, the world now faces higher population stress (with 9.2 billion people by 2050),[45] expectations of higher living standards, and stricter political boundaries (which limit migration choices and potentially fuel conflict). While climate change is a matter of adaptation for those in the developed world (requiring perhaps a 1% reduction in GDP), in the developing world it may be a question of survival (as in the Pacific and parts of Africa and Asia).

Consequently, while climate change may not inexorably lead to conflict, given the complexity of causes of conflict and the human potential for averting it, climate change can certainly be viewed as a 'threat multiplier'.[46] Climate wars are 'unlikely',[47] but there may well be increasing tensions over resource distribution and the intensification of other underlying causes of conflict, such as state failure, political instability or social violence.[48]

Paradoxically, that threat may be aggravated by certain measures taken in response to climate change. For example, if nuclear power for civilian uses replaces fossil fuels without sufficiently stringent safeguards and controls on nuclear energy, there may be a greater risk of nuclear proliferation and thus potential for conflict. Resort to biofuels has helped to increase some staple food commodity prices which may stimulate food riots and political tensions. Imposing new quarantine measures against the proliferation of climate-driven diseases could be viewed as discriminatory against migrants from developing countries, or as self-interested trade barriers and protectionism, creating further frictions.[49] The privatisation of water resources could increase water prices and — without adequate intervention to assist the poor, as is occurring in some parts of Latin America — fuel possible conflict.

Finally, developing countries may resent attempts to reduce their energy consumption without adequate compensation, since it can be seen as prejudicing their right to development which the developed world has already enjoyed and achieved. The absence of electricity in many communities in developing countries results, for instance, in adverse health effects (in turn affecting economic development) and preserves inequalities in health between developed and developing countries.[50] Too rapidly or stringently reducing emissions and generating possible energy shortages can lead to further internal instability in some societies.

A Short History of Scarcity and International Law

Historically, international law had little to say about scarcity. The colonial era was predicated on assumptions of abundance and the unlimited exploitability of natural resources, including those now regarded as non-renewable. Where powerful states encountered resource scarcity within their own territorial limits, the law prior to the *Charter of the United Nations* allowed states wide latitude in response. In the absence of a comprehensive prohibition on the use of force, hegemonic states took advantage of a sliding scale of means, from cooperative trade relations to economic coercion, unequal treaties (under which the conferral of legal personality on 'inferior' civilisations was precisely for the purpose of creating and allowing the transfer of property rights and sovereignty)[51] and exploitative concessionary contracts, to militant expansion to secure resources (including territory, labour and commodities). As Anghie writes, the colonial era from the 15th century onwards was one in which 'commercial exploitation necessitate[d] war'.[52]

Resource competition underlay many of the great historical conflicts in Europe, drove European colonial ambitions, and helps to explain the global conflagrations triggered by Nazism's quest for lebensraum and Japanese expansion in Manchuria and the Far East. The pre-1945 response to scarcity was one of normatively unconstrained unilateralism, in which the only real limits were pragmatic (such as whether a state was sufficiently powerful to take, hold and administer territory) and moral (whether political and public ethics permitted such strategies).

The post-war multilateral order was famously intended to save succeeding generations from the scourge of war, but in so aiming it nonetheless left untouched many of the structural conditions which may, given the right conditions, threaten global peace. There are certainly welcome distributive aspirations embodied in the post-war architecture: arts 55 and 56 of the *Charter of the United Nations* commit states to cooperating to achieve higher living standards and conditions of economic and social progress and development, while art 28 of the *Universal Declaration of Human Rights*[53] imagines a social and international order in which rights and freedoms can be fully realised.

Indeed, one would hope that the aspiration of law would be to even out distributive distortions in global social, political and economic structures. This is not a socialist plea for equitable redistribution of economic wealth. There is no particular virtue in levelling wealth, other than perhaps to assuage feelings of envy (and in the process, generate new envy in those whose wealth is redistributed). But there is a profound virtue in redistributing enough wealth to maintain the human dignity of all. Such a redistribution would still leave most in the developed world still very well off indeed.

International law either says little about this kind of distributive justice; or says something but ineffectively; or even works actively against it. International law has long said little about either internal or international distribution of essential resources; distributive questions were left to the market, or to politics, and are still only lighted touched upon by human rights law. Any human rights to food, water or energy security, or to development, are precarious at best and certainly poor cousins to, for instance, classically justiciable political rights of free speech and so on. The 'developmentalisation' of human rights has itself 'given rise to concerns ... that a narrow, market-oriented version of human rights is being used to promote economic liberalisation and globalization around the world',[54] resulting in a co-option of rights rather than an effective challenge to international law's distributive void.

Most pertinently, any redistribution effected by human rights law operates at the national level — that is, the zone in which states owe obligations to those within their territory or jurisdiction. Even contemporary debates about the extraterritorial application of human rights law (in the context of the right to life or freedom from arbitrary detention) do not extend to whether, for instance, a wealthy developed state such as Australia ought to also, as a matter of law, seek to guarantee the rights to food or water of the poor in developing countries, who are well within reach of the economic power of Australia, even if they are well beyond conventional concepts of jurisdiction. To the extent that human rights law provides for any redistribution, it remains conservative in its ambition only for domestic not transnational transfers.

Essentials such as food, energy and water have thus long been relegated to a discretionary field of national political decision-making and public policy choices in the exercise of foreign affairs. There is no international legal obligation on states with abundant resources to transfer food, energy or water to states which face scarcity; such decisions are consigned to the legal vacuums of foreign aid or humanitarian policy, usually guided by strategic, security and economic considerations. Indeed, benevolently forgiving Third World debt is not the same as not imposing it by law in the first place.

Thus, the MDGs may aspire to transform the lives of the poor, but a political aspiration is not the same as a legal obligation. This is not to doubt that consensus politics and 'soft law' instruments cannot deliver as much — or sometimes more — than hard law agreements. The MDGs have realised some impressive achievements, but a range of targets are unlikely to be met by 2015 — for instance at the halfway mark, 2.5 billion people still lack adequate sanitation, half a million mothers per year still die in child birth, a quarter of children in developing countries are malnourished, and a third of the urban population of developing countries still live in slums.[55] An adequate financial commitment has not followed political aspiration: in 2007, foreign aid from developed countries declined for the second consecutive year;[56] the global financial crisis from 2008–09 is likely to further set back financing for development; and despite constant reaffirmation,[57] few developed states reach the UN target of contributing 0.7% of GDP towards development assistance. An additional USD18 billion per year is needed to meet the G8 target of doubling aid by 2010.[58] As Wolfensohn notes, the USD1000 billion spent annually worldwide on military and defence (in 2000) is roughly 20 times as much as is spent on development funding.[59]

In sum, abundance can be lawfully squandered; the impoverished can lawfully starve; and the struggle for a right to development or a new economic order has never taken the law very far. Developed countries have dismissed the right to development as a mere aspiration, goal or claim, while developing countries have regarded the right to development primarily in economic terms, 'irrespective of environmental and social costs',[60] and often without consideration of inter- and intra-generational equity. There is no international law of everyday life;[61] one can rightly feel incredulous about a legal system that makes little intervention when life is at risk of a slow and hungry end. The law may not be very good at stopping genocide; but it is far worse in dealing with the more pedestrian harms of debilitating hunger or chronic waterborne disease.

Developing States and Decolonisation

Post-war decolonisation did little to dislodge the assumption of infinite exploitability in international law. As Anghie observes, '[t]he end of formal colonialism, while extremely significant, did not result in the end of colonial relations',[62] because of the persistence of international economic law in a form that preserved the economic dependency of newly decolonised states. The National International Economic Order aspired to balance the market's emphasis on efficiency with other values such as fairness and distributive equity.[63] Yet, developing countries had little success in fundamentally reorienting the logic of the international economic order, although at the margins improvements in trade, investment, finance and development policy have occurred over time. Indeed many developing countries bought into the received economic order — perhaps accepting the futility of resistance with the end of communism and the triumph of liberal economics, perhaps because many leaders of newly independent states could enrich themselves through that order.

The insistence of newly independent and developing states on securing permanent sovereignty over natural resources[64] was a distributive victory of sorts, transferring the right to exploit resources away from former colonial powers and their investors and back to the populations of decolonised states. Some arbitrators looked sympathetically upon newly independent states seeking to nationalise natural resources,[65] although the distributive promise was often tempered by the reality that newly liberated resources simply enriched new, indigenous elites, rather than realising the potential for emancipating impoverished populations.

The thrust of permanent sovereignty over natural resources was about who possessed the right to exploit, not on the equitable distribution of profits among groups within states, or 'how just the exploitation itself was'.[66] Moreover the 'international resource privilege' acquired by newly independent states brought with it 'powerful incentives towards coup attempts and civil wars in the resource-rich countries',[67] generating much instability in developing countries and often undermining development.[68]

Permanent sovereignty over natural resources was even less about the limits of exploitation; doing nothing to challenge the assumptions of abundance and exploitability in international law, or to promote a rational and equitable global distribution of resources to where they are most needed, or to pay due regard to questions of inter-generational equity. To give some examples, water shortages are often aggravated by upstream damming of transboundary river systems, accentuating shortages downstream, by states pursuing a wide view of resource sovereignty. Fisheries policies driven by 'resource nationalism' in developing Pacific countries optimise neither revenue for those countries nor the sustainability of those resources.[69] The Organisation of Petroleum Exporting Countries (OPEC) operates as a cartel which resists governance by international norms and which is premised on profiteering from unsustainable oil production.

In an era of climate change, post-war principles favourable to developing countries have become an obstacle to dealing with climate change. The emphasis on permanent national sovereignty over natural resources may exacerbate the scarcity of food, water and energy that climate change is aggravating. For example, some developing states responded to food price spikes in 2008 by prohibiting food exports — potentially safeguarding one's own population but at the same time leaving even more vulnerable populations elsewhere to face even higher food prices.

Similarly, the very low (and unreal) price of oil in producer countries such as Venezuela stimulates wasteful consumption and aggravates carbon emissions. For Venezuela and Nigeria to sell petrol at a few cents per litre in a fossil fuel dependent world is a perverse utilisation of a 'sovereign' natural resource. Faced with increasing scarcity of essential resources, stronger global governance of scarce resources becomes critical, rather than the squirreling away by sovereign states of what diminishing resources they may have left — to the detriment of those elsewhere facing even more precarious futures.

Meanwhile international economic law generally avoids distributive questions and leaves them to the market, on the assumption that economic growth will raise living standards. Sophisticated procedural rules have developed, but these do not go to the heart of substantive questions of resource distribution. For example, within the framework of the World Trade Organization (WTO), states may be required to trade in certain ways, but there is no requirement to trade at all, or at fair prices. States are perfectly entitled to opt out of the system, revert to inward-looking policies of national self-sufficiency in food or energy or other resources or commodities. So much faith has been invested in world trade law as a panacea for development, but international trade accounts for 20% of global GDP[70] and so can have only a relatively limited effect on major economic structural inequalities. Liberalisation of trade in agriculture has floundered since the Doha Round, even though it is widely agreed that it is an essential component (and indeed a plank of the MDGs) of a market-based approach to economic growth and development in developing countries.

Likewise, there are numerous bilateral investment treaty arrangements which specify the terms and safeguards of investment, but nothing at law which directs investment towards areas of most need over areas of greatest profit or least investor risk. As Philip Allott points out in Eunomia, there is an essential structural obstacle to a genuinely distributive international legal order:

> How can such a distributive effect by governmental action occur in the international economy, given that the inherent powers of individual states are powers in relation

only to the part of the international economy which is within their respective individual sovereignties and given that there are no international constitutional organs with distributive economic power?[71]

One Response to Scarcity: Unilateralism

In a world approaching absolute scarcity of key resources, distributive politics would still have a role to play, but the issue then becomes what degree of starvation or water insecurity each person should face rather than whether those problems can be eradicated, as the MDGs currently hope. Set against such an unhappily limited distributive aspiration would be more powerful forces. First, the increasing inclination of states to secure their own resource supplies by limiting exports and thus exacerbating absolute scarcity elsewhere (consider bans on rice and fuel exports in some producer countries in 2008).

Second, as scarcity becomes more acute, attempts to secure resources by force and expansion become more likely. It is no surprise that the Pentagon and the Australian Defence Forces have both identified climate change as a potential national security threats. Chinese 'resource diplomacy' in Africa and the Asia-Pacific is reconfiguring the geopolitical landscape. There is a risk of securitising the problem of scarcity and climate change — while strategically understandable to mobilise public and government attention, it risks converting concerns about the human security impacts of climate change into national security concerns, where the emphasis becomes how individual countries can secure access to resources to the exclusion of others, replacing global coordination with unilateral competition.

The Multilateral Response to Scarcity

If international law is already exceedingly weak in responding to relative scarcity, the prospects for multilateral legal responses to absolute scarcity might seem even more remote. There is, as always, hope in technology: genetically improved crop yields; renewable energy resources; and desalination plants powered by new energy technologies. Perhaps the market and political choices will drive technological solutions in time to save us.

But there is also hope in international law. First, unlike in pre-modern periods of climate-related conflict which generated conflict, today we have a greater adaptive capacity and more robust social institutions, both national and international.[72] Second, great strides have been made in constructing new regimes to deal with scarcity, particularly in international environmental law and the law of the sea.

International Environmental Law

International environmental law responds to both natural resource scarcity and, to a lesser extent, to the unequal distribution of scarce resources. On the first issue, environmental law introduces a variety of interrelated normative constraints on the commercial exploitation of scarce resources. Through the overarching principle of sustainable development, use and management of natural resources is elevated beyond exclusive sovereign, domestic jurisdiction to an international concern and subjected to international regulation and limitation, even if uncertainty remains about the scope of the principle (including the breadth of national discretion in implementation) and its relationship to human rights and international economic law.[73] The subsidiary principle of sustainable utilisation, use and conservation of natural resources, and the related precautionary principle, aim to preserve scarce resources and challenge the classical approach to such resources as infinitely exploitable.

While there is no systematic obligation in international environmental law to conserve and sustainably use all natural resources,[74] numerous specialised regimes have developed to protect particular resources: special ecosystems, endangered species, wildlife and habitat, biological diversity, natural heritage, international water-courses and shared groundwater, and atmospheric and marine resources (including fisheries).[75] There is a concern to balance commercial and developmental imperatives against the conservation and sustainable usage of scarce resources.

Preserving or sustainably developing scarce resources is not the same as redistributing them to those who need them most. Principles of sustainability and conservation limit how quickly and in what manner resources may be exploited, but the core of those principles do not challenge who may exploit them and benefit from the profits. Related principles of environmental law do, however, begin to address distributive questions.

First, a nascent concept of intra-generational equity is evident in some areas of environmental law, including cooperation against poverty (*Rio Declaration on Environment and Development Principle 576*) and special concessions for developing countries in the Vienna Convention for the Protection of the Ozone Layer[77] and *United Nations Framework Convention on Climate Change*[78] (such as financial assistance, capacity building, and 'common but differentiated responsibilities').[79] The *Kyoto Protocol's*[80] climate change adaptation fund within the Global Environment Facility is a limited distributive mechanism, being substantially underfunded, and adaptation costs in developing countries are still not met even when supplemented by World Bank investment initiatives.[81] As others have noted, those responsible for emissions and those capable of providing assistance should bear the costs of financing adaptation in developing countries.[82]

In another area of environmental law, the Biological Diversity Convention provides for a 'fair and equitable' distribution to developing countries of the benefits of the exploitation of genetic resources, despite uncertainty about the interface with international intellectual property law.[83] The principle of intra-generational equity aims at 'redressing the imbalance between the developed and developing worlds and giving priority to the needs of the poor'.[84] It is not only concerned with equity within communities but also across them at the international level.[85]

Transboundary freshwater governance provides a good example of the progressive development of a distributive approach to resource scarcity. Until the mid-20th century, '[b]ecause water was generally abundant, the utilisation of water did not often create a conflict over use that extended beyond national borders'[86] and water was generally treated as subject to exclusive sovereign regulation. Increasing post-war economic and nationalist competition among water users, short term and unsustainable economic development,[87] demographic growth and pollution increasingly generated water scarcity in various regions. As Elver notes:

> ... freshwater scarcity and management is one of the most difficult challenges facing international law. Solving freshwater scarcity in an equitable manner contradicts the vital principles of international law, such as "absolute state sovereignty" over natural resources.[88]

Regimes on international watercourses and shared groundwater have developed which seek to limit national sovereignty over water resources by recognising the legitimate interests of other water users. In particular, the principle of equitable sharing in the beneficial use of water, and procedural norms enabling cooperation among water users, are distributive mechanisms in respect of current users, although there is less clarity in balancing present users against future users,[89] and equitable utilisation of watercourses chiefly concerns water quantity rather than quality.[90]

International environmental law pays less attention to water resources wholly within national territory, although the human 'right to water' is arguably engaged within national jurisdictions to ensure at least a modicum of redistribution to meet basic human needs. Here the conceptualisation of water as a human right rubs up against other ways of conceiving of water: as rights, as common heritage, as a tradeable commodity, as a sovereign resource, and so on. There is indeed contestation about how to best achieve the related goals of conservation and redistribution. On one view, pricing water as a commodity according to its scarcity is necessary to preserve it by preventing wastage.[91] But the distributive impacts of privatising and/or commodifying water (including at the insistence of donors imposing reform conditionalities) may be uneven: on the one hand, it may deprive the poor of access to a fair share of essential water resources, while on the other hand it may prevent the depletion of those water resources for everyone, as well as future generations. Even in developed economies such as Australia, efficiency gains brought by market pricing of water may come with social and equity costs.[92]

Second, a concept of inter-generational equity is an emerging, if imprecise, subsidiary principle of sustainable development,[93] which aims to ensure that future generations can benefit from the natural environment and resources at least to the same level as the present generation. Notions of inter-generational equity are implicit in environmental law regimes governing whaling, dumping at sea, Antarctica, climate change, the ozone layer, biological diversity and some fisheries.[94] Whether inter-generational equity is a right as such, whether it is justiciable, and who might have standing to enforce it in particular fora[95] are distinct questions from whether a concept or principle exists: they are, rather, matters of legal form, procedure or jurisdiction. Nonetheless, the principle remains ill-defined in relation to determining matters such as the appropriate balancing between the interests of current and future generations.

International Law of the Sea

The international law of the sea includes a number of rules with distributive ambitions, although in practice some of them have limited effect. The *United Nations Convention on the Law of the Sea*[96] recognises the rights of geographically disadvantaged states (art 70), provides a right of landlocked states to share in exploiting any surplus of fishing resources in Exclusive Economic Zones (art 69), and acknowledges the interests of developing states in a given region in the utilisation and conservation of living resources (art 62). UNCLOS also provides for the transfer of marine science and technology (art 266) and the equitable sharing of benefits of the exploitation of continental shelf resources beyond 200 nautical miles (art 82).

Most prominently, the deep sea bed (that is, an area beyond national jurisdiction) is recognised as 'common heritage' of mankind and its governing principle is the equitable sharing of any benefits of its exploitation (art 140). The designation of the deep seabed as common heritage was deeply contested by certain developed states and produced a 'legal stand-off' at the time.[97] Since the advanced industrial economies are most likely able to afford to invest in the technology and infrastructure to extract resources from the relatively inaccessible deep seabed, from a purely economic point of view it is understandable that they were unhappy about having to share the proceeds with others not involved in its exploitation. No equivalent distributive principle applied to the high seas, which was ripe for exploitation by any state able to undertake it, at least until agreements on specific fisheries or marine resources were concluded. Since little exploitation of the deep seabed is currently underway, disagreement about equitable sharing of benefits presently is moot, though likely to be reignited as increasing resource scarcity globally decreases the relative cost of its exploitation.

Assuming the principle does gain traction in practice, it remains a crude distributive mechanism in one sense: it is foremost a government-to-government sharing of benefits. That does not necessarily serve the purpose of 'common heritage' — intended to benefit humanity as a whole — if it simply results in developed countries transferring benefits to corrupt, undemocratic leaders in developing countries, where those leaders are unlikely to deploy the additional resources for the benefit of their peoples. As one writer warns, the common heritage principle should rather be used as a means 'to ensure that the resources of the global South are prioritised to address questions of hunger, disease and poverty, and not simply to further enrich the wealthy'.[98]

An International Law of Distributive Justice

What might a distributive international law look like, in the face of growing resource scarcity? A more rational and equitable approach to the utilisation of resources and the benefits of their exploitation must replace the old view of permanent national sovereignty over resources. Stronger international intervention needs to cut across many regulatory fields.

In international economic law, there is need for stronger global or multilateral governance and management of scarce resources that are essential to human dignity, to ensure a more rational, equitable and timely global distribution of benefits. Such intervention is needed across the resource spectrum. To give one example, transnational governance of freshwater resources needs to improve on the best practice experiences in some river management contexts, building on the lessons learned and successes achieved in those specific contexts. Emerging regimes such as that governing transboundary watercourses and groundwater aquifers will need to factor climate change impacts into the implementation of key norms such as the principle of equitable and reasonable utilisation, the obligation not to cause significant harm, the protection and preservation of ecosystems, and the 'precautionary approach'.[99]

In international trade law, radical steps forward in the regulation of global agricultural trade are needed, including the expedited removal of protectionist distortions in developed countries and enhanced market access for developing countries.[100] Agricultural subsidies in developed states run at more than $300 billion per year,[101] with each European cow subsidised by €2.50 per day while 300 million Africans live on less than €1 a day.[102] Rather than subsidising European farmers, international law should safeguard the absolute poor in developing countries, where the market is failing to provide affordable food for vulnerable populations. At the same time, international law needs to become sensitive to climate change impacts of trade flows, such as emissions from transporting high volume or heavy goods to distant places when local substitutes are readily available.

In international intellectual property law, better sharing and improved transfer of technology is needed across a range of areas essential to human development, whether in the patenting of medicines against disease, new energy technologies, or genetically modified organisms and gene technology. In international investment law, there are a number of impediments to the implementation of climate change mitigation measures[103] which require attention if private investment patterns are not to aggravate the risks associated with climate change. In the area of labour migration, international law needs to think seriously about how to better open domestic labour markets in developed countries to foreign labour from developing countries on more equitable terms, as another form of redistribution.

In international environmental law, the principle of sustainable development needs to be modified to include the long-term and global effects of climate change in judgments about the sustainability of particular development projects, so as to defuse potential conflicts arising over unsustainable or ill-devised projects. Those international institutions and donor states which are involved in sponsoring development also need to be mindful of both the climate change impacts of development initiatives and the impact of climate change on such initiatives. The World Bank, for instance, estimates that 40% of its development projects are affected by climate change.[104] At the same time, the World Bank has been increasing its lending for fossil fuel based energy relative to renewable energies.[105] From an economic and developmental perspective, it makes little sense to invest in projects maladapted to dealing with the effects of climate change, and even less sense to use international money to aggravate climate change where it can be avoided.

Climate change impacts should consequently be factored in the approvals process for international loans and assistance, and multilateral or bilateral foreign aid, while climate change adaptation and mitigation measures could be made conditions of finance or assistance. At the same time, such conditionality must be sensitive to the risk of forcing premature energy reductions onto developing states, which itself risks generating conflict if developing states feel aggrieved that their development process is being thwarted by already developed donors. Ultimately, climate change adaptation cannot be regarded merely as a technical and developmental activity, but will require multilateral governance and norms to intervene 'in difficult and highly charged internal resource management issues'.[106]

Nothwithstanding distributive reforms in specific branches of international law, there may still be a need to reconsider creating an overarching distributive mechanism that can secure consistent financing for development over time. If there is emerging a nascent area of 'global public finance',[107] then an international law of global social welfare might logically follow, and is supported by the overall thickening of social relations at the global level. A legally binding finance mechanism would immunise the global poor from the vagaries of national political discretion in foreign aid funding and project priorities, and enable a more rational and sustainable redistribution to meet basic human needs. There are numerous models available, such as a global currency flow or transaction tax ('Tobin tax'), new Special Drawing Rights within the IMF or an International Finance Facility,[108] global environmental or carbon taxes, or a Global Resources Dividend.[109] The latter proposal 'envisions that states and their governments shall not have full libertarian property rights with respect to the natural resources in their territory, but can be required to share a small part of the value of any resources they decide to sell', on the basis that 'the global poor own an inalienable stake in all limited natural resources'.[110]

In the international law of the sea, where maritime rights concern vital questions of resource allocation and distribution, the European Commission has suggested that '[t]here might be a need to revisit existing rules ... as regards the resolution of territorial and border disputes'.[111] What that specifically means is less clear. The existing normative framework might be capable of accommodating the impacts of climate change: it is simply a matter of patiently negotiating disputes in the Arctic or over maritime or continental shelf delimitation through the usual processes. Paradoxically, climate change may strengthen the existing law of the sea regime in unexpected ways: resource competition in the Arctic has stimulated United States interest in finally becoming a party to UNCLOS, which would enable the US to participate more fully in maritime governance.[112] Issues presently suspended for pragmatic reasons — such as whether certain Arctic waters are Canadian sovereign territory or international straits — may be pressed for an ultimate resolution according to law, given that the stakes are increasingly raised over resource competition.

There are, however, arguments that the existing regime may not be entirely adequate to respond to the challenges of climate change. For instance, 'permanently fixing ocean boundaries' may be preferable to the increasing uncertainty (and potential for conflict), following sea-level rises, generated by the existing 'ambulatory' nature of baselines and the maritime boundaries and resource rights which flow from them.[113] This is particularly important in areas where sea rises spell the end of permanent habitation and thus the ability to claim maritime areas and attendant economic rights could otherwise disappear — which may particularly harm (or even extinguish) small, underdeveloped island states and their economies. Declining global fish stocks may also require more stringent regulation, as well as a shift away from 'resource nationalism' towards more sustainable collective management of an increasingly scarce resource.[114] Unresolved legal issues in the Arctic and Antarctic polar regions may also benefit from clarification, including the determination of baselines and maritime zones, the legal status of waters formerly covered by ice, extended continental shelf claims, and areas beyond national jurisdiction, along with marine environmental protection and biodiversity concerns.[115]

In the field of global migration, international law is currently poorly equipped to respond to the challenge of climate-induced displacement and the potential security risks it may bring. Migration as a whole is generally left to national legal discretion rather than being subject to comprehensive international legal control that regulates the redistribution of people compelled to move; international law is limited to narrow and specialised interventions in areas such as refugees or migrant workers. The climate-induced displaced are very unlikely to meet the legal definition of a refugee and therefore generally fall outside of the parameters of international legal protection, and complementary human rights-based protection is currently underdeveloped beyond agreed areas (such as non-return to torture).[116] They are likely to be seen as another manifestation of mere 'economic migrants' — much maligned and seldom admitted to developed countries, particularly given increasing restrictions on migration in the wake of the global financial crisis of 2008–2009.

As argued elsewhere,[117] there is accordingly a need to develop a comprehensive international framework for managing likely new flows of displaced persons, based on burden-sharing principles, a human-rights oriented approach, and the allocation of well-defined institutional responsibilities for humanitarian relief, legal protection, and the forging of durable, permanent solutions. This includes developing innovative ways for dealing with the disappearance of island states and the consequent extinction of their legal personality (and allied nationality) — which may carry with it new problems of statelessness and instability in neighbouring countries receiving mass influxes.

For pragmatic reasons, a comprehensive migration strategy need not necessarily involve the formulation of a new international treaty covering, for instance, environmental migration,[118] but it would require a commitment by developed states to actively manage the problem, including by providing development assistance for internal relocation where appropriate. In normative terms it would also demand greater attention to the human rights implications of climate change.[119] It is doubtful that major carbon emitters can be causally characterised as legally responsible (and thus obliged to make reparation) for human rights violations resulting from climate change. Climate change is, however, likely to weaken some states and render them unable to fulfil their obligations to protect and ensure rights.[120] International, regional and national responses to climate-induced displacement (including as a result of conflict) must accordingly be attentive to the fundamental rights of those forced to relocate.

Conclusion: The Limits of Toleration

There are, of course, potential political limits on any distributive international law: any such project immediately grinds up against the limits of toleration of national political communities, which remain the primary organising principle of a legal system of sovereign states. Why, for instance, should international law compel Australians to surrender part of their national income, and thus diminish their standard of living (even slightly), to help distant foreigners? Many would see moving development beyond charity and into law as an unreal enterprise. It may, indeed, be electoral suicide to force a nation's people to give up too much to benefit distant and unknown others; the Kantian vision of a cosmopolitan international order of citizens remains far off — and has taken a beating by moves in recent years to constitution-alise a pan-European community.

But the real possibility of absolute global food and water scarcity induced by climate change might just be electric enough to defibrillate the global conscience into motion. As Pogge notes, 'a future that is pervaded by radical inequality and hence is unstable would endanger not only the security of ourselves and our progeny, but also the long-term survival of our society, values, and culture'.[121] The scarcity wrought by climate change, in combination with other global stresses, might provide enough of a systemic shock that we will finally begin to tentatively configure a distributive inter-national legal architecture — one that places human dignity at its centre, rather than at its periphery as a byproduct of economic forces.

A distributive approach would not discount other means of benefiting the poor: trade, investment, economic growth, the rule of law, good governance, human rights and so on are all necessarily part of any comprehensive developmental strategy for enlarging human freedom. But when 'aspects of the [present] global economic order … contribute substantially to the persistence of severe poverty',[122] and do not look set to soon change, reliance upon the market, discretionary aid or ad hoc political responses is not enough. There may well be costs in a new international juridification of politics, but they are arguably outweighed by the ethical thickening that comes with cementing in law a global obligation of redistribution to the poor.

Endnotes

1   Rajagopal critiques the dominant concept of scarcity in development policy and in human rights law as constructing a 'modern market being … whose attempt to realize his/her full potentialities are confined within the moral possibilities of the state and the material conditions of the global market': Balakrishnan Rajagopal, *International Law from Below: Development, social movements and third world resistance* (2003), 199.

2   For example, on the side of government expenditure, distributive justice commonly manifests in social security and welfare programs (including those aimed at the unemployed, pensioners, students, war veterans and the disabled), housing assistance, maternity and parental leave, new baby payments. Some such payments also recognise the vulnerabilities of particular groups; others recognise wider social contributions by the targeted groups (new mothers, veterans), and some pursue long-term economic objectives (student support which raises education and skills levels and usually translates into stronger national economic outputs). Government regulation of the labour market, such as minimum wage requirements, and of taxation (including progressive sliding tax scales, tax cuts and tax exemptions), are also part of the distributive project.

3   See Amartya Sen, *Development as Freedom* (2000).

4   Philip Alston, Remarks on Professor BS Chimni's A Just World Under Law: A view from the south' (2007) 22 *American University International Law Review* 221, 230.

5   Bhupinder Chimni, The Sen Conception of Development and Contemporary International Law Discourse: Some parallels (2008) 1 *Law and Development Review* 1, 18.

6   Thomas Pogge, *World Poverty and Human Rights* (Polity Press, Cambridge, 2002), 211.

7   David D Zhang et al, Global Climate Change, War, and Population Decline in Recent Human History (2007) 104(49) *Proceedings of the National Academy of Sciences of the United States of America* 19214.

8   Ibid.

9   Ibid.

10   Ibid.

11   According to the recent International Assessment of Agricultural Science and Technology for Development.

12   See, eg, John Podesta and Peter Odgen, 'The Security Implications of Climate Change' (2008) 31 *The Washington Quarterly* 115; Gwynne Dyer, *Climate Wars* (2008); Dan Smith and Janani Vivekananda, *A Climate of Conflict: The links between climate change, peace and war* (International Alert: 2007); Nick Mabey, 'Delivering Climate Security: International Security Responses to a Climate Changed World' (2007) 69 *Whitehall Papers* 1; Kurt Campbell et al, *The Age of Consequences: The foreign policy and national security implications of global climate change* (Center for Strategic and International Studies: 2007); German Advisory Council on Global Change, *World In Transition: Climate change as a security risk* (2007) <http://www.wbgu.de/wbgu_jg2007_engl.html>; Cleo Paskal, *How Climate Change Is Pushing the Boundaries of Security and Foreign Policy* (2007) (Chatham House Briefing Paper, EEDP CC BP 07/01); Ben Wisner et al, *Climate Change and Human Security* (Peace Research and European Security Studies: 2007); International Institute for Strategic Studies, *Strategic Survey* 2007 (2007); Alan Dupont and Graeme Pearman, *Heating up the Planet: Climate change and security* (2006) (Lowy Institute Paper 12).

13   Mabey, above n 12, 2.

14   Smith and Vivekananda, above n 12.

15   UN SCOR, 62nd sess, 5663rd meeting, New York, UN Doc S/PV 5663 (17 April 2007), 13 and 14.

16   Wisner et al, above n 12, 2.

17   David Adam, 'Food Prices Threaten Global Security — UN', The Guardian (London), 9 April 2008.

18   Ibid.

19   Rafael Epstein, 'UN Chief Warns of Civil Unrest Amid World Food Shortage', ABC News Online, 30 April 2008 <http://www.abc.net.au/news/stories/2008/04/30/2231223.htm>.

20   John Vidal, 'Nations Split on Ways to Tackle Hunger Alert', *Sydney Morning Herald*, 17 April 2008.

21   US National Intelligence Council, Global Trends 2025: A Transformed World (USGPO, Washington DC, 2008), 51.

22   Intergovernmental Panel on Climate Change, *IPCC Fourth Assessment Report* (AR4) (2007) <http://www.ipcc.ch/> at 31 July 2009.

23   IPCC, 'Statement on the melting of Himalayan glaciers', Geneva, 20 January 2010, <http://www.ipcc.ch/pdf/presentations/himalaya-statement-20january2010.pdf> (noting that errors concerning the speed of Himalayan glacial melting do not affect the conclusion that melting is ??.

24   World Health Organization-UNICEF, *Meeting the MDG Drinking Water and Sanitation Target: The urban and rural challenge of the decade* (2006).

25   UNESCO, *Facts and Figures from the United Nations World Water Development Report 2*. Available at www.unesco.org/bpi/wwap/press/pdf/wwdr2_facts_and_figures.pdf

26   United Nations Secretary-General Ban Ki Moon, 'Scarcity Threatens Socio-Economic Gains, Could Fuel Conflicts' (Speech delivered at the inaugural Asia-Pacific Water Summit, Japan, 3 December 2007) <http://www.un.org/News/Press/docs/2007/sgsm11311.doc.htm>.

27   World Health Organization, Fact Sheet: Climate and health (2005) <http://www.who.int/globalchange/news/fsclimandhealth/en/index.html>. See also World Health Organisation (Executive Board), Climate Change and Health: Report by the Secretariat, 16 January 2008, EB122/4; 61st World Health Assembly, Resolutions WHA51.29 and WHA61.19.

28   Ibid. Asia alone accounts for 80% of world's total deaths due to water-related disasters (2001–2005); 1960–2006 over 600 000 water-related disasters (floods, droughts, tsunamis, windstorms, landslides, storm surges, water-borne diseases, epidemics).

29   Ibid.

30   Ibid.

31   Atul A Khasnis and Mary D Nettleman, Global Warming and Infectious Disease (2005) 36 *Archives of Medical Research* 689; see also Kathryn Senior, Climate Change and Infectious Disease: A dangerous liaison? (2008) 8 *The Lancet Infectious Diseases* 92. For a more cautious assessment, see Roland Zell, Global Climate Change and the Emergence/Re-emergence of Infectious Diseases (2004) 293(37) *International Journal of Medical Microbiology Supplements* 16.

32   *Lancet* and University College London Institute for Global Health Commission, Managing the Health Effects of Climate Change (2009) 373 *The Lancet* 1693, 1694.

33   Smith and Vivekananda, above n 12, 3.

34   Ibid.

35   According to the US National Snow and Ice Data Center, cited in US National Intelligence Council, op cit, 53.

36   See, e.g., Chatham House International Law Discussion Group, The Arctic and Climate Change (Meeting Summary, 14 February 2008) Available at http://www.un.org/News/Press/docs/2007/sgsm11311.doc.htm

37   Security Council debate, op cit, 12.

38   Jon Barnett, *Security and Climate Change* (Working Paper 7, Tyndall Centre for Climate Change Research, 2001) 4.

39   Paskal, above n 12, 3.

40   IPCC, above n 22; Sir Nicholas Stern, *Stern Review on the Economics of Climate Change* (2006).

41   India is currently building a border wall in West Bengal to stem the flow of millions of unauthorised Bangladeshi migrants who enter India each year.

42   See, eg, Barnett, above n 38, 5.

43   Ibid.

44   Zhang et al, above n 7.

45   Associated Press, World Population May Reach 9.2 Billion by 2050, *MSNBC*, 14 March 2007 Available at http://www.msnbc.msn.com/id/17605186

46   *Climate Change and International Security* (Paper from the High Representative and the European Commission to the European Council, S113/08, 14 March 2008) 2.

47   German Advisory Council on Global Change: Summary for Policy Makers, 'World In Transition: Climate Change as a Security Risk', May 2007.

48   Ibid.

49   See, eg, Podesta, op cit, 123–4.

50   Lancet and University College London Institute for Global Health Commission, above n 32, 1706.

51   Antony Anghie, The Evolution of International Law: Colonial and postcolonial realities. In Richard Falk, Balakrishnan Rajagopal and Jacqueline Stevens (Eds), *International Law and the Third World: Reshaping justice* (2008) 35, 41.

52   Ibid 40.

53   *Universal Declaration of Human Rights*, GA Res 217A (III), UN GOAR, 3rd sess, 183rd plen mtg, UN Doc A/RES/217A (III) (10 December 1948).

54   Rajagopal, 218.

55   United Nations, *The Millennium Development Goals Report 2008* (2008) 4.

56   Ibid.

57   Most recently, at the Follow-up International Conference on Financing for Development to Review Implementation of the Monterrey Consensus, Doha, 29 November — 2 December 2008.

58   UN, High Level Event on the MDGs, UN Headquarters, New York, 25 September 2008 (based on the promise to increase foreign aid from $80 billion in 2004 to $130 billion by 2010): see UN, *Fact Sheet: Goal 8: Develop a global partnership for development*, DPI/2517 N (2008).

59   James Wolfensohn, 'Some Reflections on Human Rights and Development' in Philip Alston and Mary Robinson (eds), *Human Rights and Development: Towards mutual reinforcement* (2005) 19, 24.

60   Rajagopal, above n 1, 220.

61   See Hilary Charlesworth, 'International Law: A Discipline of Crisis' (2002) 65 *Modern Law Review* 377–392.

62   Anghie, above n 50, 44–5.

63   James Thuo Gathii, 'Third World Approaches to International Economic Governance' in Richard Falk, Balakrishnan Rajagopal and Jacqueline Stevens (eds), *International Law and the Third World: Reshaping justice* (2008) 255, 255.

64   UN General Assembly resolution 1803 (XVII) of 14 December 1962, 'Permanent Sovereignty over Natural Resources'.

65   Rosalyn Higgins, *Problems and Processes: International law and how we use it* (1994) 141–2.

66   Rajagopal, above n 1, 199.

67   Pogge, above n 6, 113.

68   Particularly when coupled with the 'the international borrowing privilege' which 'helps [corrupt, violent] rulers to maintain themselves in power even against near-universal popular opposition': Pogge, above n 6, 115.

69   See Tim Stephens, 'Fisheries-Led Development in the South Pacific: Charting a "Pacific Way" to a sustainable future' (2008) 39 *Ocean Development & International Law* 257.

70   Pankaj Ghemawat, 'The Myth of Globalisation', *Australian Financial Review* (Sydney), 16 March 2007, 3.

71   Philip Allot, Eunomia: *New order for a new world* (1990) 360.

72   Zhang, above n 7, 19218.

73   Patricia W Birnie and Alan E Boyle, *International Law and the Environment* 85.

74   Ibid 89, noting gaps in coverage such as tropical forests.

75   See, e.g., *World Charter for Nature*, GA Res 37/7, UN GAOR, 48th plen mtg (28 October 1982); *United Nations Framework Convention on Climate Change*, opened for signature 4 June 1992, 1771 UNTS 107 (entered into force 21 March 1994) (*'UNFCCC'*); *United Nations Convention on Biological Diversity*, opened for signature 5 June 1992, 1760 UNTS 79 (entered into force 29 December 1993) (*Biological Diversity Convention*); *United Nations Convention to Combat Desertification*, opened for signature 14 October 1994, 1954 UNTS 3 (entered into force 26 December 1996) (*'Desertification Convention'*); fisheries (*Icelandic Fisheries* case), international watercourses.

76   United Nations Environment Programme, *Rio Declaration on Environment and Development*, UN Doc A/CONF.151/26 (Vol I) (12 August 1992) (*'Rio Declaration'*).

77   *Vienna Convention for the Protection of the Ozone Layer*, opened for signature 22 March 1985, 1513 UNTS 293 (entered into force 22 September 1988) (*'Ozone Convention'*).

78   *United Nations Framework Convention on Climate Change*, opened for signature 4 June 1992, 1771 UNTS 107 (entered into force 21 March 1994) ('UNFCCC').

79   Birnie and Boyle, above n 71, 91–2.

80   *Kyoto Protocol to the United Nations Framework Convention on Climate Change*, opened for signature 16 March 1998, 2303 UNTS 148 (entered into force 16 February 2005) ('Kyoto Protocol').

81   *Lancet* and University College London Institute for Global Health Commission, above n 32, 1712.

82   See, e.g., Oxfam's *Adaptation Financing Index: Oxfam International, Adapting to Climate Change: What's needed in poor countries, and who should pay*', Briefing Paper, 29 May 2007. But note (problematic) objections that that the response of developed countries to climate change cannot be easily justified by distributive or corrective justice theories: Eric Posner and Cass Sunstein, 'Climate Change Justice' (John M Olin *Law & Economics Working Paper* No 354 (2nd series), University of Chicago Law School, August 2007).

83   Birnie and Boyle, above n 73, 92.

84   Ibid 91.

85   Jacqueline Peel, *The Precautionary Principle in Practice: Environmental decision-making and scientific uncertainty* (2005) 26.

86   Hilal Elver, 'International Environmental Law, Water and the Future' in Richard Falk, Balakrishnan Rajagopal and Jacqueline Stevens (eds), *International Law and the Third World: Reshaping justice* (2008) 181, 183.

87   Ibid 185.

88   Ibid 182.

89   Higgins, above n 65, 134–5.

90   Stephen C McCaffrey, *The Law of International Watercourses: Non-navigational uses* (2001) 325.

91 Elver, above n 86, 191.

92 See, e.g., Karen Hussey, Recognising and Reconciling Social Equity Issues Contemporary Water Policy. In R Quentin Grafton, Jeff Bennett and Karen Hussey, *Dry water* (Policy Briefs 3, Crawford School of Economics and Government, ANU, 2007) 8.

93 See, eg, *Rio Declaration* Principle 3; *United Nations Framework Convention on Climate Change*, opened for signature 4 June 1992, 1771 UNTS 107 (entered into force 21 March 1994) (UNFCCC) art 3(1).

94 Birnie and Boyle, above n 73, 90.

95 Ibid 90–1.

96 *United Nations Convention on the Law of the Sea*, opened for signature 10 December 1982, 1833 UNTS 3 (entered into force 16 November 1994) ('UNCLOS').

97 Higgins, above n 65, 132.

98 Thuo Gathii, above n 63, 261.

99 *Resolution on the Law of Transboundary Aquifers*, GA Res 63/124, UN GAOR, 63rd sess, 67th plen mtg, UN Doc A/RES/63/124 (2008) arts 4, 6, 10 and 12 respectively.

100 See, e.g., German Advisory Council, above n 12, 10.

101 Wolfensohn, above n 59, 19, 24.

102 Giles Bolton, *Aid and Other Dirty Business* (2007) 7.

103 Kate Miles, 'International Investment Law and Climate Change: Issues in the transition to a low carbon world' (Paper presented at *New Horizons of International Economic Law*, the inaugural conference of the Society of International Economic Law, Geneva, Switzerland, 15–17 July 2008); see also Kate Miles, 'Sustainable Development, National Treatment and Like Circumstances in Investment Law' in Marie-Claire Cordonier Segger, Andrew Newcombe and Markus Gehring (eds), *Sustainable Development in World Investment Law* (forthcoming).

104 World Bank, *Clean energy and development: Towards an investment framework* (DC2006-0002, 5 April 2006) annex K.

105 Heike Mainhardt-Gibbs, *World Bank energy sector lending: Encouraging the world's addiction to fossil fuels*, Bank Information Centre Brief, February 2009.

106 RUSI/Mabey, 6.

107 AB Atkinson, Funding the Millennium Development Goals: A challenge for global public finance (2006) 14 *European Review* 555, 558.

108 Ibid 557.

109 Pogge, above n 6, ch 8.

110 Ibid, 196.

111 European Commission, above n 46, 4.

112 Associated Press, As Arctic Ice Melts, US Moves to OK Sea Treaty, *MSNBC*, 4 October 2007 <www.msnbc.msn.com/id/21131181>.

113 David Caron, 'Climate Change, Sea Level Rise and the Coming Uncertainty in Ocean Boundaries: A proposal to avoid conflict' in Seoung-Yong Hong and Jon M Van Dyke (eds), *Maritime Boundary Disputes, Settlement Processes, and the Law of the Sea* (2009), 14.

114 See Stephens, above n 69.

115 See, eg, Rosemary Rayfuse, 'Melting Moments: The future of polar oceans governance in a warming world' (2007) 16 *Review of European Community and International Environmental Law* 196.

116 Jane McAdam and Ben Saul, 'Weathering Insecurity: Climate-induced displacement and international law' in A Edwards and C Ferstman (eds), *Human Security and Non-Citizens: Law, Policy and International Affairs* (2009), ch 13.

117 Ibid.

118 As suggested by the German Advisory Council, above n 12, 11.

119 See above n 7.

120 McAdam and Saul, above n 116.

121 Pogge, above n 6, 213.

122 Ibid 115.

PART III
Emissions Trading

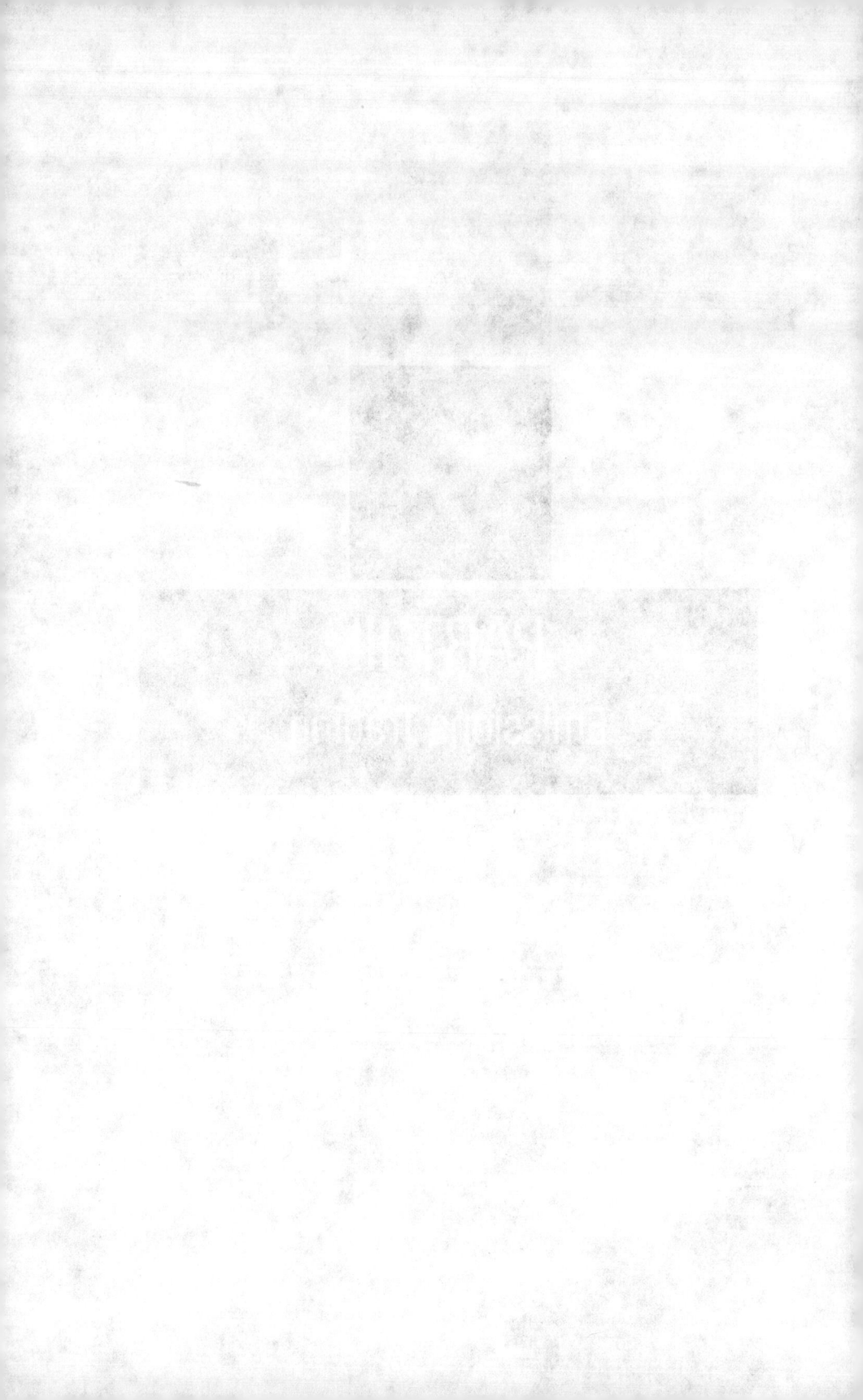

Reducing Emissions from Deforestation and Degradation: The Road to Copenhagen

Rosemary Lyster

To date, the vast majority of research and writing on the issue of climate change by environmental lawyers has focused on greenhouse gas (GHG) mitigation measures, such as emissions trading schemes and complementary measures including renewable energy and energy efficiency. Yet the issue of deforestation in tropical rainforest countries is a significant contributor to the phenomenon of global climate change and deserves the attention of legal scholars. Deforestation, especially in the tropics, contributes around 20% of annual global GHG emissions and in the case of Indonesia amounts to 85% of its annual emissions from human activities,[1] making Indonesia the third highest emitter of GHGs in the world.

International attention is now turning to the incentives that may be offered to tropical rainforest countries for Reducing Emissions from Deforestation (and Degradation) (REDD). As well, jurisdictions which have established emissions trading schemes, or are proposing to do so, have had to consider whether to allow liable entities to rely on forestry offsets to meet their legally binding carbon abatement obligations. As this article will demonstrate, there are a number of pressing legal issues associated with REDD offsets that need greater clarification and resolution in order for REDD schemes to be implemented in a way that is equitable while achieving the stated goals of REDD programs.

Forest Sequestration

Standing forests[2] are the most important reservoir of carbon dioxide. As mentioned above, deforestation, especially in the tropics, contributes around 20% of annual GHG emissions while conserved rainforests sequester almost 17% of atmospheric carbon each year.[3] Given the Intergovernmental Panel on Climate Change's (IPCC) call to keep the concentration of GHGs in the atmosphere at between $445^1/n$ parts per million (ppm),[4] attention has turned to the opportunities for significant carbon abatement offered by the forest sector, particularly in tropical rainforest countries. Stern regards REDD as an opportunity to significantly reduce emissions without the need for new technology, except perhaps for important monitoring purposes.[5] Stern warns that without prompt action by the international community, emissions from deforestation are expected to total 40Gt CO_2 between 2008–2012 that will result in an increase in GHG concentrations in the atmosphere of 2 ppm. These emissions are greater than the cumulative total in aviation emissions since the invention of aircraft until at least 2025.[6]

While much attention has been placed on the sequestration benefits of tropical forests, it is also important to remember that they have the potential to deliver far more than carbon sequestration services. Forests also provide valuable local, regional and global ecosystem services including water quality, flood control, soil stability and biodiversity.[7] If properly implemented, REDD projects also have the potential to contribute to the protection of forest livelihoods among forest dependent populations.[8]

The International Law Framework for REDD

Prior to the climate change negotiations in Bali in December 2007, REDD was not incorporated in the international legal framework for dealing with climate change. However, given the scientific evidence, there is growing consensus by the parties to the *United Nations Framework Convention on Climate Change*[9] that REDD must be a priority in negotiations on a successor to the *Kyoto Protocol*.[10]

The Bali Action Plan

Under the Bali Action Plan, negotiated at the 13th Conference of the Parties to the *United Nations Framework Convention on Climate Change* (COP13), the parties decided to begin a process immediately to allow them to adopt a decision at COP15 on a shared vision for long-term cooperative action on climate change. COP15 takes place in Copenhagen in December 2009. This vision will include a long-term global goal for emission reductions, based on the principle of common but differentiated responsibilities.

Significantly, the Action Plan requires 'enhanced consideration of policy approaches and positive incentives on issues relating to reducing emissions from deforestation and forest degradation in developing countries; *and the role of conservation, sustainable management of forests and enhancement of forest carbon stocks in developing countries*.'[11] The addition of sustainable forest management to REDD results in REDD now being referred to as 'REDD+'. This builds on a decision taken at COP11, in December 2005, to establish a two-year review of relevant scientific and methodological issues, and to consider policy approaches and incentives for reducing emissions from deforestation in developing countries.[12]

The incorporation of REDD+ in the Bali Action Plan is highly significant as prior to this there has been no mention of it in international agreements. Under art 3.3 of the *Kyoto Protocol*, Annex I Parties[13] may rely on *domestic* reductions in GHG emissions resulting from forestry activities, limited to afforestation[14] and reforestation[15] since 1990, to meet their emissions reduction targets under the *Protocol*.[16] Similarly, afforestation and reforestation projects undertaken by Annex I Parties in developing countries may be relied upon, under the Clean Development Mechanism (CDM),[17] to satisfy their *Kyoto* commitments.

There are a number of reasons why REDD+ has been excluded previously from the provisions of the *Kyoto Protocol*, including from the CDM. These include concerns about: the risk of leakage,[18] non-permanence,[19] establishing baselines,[20] additionality,[21] and difficulties associated with monitoring and measurement. During the two-year review of REDD established at COP11, considerable advances have been made in addressing these problems, particularly with respect to monitoring and measurement.[22] It should be noted also, that many of the problems relating to REDD also arise with respect to afforestation and reforestation CDM projects. For this reason, the CDM Executive Board has developed unique rules governing these types of projects.[23]

Given these technological advances and the pre-existing afforestation and reforestation CDM Rules, an international project to develop rules for REDD+ is not impossible. However, one of the purposes of this article is to draw attention to some of the ongoing dilemmas surrounding REDD+ and to illuminate the potential for a range of legal academics to engage in this globally significant project. To date, there has been very little legal scholarship in this area.[24] The areas requiring the engagement of lawyers include: an analysis of the legal reforms needed in discrete jurisdictions to implement any international agreement emerging from COP15; the structuring of public and private sector financing for REDD+, whether on a national or a project basis; necessary institutional and forestry governance arrangements; questions of land and/or resource tenure; benefit sharing arrangements under Payment for Environmental Services schemes; and monitoring, reporting and verification regimes necessary for compliance and enforcement of REDD+ targets.

The Negotiating Text in the Lead-Up to the Fifteenth Conference of the Parties

On 19 May 2009, the UNFCCC secretariat released the Negotiating Text[25] developed by the Ad Hoc Working Group on Long-Term Cooperative Action under the Convention for the Group's sixth session in Bonn from 1–12 June 2009. It provided a starting point for negotiations and took account of the proposals contained in the recent submissions from parties to the AWG-LCA following its fifth session. The text, which has been substantially revised following the Bonn negotiations,[26] is intended to be a 'living document'.[27] The text comprises the following chapters: Chapter I: A shared vision for long-term cooperative action;[28] Chapter II: Enhanced action on adaptation; Chapter III: Enhanced action on mitigation; Chapter IV: Enhanced action on financing, technology and capacity-building. In this article, Chapters III and IV will be discussed because of their relevance for REDD+.

At the outset, it is important to note that the Negotiating Text proposes a number of options with regard to the legal form that a post-Kyoto agreement will take. The following views have emerged:

- That a new agreement should be based on a set of COP decisions which would be legally binding in nature, emanating from the obligations and commitments of the parties under the UNFCCC.
- That a new legal instrument or instruments should be established within the framework of the UNFCCC:
  - perhaps by way of amendment to the Convention; or
  - by way of a new protocol that may be separate from the *Kyoto Protocol*; or
  - which would build on the *Kyoto Protocol* and incorporate its commitments; or
  - where its relationship with the *Kyoto Protocol* is not expressly addressed.[29]

Consequently, the language of the text does not prejudge an outcome and careful use is made of the verbs 'shall' and 'should' to indicate that at this stage various options are simply being proposed. One of the important implications of these options is that the United States has ratified the *UNFCCC* which means that any decisions made by the COP are automatically binding on it. However, given that the United States has not ratified the *Kyoto Protocol*, a two-thirds majority vote in the Senate would have to be obtained to authorise ratification.

Nationally Appropriate Mitigation Action by Developing Countries

The Negotiating Text on mitigation by developing countries comprises five parts, which are: nationally appropriate mitigation actions by developing countries; means of implementation; measurement, reporting and verification of actions; measurement, reporting and verification of support; and institutional arrangements.

According to the revised Text, developing countries contribute to enhanced mitigation by undertaking nationally appropriate mitigation actions (NAMA). These should be:

- country-driven;
- voluntary and undertaken in the context of sustainable development;
- in conformity with prior needs of sustainable development and eradication of poverty; and
- determined and formulated in accordance with the principles equity and common but differentiated responsibilities and capabilities.

NAMAs are contingent on financial and technical support by developed countries and only those that are supported will be subject to the requirements of monitoring, reporting and verification (MRV).[30] The wording in the revised Text is different from the original Text which simply stated that NAMAs should be supported and enabled by technology, financing and capacity-building in accordance with the UNFCCC since mitigation actions will depend on the effective provision of financial and technical support by developed countries. NAMAs *shall* be undertaken in a measurable, reportable and verifiable manner.[31]

NAMAs may be: enabled and supported by finance, technology and capacity-building from developed countries; or be undertaken unilaterally by a developing country; or be pursued with the intention of generating carbon credits for participation in the carbon market.[32] A number of different NAMA activities are mentioned in the text including: the development of a national action plan, low carbon development strategies and plans, renewable energy strategies and plans; energy efficiency or energy pricing measures CDM projects, cap-and-trade schemes and carbon taxes, economy wide or sectoral intensity targets; mitigation actions at subnational and local levels; and, for our purposes, REDD+.[33]

There is clearly a difference of opinion among the parties to the Convention as to whether the NAMAs should constitute binding obligations or targets for developing country parties, since eleven different permutations are proposed. The proposals include: whether targets should be different for different groups of developing countries depending on their levels of development; whether the NAMAs should be in the form of sectoral and economy-wide GHG emissions intensity targets; whether the NAMAs should be incorporated in national schedules; and whether they should evolve over time to reflect changes in national circumstances assessed in accordance with objective criteria of economic development.[34]

Importantly, it is proposed that developing countries *should* register their NAMAs in order to gain international recognition and financial and technical support for their proposed actions.[35]

As mentioned above, the Text provides that MRV of NAMAs will only apply to the extent that they are enabled and supported financially and technologically by developed countries through an agreed financing mechanism. Guidelines for MRV will be established and approved by the COP, and developing countries may establish national coordinating bodies to estimate the full incremental costs of implementing the NAMA to be met by the financial mechanism. The body should also build institutional capacity, and facilitate the submission of proposals for NAMAs which indicate the level of support required to implement them.[36]

It is also proposed that developing countries should submit their national inventories to the *UNFCCC* on a more frequent and regular basis.[37]

The revised Text also contains extensive new provisions on methodological issues for MRV. These include obligations to have in place national systems for the estimation of anthropogenic emissions by sources and removals by sinks of all GHGs no later than one-year prior to the start of any new commitment period. Methodologies for such estimations must be accepted by the IPCC and agreed upon by the COP at its third session following the conclusion of a post-*Kyoto* agreement. The Text also includes provisions on annual GHG inventories, the review of information contained there by expert review teams and the preparation by the teams of reports assessing the implementation of commitments under a new agreement.[38]

The REDD Plus Mechanism

The revised Text envisages the establishment of a REDD+ mechanism (the Mechanism) the objectives of which are to assist developing countries to reorganise their forestry sectors in a way that contributes to the global effort to stabilise and reduce GHG concentration in the atmosphere.[39] The goal is to halt forest cover loss in developing countries by 2030 at the latest and reduce gross deforestation in developing countries by at least 50% by 2020 compared with current levels.[40] For the Mechanism to be effective there should be a phased-in approach. Here, Phase 1 would entail national REDD+ strategy development and core capacity-building; Phase 2 would see the REDD+ policies implemented with compensation for proxy-based results for emission reductions and removals from selected forest activities and land-use and land-use change categories; and Phase 3 would see the policies finally evolving into a result-based compensated mechanism in accordance with emissions reductions that are fully measured, reported and verified.[41]

The Text posits that in time the scope of the Mechanism may be expanded to cover the full land sector in developing countries. Initially, however, it will encompass activities that reduce GHG emissions from deforestation and degradation, and increased removals from afforestation, reforestation and enhancement of forest carbon in developing countries.

Nationally Appropriate Activities

The nationally appropriate activities that will qualify for receiving positive incentives under the Mechanism include:

- stabilisation of forest cover, and thereby forest carbon stocks
- sustainable management of forests leading to the conservation and maintenance of forest carbon stocks
- reduction in deforestation rates
- reduction in forest degradation
- enhancement of forest carbon stocks due to conservation and sustainable management of forests, and/or increase in forest cover due to afforestation and reforestation.[42]

REDD+ activities should be country-driven and voluntary taking into account the need to ensure equitable access to financial and technological support for these actions.[43]

Biodiversity and Ecosystem Protection and Participation

There should be national level participation in REDD+ activities while forest governance, permanence and co-benefits such as biodiversity and ecosystem services should be promoted. Precautionary measures should be taken to protect biological diversity including by safeguarding against the conversion of natural forests to forest plantations.[44] There is an acknowledgement of the necessity for good local government in which indigenous peoples and local communities participate and are able to adopt customary patterns of government. Indeed they should be involved in the design of plans and actions as well as the implementation of REDD+ activities and their land and their rights should be respected.[45] Leakage should be avoided as much as possible.[46]

REDD Obligations

The obligations on developing countries that choose to implement REDD+ actions are: to develop national or subnational plans covering different phases of implementation; to designate a national authority to implement the actions; to establish national reference emissions levels in accordance with national circumstances; to collaborate closely with the independent review authority; and have in place a process that promotes the full and effective participation of all relevant stakeholders including indigenous peoples, forest dwellers and local communities.[47] National accounting and monitoring systems of emissions and removals from the forestry sector should be established, although sub-national accounting is acceptable as an interim measure.[48] At COP15, modalities and procedures must be elaborated to ensure that REDD+ activities carried out at the sub-national level account for, and address, leakage in a comprehensive and conservative way.[49]

The Text provides, perhaps rather idealistically given capacity constraints, that if developing countries decide to engage in REDD+ activities they must ensure that:

- national and international emissions displacement is avoided;
- REDD actions are permanent and do not result in increased emissions from defor-estation and forest degradation in the future;
- appropriate governance structures are in place to facilitate the appropriate use of funds provided for REDD;
- consultative mechanisms are established and the rights of indigenous people and communities are protected in domestic legislation;
- land tenure systems are recognised; and
- actions are consistent with biodiversity conservation.[50]

Meanwhile, developed countries should ensure that the importation of forest products and other commodities from developing countries does not exacerbate the problem of deforestation and degradation.[51]

A Phased Approach to Implementation

With regard to implementation of REDD+ activities, the Text recognises that developing countries have different capabilities and capacities. A phased approach is therefore envis-aged whereby there will be an initial readiness phase, which can be followed by a policy implementation and demonstration activities phase, and a full implementation phase. It is clear that funding from developed countries is integral to the implementation of REDD+ activities. They are required to provide 'adequate, predictable and long-term sus-tainable funding' on an appropriate and effective international funding mechanism.[52] Various options for the structuring of the funding mechanisms have been proposed both by the Text and other organisations. These are discussed in more detail below under a separate section dealing with the funding mechanism.

Monitoring, Reporting and Verification

As with all NAMAs, MRV is absolutely crucial for the successful implementation of REDD programs. Yet it remains one of the most contentious aspects of REDD given concerns about whether or not the technology for monitoring emissions from defor-estation is currently available, and whether developing countries have the capacity to engage accurately in MRV processes.

The Text illustrates this dilemma by stating *either* that developing countries partici-pating in the forest carbon mechanism must register their national forest emissions level in their National Schedules, with their REDD activities properly monitored, reported and verified against their agreed national forest emissions levels, *or* that those countries requesting support for REDD+ actions shall/should record under the NAMA registry the actions undertaken, including information on the extent and type of support requested, the nature of the action, and any information received on MRV of actions.[53] To assist the COP with this, developing countries must develop: a national capacity needs assessment; national forest inventories; national and where appropriate subnational baselines for calculating emissions from deforestation and degradation; strategic plans for REDD; and regulations relating to quality assurance and quality control to ensure that REDD funding is not diverted for other purposes.[54]

Establishing National Reference Emission Levels

When establishing national reference emission levels for emissions, removals, conservation areas, and existing forest carbon stocks, developing countries that are requesting funding support must follow any guidance developed by the COP on how to establish the levels, including ways to address domestic leakage.[55] The Text provides that in order to avoid carbon leakage and to ensure the environmental integrity of the REDD+ mechanism, a global reference level for future emissions and removals from the forestry sector, and other selected land-use and land-use change categories and activities in developing countries, must be established.[56] The methodology for establishing the global reference level must: be robust and based on objective, measurable and verifiable criteria; and ensure additionality both at the national and global level compared with business as usual.[57]

Developing countries that aim to implement REDD+ actions must report on the implementation of any REDD+ readiness activities, including policy implementation and demonstration activities as well as identified co-benefits such as biodiversity. Any quantitative reductions in GHG emissions must be measured and reported on in relation to the reference emissions level. Consideration will be given to credit for early actions which are taken by developing countries up to 2012. If any correction factor to the relevant national reference levels has been applied taking into account national circumstances, historically low rates of deforestation, developmental divergence and respective capabilities and capacities, this must be reported.[58]

Robust National Monitoring Systems

The Text requires developing country parties to develop robust national monitoring systems that take into account relevant methodological guidance provided by the COP on the advice of the Subsidiary Body for Scientific and Technological Advice (SBSTA) and relevant IPCC guidelines and methodologies for GHG inventories. The monitoring system should include above and below ground carbon stocks, subject to the availability of technology, finance and capacity-building. They should also take into account ancient indigenous and local community knowledge.[59]

Financing REDD

A recent report has estimated that developing countries will need international financing flows of between USD22–44 billion per year between 2010–12 to implement wide ranging emissions and adaptation programs, rising to USD95–147 billion per year for the next decade. The financing is likely to be provided through direct offset purchases (15%), auction revenues (30–50%) and public and international finance (50–70%) through international maritime and aviation levies, concessional debt and direct public fiscal revenues.[60]

It is difficult to estimate the specific funding needs for REDD; however, one of the most recent estimates has been provided by the Meridian Institute which, in March 2009, prepared An Options Assessment Report (OAR) for the Norwegian government.[61] The OAR estimates that, to achieve 50% global reductions in forest emissions, funding in the order of USD15–35 billion per year will be needed, while at present only USD2 billion is available.[62]

There has been ongoing philosophical debate about the best financing approach to REDD. It has been suggested, for example, that public funding schemes are not an adequate response to deforestation and degradation and that, to remedy the missing or incomplete market for forest ecosystem services, a market-based instrument to capture the carbon, and other, values of forests should be developed.[63] Market advocates maintain that public funding schemes will not be sufficient to generate the required volume of funds to provide attractive and sustained economic incentives for REDD.[64]

Part IV of the Negotiating Text deals with 'enhanced action on financing, technology and capacity building' and states that developed countries agree to establish in accordance with the UNFCCC the Financial Mechanism to enable, enhance and support mitigation and adaptation by developing countries.[65] However, there are special provisions for the financing of REDD. These seem to address the philosophical questions alluded to above.

According to the Text, a REDD+ funding mechanism must be established at COP15 under which developed countries commit to providing adequate, predictable and long-term sustainable funding for REDD, and other select land-use and land-use change sectors in developing countries. Funding will be allocated according to three phases: an initial readiness phase; a policy implementation and demonstration activities phase; and a full implementation phase. The provision of funding should be transparent, efficient and equitable while ensuring that all relevant stakeholders and indigenous people and local communities derive a fair distribution of benefits.[66]

The Readiness Phase

The Text includes five different options for establishing a fund that will support the readiness, policy implementation and demonstration activity phases. The first option is for a readiness fund to be established under the COP using funds from the auctioning of allowances in domestic compliance markets. There are four alternatives to this option. One envisages a special REDD funding window within a single consolidated fund that might be established for mitigation in developing countries by amalgamating the Global Environment Facility (GEF) with the two funds established under the *Kyoto Protocol* — the Least Developed Countries Fund and the Special Climate Change Fund.[67] The Negotiating Text then gives five different alternatives to these funding options. Clearly it will be difficult to achieve consensus at COP15 as to the appropriate REDD funding mechanism for these phases.[68]

The Full Implementation Phase

With regard to the full implementation phase, three quite distinct options are provided. These are: the use of public funds; the use of markets; and a combination of both. Where public funds are decided upon, these could be established through one or more of the following approaches: a specialised fund established under the COP for REDD; or trust funds for community forestry accounts; a Convention adaptation fund, by which conservation and sustainable forest management could be supported as adaptation measures; and/or a forest reserve fund for conservation and sustainable forest management under an amalgamated mitigation fund described above. Clearly, where developing countries receive international funding for REDD projects, governments will have to devise schemes whereby financial incentives to preserve forests are passed on to those who own, or control, the land on which the forests are situated, or who have resource rights over the forests. Land tenure, property rights and the contractual arrangements between government and landholders have emerged as crucial legal issues in this regard.[69]

Under the markets approach, developing countries would have access to the carbon market through the issuance of carbon credits for REDD activities, or for REDD activities and for conservation and enhancement of carbon stocks in existing forests.

Under the combined approach one option would be to establish a special climate change fund complementary to the GEF and bilateral and multilateral funding for forest conservation efforts that enhance carbon stocks, while using certified emission reductions from CDM project activities. Both the public and private sectors could participate in these projects to contribute to compliance with part of the countries' quantified emission limitation and reduction obligations under the UNFCCC.[70]

Existing Public Funding Schemes

An example of a public funding scheme is the Forest Carbon Partnership Facility (FCPF)[71] launched at COP13 by the World Bank in response to a request by developing and industrialised countries to explore a framework for piloting REDD activities. Two separate mechanisms have been established: the readiness mechanism and the carbon finance mechanism. The readiness mechanism assists developing countries to calculate a credible estimate of their national forest carbon stocks and sources of forest emissions, as well as assisting the country in identifying its reference scenario based on past emission rates for future emissions estimates. Technical assistance is offered in calculating the opportunity costs of possible REDD schemes, and designing an adapted REDD strategy that takes into account country priorities and constraints. The readiness mechanisms are carried out in accordance with a Charter Establishing the Forest Carbon Partnership Facility.[72]

Under the carbon finance mechanism, a few countries will be selected to participate in pilot incentive programs for REDD based on a system of compensated reductions. The selected countries: (a) must have demonstrated a commitment to REDD and have adequate monitoring capacity; (b) must have established a credible reference scenario and options for reducing emissions; and (c) will receive payments for reducing emissions below the reference scenario. Payments will only be made to countries that achieve measurable and verifiable emission reductions.[73] The World Bank believes that fears about the future supply of carbon credits could be allayed by ensuring that REDD CDMs are incorporated into the post-2012 regime.[74]

As of 13 March 2009, 37 countries from Asia, Latin and Central America, and Africa are participating in the Readiness Mechanism based on Readiness Plan Idea Notes ('R-PIN'). The R-PINs have to be reviewed by the Participants Committee and independent reviews by a Technical Advisory Panel. Once selected, REDD Country Participants receive grant support to develop a Readiness Plan, which contains a detailed assessment of the drivers of deforestation and degradation, terms of reference for defining their emissions reference level based on past emission rates and future emissions estimates, establishing a MRV system for REDD, and adopting or complementing their national REDD strategy. A consultation plan is also part of the Readiness Plan. To date, approximately USD107 million has been contributed by 11 donor countries to the Readiness Fund, the target being to raise USD185 million to support the REDD Readiness efforts of the 37 countries selected into the FCPF.[75]

Is an Emergency Package for Tropical Forests Needed?

The Prince's Rainforest Project (PRP) has taken the view that the world cannot afford to wait for a funding mechanism for REDD to be established by the international community under a post-*Kyoto* agreement. For this reason, it proposes the creation of a 'light temporary global institutional framework' for funding REDD as a bridge to a long-term *UNFCCC* solution.[76] The proposal is to establish a Tropical Forests Facility with which rainforest nations would negotiate multi-year agreements offering to achieve defined performance targets in return for annual payments. In the early years, payments would be made for progress towards policy and institutional reform, the establishment of monitoring and enforcement systems, and the implementation of an alternative economic development plan. As soon as possible thereafter, the payments would be made on the basis of actual emissions reductions which have been verified *ex post*. Rainforest nations would need to commit to national deforestation, as opposed to project-level, targets. National monitoring would be undertaken by national systems owned and operated by rainforest nations and there would also be a global verification system.[77]

Although the PRP is aware that the facility could obtain long-term annual funding commitments from developed country governments, it contends that private capital markets in the form of fixed income securities — 'Rainforest Bonds' — could be issued by an international institution like the World Bank while being backed by developed country governments. Alternatively, the facility might be able to issue bonds itself, through a Special Purpose Vehicle (SPV). The bond would offer investors a return that is competitive to other AAA-rated fixed income securities.[78] On 1 April 2009, the Prince of Wales hosted a meeting of world leaders to discuss ways to reduce tropical deforestation. They agreed to form an international working group to study the PRP proposal, along with other suggestions from governments and international organisations, before making recommendations for a co-ordinated, global plan of action.[79]

Establishing a Market for REDD Credits

As discussed above, the Negotiating Text proposes a phased-in implementation approach to REDD with Phases 1 and 2 being publicly funded with a market-based approach to REDD being introduced in Phase 3. Although the Negotiating Text does not provide any details on how a Phase 3 market-based implementation approach to REDD might operate, a number of proposals have been put forward for developing a market in REDD carbon credits.[80] Prior to the release of the Negotiating Text, it was proposed that a separate REDD instrument be incorporated into the post-2012 climate change regime by way of a new protocol,[81] and under which the ability of Annex I Parties to rely on REDD credits is restricted.[82] Here, the COP would set a maximum on the percentage of emissions reductions Annex I Parties can achieve through overseas REDD. This satisfies the principle of 'supplementarity' as well as the inherent concerns about REDD programs, mentioned above. Also, it avoids the prospect of an oversupply of REDD credits disrupting a well-established carbon market[83] and lowering carbon prices in the post-2012 era. In essence, this approach 'keeps separate an emerging market (REDD) from the more mature carbon market until questions of volatility have been resolved.'[84]

A National or Project-Based Approach to REDD Credits?

Irrespective of how a REDD carbon credits approach is integrated into the international climate change regime, the more fundamental question is whether credits should be granted to national governments, or whether a project-based/CDM type approach should be adopted.

A National Approach

Under a national approach, credits could be distributed/auctioned to Annex I Parties by developing countries which accept nationally binding caps, or deforestation targets.

There is some concern,[85] that where governments generate REDD credits for the international compliance market, agents of deforestation,[86] whose property rights are not officially recognised, might be subject to coercive activities to halt deforestation; yet, if they practice subsistence agriculture they are particularly vulnerable. Conversely, private companies to which the government grants logging licences would be able to negotiate favourable compensation arrangements based on their lost opportunity costs. This leads to the risk of an inequitable distribution of wealth where a distinction is drawn between the implementation costs, borne by those without clearly recognised property rights, and the opportunity costs of deforestation reduction.[87]

A Project-Based Approach to REDD Credits

Under a project-based approach to REDD credits, private or public entities would be authorised to engage in REDD activities at the project level, irrespective of whether or not a host country has negotiated a national emissions reference level.[88] The rationale for a project-based approach is that developing countries with capacity constraints to implement forest protection measures, may not be able to implement the necessary policy, legal and institutional reforms to meet a nationwide REDD target. There has also been some doubt that the private sector would participate in a REDD mechanism that links investment risk to government performance.[89] In this regard, the temporary Certified Emissions Reductions (tCER) arising from afforestation and reforestation CDM (A/R CDM) projects have been suggested as a basis for a CDM approach to REDD.[90]

However, there is a view that a project-based approach would not provide for national coverage and would be more likely to cause 'leakage'. Here 'leakage' might counteract any emissions reduction in the project area so participants might find it difficult to claim the expected carbon credits. In practice then, because it is virtually impossible for project-based mechanisms to guarantee an overall reduction of emissions from a country's forests, it may be ultimately unsuitable as a REDD instrument. For this reason, it has been suggested that perhaps a national approach is preferable to a project approach, although it should also be acknowledged that the national approach does not solve the problem of leakage from one country to another.[91]

Developing an Internationally Recognised Legal Standard for REDD Credits

Clearly, the UNFCCC in association with the IPCC will need to develop the standards by which carbon credits from tropical forests can be generated. The interest in the voluntary carbon market for REDD credits, discussed below, has given rise to the development of a wide variety of REDD standards. These include: the American Carbon Registry Forest Carbon Project Standard;[92] the Carbon, Community and Biodiversity Project Design Standards;[93] the Chicago Climate Exchange Standard; Plan Vivo Standards;[94] the Social Carbon Guidelines;[95] and the Voluntary Carbon Standard.[96] It seems likely that in developing legally-binding standards for REDD offset credits post-COP15, the UNFCCC will be guided to some extent by these standards. It should also be remembered that the CDM Executive Board has developed methodologies for afforestation and reforestation CDM projects[97] which will provide a comparative baseline for developing methodologies and standards for REDD. Finally, the IPCC has developed the IPCC 2006 Guidelines for Agriculture, Forestry and Other Land Uses (AFOLU).[98]

An analysis of the carbon offset standards shows that a complete and fully-fledged carbon offset standard comprises the following three components: accounting standards; monitoring, verification and certification standards; and registration and enforcement standards.[99] The accounting standards ensure that offsets are 'real, additional and permanent' while the monitoring, verification and certification standards ensure that offset projects perform against key project design criteria.[100] Verification and certification occur *ex post facto* and are distinguishable from validation of the project activity. Validation is usually undertaken by an independent auditor which reviews and validates the project design documents and other related documentation such as environmental impact assessments and stakeholder consultation processes.[101] Finally, registration and enforcement systems ensure that offset credits are sold only once with the registry containing publicly accessible information which identifies individual offset projects and tracks the ownership of offset credits. Contractual or legal standards should also stipulate the distribution of risk in the case of project failure, and perhaps insurance arrangements for insuring against the risk.[102] Together, these aspects of the standards enable the trading of offset credits.[103]

According to the author's assessment of these standards it appears that the Voluntary Carbon Standards' *Tool for AFOLU Methodological Issues*[104] is relied upon heavily to provide the methodology for generating REDD credits. To be compliant with this standard the project proponent must follow the following steps: follow the general methodological guidance (for example determine baselines and assess 'leakage'); determine the land eligibility; determine the project boundary; determine the carbon pools; establish a project baseline; assess and manage leakage; and estimate and monitor net project GHG benefits.

'Additionality' is probably the most controversial aspect of all offset projects, including REDD projects. Consequently, all of the standards require adherence to the *A/R CDM additionality tool*.[105] Under this tool, project developers must:

- Provide evidence that, in order to implement the project, the generation and sale of credits is crucial.
- The identification and description of a minimum of two land-use scenarios — one 'project scenario' and one 'without-project scenarios'.
- The undertaking of an investment analysis which compares these two scenarios and proves that additional financial means from the sale of carbon credits are required to implement the project. If this cannot be proved then a barrier analysis must be done to indicate that the sales of carbon credits will result in overcoming barriers which are decisive for project implementation, such as the lack of technical expertise or the lack of law enforcement.
- A 'common practice' test must be applied indicating that the project is not common practice compared with projects that do not generate carbon credits.[106]

Clearly, the design of an international offset standard for REDD credits by the UNFCCC will be an essential part of establishing a REDD carbon market.

Likely Demand for REDD Credits

While it is difficult to predict the likely demand for REDD credits following Copenhagen, the existing demands for compliance offsets grows exponentially each year, while corporate investors and investment banks are already investing in REDD projects in developing countries.[107]

Demand for Offsets in Compliance Markets

The World Bank regularly releases updates on the world carbon market, which essentially comprises the compliance markets established under the European Union Emissions Trading Scheme, the New South Wales Greenhouse Gas Abatement Scheme and the Regional Greenhouse Gas Initiative, established by north-eastern states in the US.[108] *The World Bank State and Trends of the World Carbon Market 2009*[109] reports that in 2008 the carbon market reached a total value of about USD126 billion, double its 2007 value in spite of the constraints of the global financial crisis.[110] Importantly, the second largest segment of the carbon market was the secondary market for Certified Emission Reductions ('sCERs') representing the resale of credits derived from carbon offset projects under the CDM. This market comprising spot, futures and options transactions was valued in excess of $26 billion, representing a five-fold increase in both value and volume over 2007.[111] One of the reasons for an increase in value for sCERs is that the primary CDM market has faced regulatory delays, difficulty in obtaining financing for projects during the global financial crisis, and the renegotiation or cancellation of some carbon contracts.

While the report provides interesting insights into the state of the CDM offsets market, it also suggests that if the difficulties associated with the CDM market are not adequately resolved at COP15, the REDD offsets market might provide a more attractive option for liable entities in compliance markets. As well, on the REDD demand side, the outcomes of COP15 are likely to be significant from two perspectives. First, developed countries are likely to accept new emissions reduction targets at COP15 along with new rules regarding 'supplementarity'. Also, as discussed above, COP15 will establish the international legal framework for REDD. Taken together, these two factors will shape the demand for REDD credits. Yet even prior to COP15, the European Union has indicated that if there is a satisfactory international agreement, then it will adopt a stronger target of 30% below 1990 levels by 2020, instead of its current commitment of 25% below, so creating an additional demand for emissions reductions of about 300 MtCO2e per year over 2013–2020.[112] This, together with the REDD credits provisions under the United States *American Clean Energy and Security Act of 2009*,[113] discussed below, indicates that the demand for REDD credits in compliance markets is likely to be significant in the post-2012 era.

Demand for REDD Credits in the Voluntary Market

Despite the lack of an existing international legal framework for REDD, the voluntary market for REDD credits is emerging. For example, in 2007, a Utah-based environmental foundation, signed an agreement with the Costa Rican government to buy credits generated under the country's Payment for Environmental Services scheme, discussed above. Pax Natura, the foundation, will pay $10 million for the credits which it will sell into the voluntary carbon market in North America.[114] Then on 7 April 2008, Marriott International signed the first REDD agreement in the Brazilian state of Amazonas committing $2 million, with an additional $4 million to be added over three years. This is indicative of a pipeline of new projects, which incorporate sustainable development and biodiversity benefits, seeking certification under the Climate, Community and Biodiversity Project Design Standard.

In another development, US Investment Bank, Merrill Lynch, has signed a six-year, $9 million agreement with Carbon Conservation, an Australian-based project developer, and UK-based NGO Flora and Fauna International (FFI) to buy voluntary emissions reductions (VERs) from a REDD project in Indonesia's Aceh province. Merrill Lynch has the option to expand the deal to $400 million. The project, which is regarded as the largest carbon offset project in the world, seeks to avoid the emission of 100 million tonnes of CO_2 over 30 years. Named Aceh Green, the project hopes to achieve: avoided deforestation; improved forestry management; small-holder estate crop development and land reform; the development of public infrastructure; and 'green' soft commodity production marketed with Aceh Green branding.[115]

On 24 June 2008, Australia's Macquarie Group and FFI announced the formation of a task force to invest in the management of tropical forests and generate carbon credits for sale. Over three years, the partnership will work with local communities and governments to protect six forests at risk from deforestation in South East Asia, South America and Africa. The drivers of deforestation will be addressed by developing new economic opportunities for forest-dependent communities. Capital and financial services for the forest projects will be provided by Macquarie Group which will also take responsibility for ensuring compliance with carbon standards. Macquarie has reserved the right to sell the carbon credits internationally. FFI will draw on its conservation experience to work with local governments and communities to implement the projects.[116]

The United States Moves Ahead on REDD

Most countries are awaiting the REDD outcomes at COP15 before incorporating REDD offsets in their emissions trading schemes, while at the same time devoting considerable amounts of money towards capacity building in developing countries.[117] The United States, however, has incorporated REDD provisions in the *American Clean Energy and Security Act of 2009*,[118] which was passed by the House of Representatives on 26 June with a vote of 219–212 prior to any outcomes at COP15.

The Act allows covered entities[119] to rely on REDD offsets. It also provides for the allocation of US emissions allowances to developing countries to secure Supplemental Emissions Reductions for the United States through reduced deforestation programs, over and above its domestic emissions reduction target. Irrespective of anything which is agreed upon at Copenhagen, if the Act manages to win Senate approval, it will have an important bearing on REDD projects in developing countries.

Offset Credits from REDD

The Act provides a compliance option whereby a certain percentage of a liable entity's liability may be satisfied by surrendering both domestic and international offset credits on the ratio of 1.25 offset credit: 1 emission allowance.[120] Covered entities may collectively use offset credits to demonstrate compliance for up to a maximum of two billion tons of greenhouse gas emissions annually. The ability to demonstrate compliance with offset credits will be divided pro rata among covered entities by allowing each covered entity to satisfy a percentage of the number of allowances required to be held. The percentage is calculated on an annual basis according to a formula set out in the Act, with half of that percentage coming from domestic offsets and the other half from international offsets.[121]

Eligible Countries, States or Provinces

Offset credits from REDD must be compliant with the general provisions for international offsets but are subject to additional requirements.[122] The Act requires the Administrator, in consultation with the Secretary of State, the Administrator of USAID and any other appropriate federal agency, and taking into consideration any recommendations of the Advisory Board, to make regulations for international offsets. These offsets can only be issued if the US is a party to a bilateral or multilateral agreement with the country in which the offset activity will occur. In addition, the regulators should seek to ensure that these countries have in place enforceable legal regimes which give due regard to the rights and interests of local communities, indigenous peoples, forest-dependent communities and vulnerable social groups and that full consultation with, and the participation of, these groups will occur during the design, planning, implementation, monitoring and evaluation of REDD activities. There should be an equitable sharing of profits and benefits derived from REDD credits with these people.[123] Offset credits may be issued for REDD activities only in countries, or in states and provinces of a country, if they are listed by the Administrator.

For a country (for sake of clarity to be known as a 'Category A' country) to be listed it must have:

- The technical capacity to monitor, measure, report and verify forest carbon fluxes for all significant sources of GHG emissions from deforestation with an acceptable level of certainty. This should be done in accordance with internationally accepted methodologies established by the IPCC.

- The institutional capacity to reduce emissions from deforestation, including strong governance mechanisms and ways of distributing REDD resources equitably at the local level.

- A land-use or forest sector strategic plan that: assesses national or local drivers of deforestation and identifies national policy reforms to address them; estimates the country's emissions from deforestation and degradation; identifies improvements and institutional capacity needed to implement a national REDD program; and establishes a timeline for implementing the program.[124]

There is another category under which a country (for sake of clarity to be known as a 'Category B' country) may be listed and that is where, based on recent, credible and reliable emissions data, the Administrator determines that it accounts for less that 1% of global GHG emissions and less than 3% of global forest-sector and landuse change GHG emissions. In addition, the Administrator must determine that the country is making a good faith effort to develop a land-use or forest sector strategic plan as described above.[125] In these countries REDD credits may be issued for projects or programs that reduce deforestation.

For a state or province to be listed by the Administrator, it must itself be a major emitter of GHGs from tropical deforestation on a scale that is commensurate to the emissions of other countries. It must meet the same capacity criteria as a country, mentioned above, and have the same regard for local communities, indigenous people, forest-dependent communities and vulnerable social groups.[126]

Issuing Offset Credits for National REDD Activities Carried Out in a Category A Country

Subject to a Category A country being an eligible country, the Administrator will issue a quantity of REDD credits by comparing the national emissions from deforestation relative to a national deforestation baseline for that country.[127] The baseline must be: national in scope; consistent with REDD NAMAs taking into account the average annual rates of historical deforestation during the previous five years, the applicable drivers of deforestation and other factors to ensure additionality; establish a trajectory that will result in zero net deforestation within 20 years from the date that the baseline is established; be adjusted over time due to changing national circumstances; and account for all significant sources of GHG emissions from deforestation in that country.[128]

The reductions in emissions from deforestation must have occurred before the issuance of the REDD offset credits, and taking into account relevant international standards, must have been demonstrated through the use of ground-based inventories, remote sensing technology, and other methodologies so that all relevant carbon stocks have been accounted for.[129] In issuing the credits, the Administrator must have made adjustments to discount for any additional uncertainty.[130] The Administrator must ensure that activities, for which REDD offset credits are issued have been carried out and managed: in accordance with the practices of environmentally sustainable forest management; in a way that promotes or restores native forest species and ecosystems and avoids the introduction of invasive species, and gives regard to the rights and interest of local communities, indigenous people, forest-dependent communities and vulnerable social groups; and consults with all of these stakeholders while ensuring the equitable distribution of profits.[131]

Issuing Offset Credits for REDD Projects or Programs Carried Out in a Category B Country

REDD offset credits can also be issued by the Administrator for project or program level activities in Category B countries. Here offset credits will be issued where emissions are below the established project or program level baseline.[132] The baseline must be: consistent with REDD NAMAs taking into account the average annual rates of historical deforestation during the previous five years, the applicable drivers of deforestation and other factors to ensure additionality; be designed to account for all significant sources of GHG emissions from deforestation in the project or program boundary; and be adjusted to account for emissions leakage outside the project or program boundary.[133] These offset credits should be phased out five years after the Act's compliance period begins but the deadline can be extended by the Administrator by up to eight years where the project and program activities are occurring in least developed countries. It must be determined that the country: lacks sufficient capacity to achieve reductions in deforestation measured against national baselines; is receiving support to develop such capacity under the Supplemental Emissions Reduction provisions (see below); and has developed and is working to implement a credible national strategy or plan to reduce deforestation. Forest degradation, or soil carbon losses associated with forested wetlands or peat-lands may be included in the meaning of deforestation.[134] Where the Administrator, in consultation with the Secretary of State and the Administrator of USAID, determines that any of these requirements are not feasible in a given country, they can be modified or omitted. The Administrator is still required to ensure, with an adequate margin of safety, the integrity of REDD credits issued.[135]

Issuing Offset Credits for REDD Activities Carried Out in a State or Province

Provided that a state or province is eligible, the Administrator will issue REDD offset credits for GHG emissions reductions achieved at a state or provincial level by comparing the reductions with the deforestation baseline for that state or province. The baseline must be: consistent with REDD NAMAs taking into account the average annual rates of historical deforestation during the previous five years, the applicable drivers of deforestation and other factors to ensure additionality; establish a trajectory that will result in zero net deforestation within 20 years from the date that the baseline is established; and be designed to account for all significant sources of GHG emissions from deforestation within the state or province and be adjusted to fully account for emissions leakage outside the state or province.[136] REDD offset credits from state or provincial activities will be phased out five years after the start of the compliance period under the Act.[137]

To avoid double counting, the Administrator must ensure that any REDD offset credits issued under the US Act are not also used for compliance purposes under a foreign domestic or international regulatory system.[138]

Supplemental Emissions Reductions From REDD

The supplemental provisions are designed to secure additional US reductions in GHG emissions over and above those that will be achieved under a domestic emissions trading scheme. It is important to note that activities which receive support under these supplemental provisions will not generate offset credits for the GHG emissions reductions or avoidance, or GHG sequestration, produced by the activities.[139] They are essentially a funding mechanism for REDD.

The provisions are based on Congress' findings that: deforestation is one of the largest sources of GHG emissions in developing countries; without reductions in these emissions it will be difficult to limit global temperature increases to less than two degrees centigrade; it is in the US national interest to assist developing countries to halt deforestation thereby also delivering environmental and social co-benefits; and under the Bali Action Plan developed countries had committed to provide financial resources to support mitigation actions in developing countries.[140]

Objectives of the Act and disposition of allowances

This part of the Act requires the Administrator, in consultation with the USAID Administrator, to make regulations establishing a program to use emissions allowances to achieve supplemental reductions in GHG emissions from deforestation in developing countries.[141] The stated objectives of the Act are to:

- Achieve supplemental emissions reductions from deforestation of at least 720,000,000 tons of CO_{2-e} in 2020 and a cumulative amounts of at least 6,000,000,000 tons by 31 December 2025 and additional supplemental emissions reductions in subsequent years;
- To build capacity in developing countries for REDD and to prepare them to participate in international markets for REDD; and

- To preserve existing forest carbon stocks in countries where they may be vulnerable to international leakage, especially countries with largely intact native forests.[142]

In accordance with these objectives, each year the Administrator must allocate the following percentages of emissions allowances to recipient developing countries to support their REDD capacity building activities:
- For the years 2012–2025, 5%.
- For the years 2026–2030, 3%.
- For the years 2031–2050, 2%.

These percentages may be modified by the Administrator to ensure that the reductions objectives for 2020 and 2025 are met. Any allowances that have not been distributed in any given year by the end of that year can be carried over to the following vintage year.[143]

Activities carried out consistently with the objectives of the Act must, to the extent practicable, be aligned with broader development, poverty alleviation, or natural resource management objectives and initiatives in the recipient country.[144]

Eligible Countries

The Administrator will identify eligible countries to participate in the scheme in consultation with the Administrator of USAID. Such countries must be, or must be at risk of, experiencing deforestation and forest degradation and must have entered into a bilateral or multilateral agreement with the US establishing the conditions of its participation in the scheme. Beginning five years after the date that the country entered into such an agreement, no further compensation through emissions allowances will be granted to that country for any subnational deforestation reduction activities. However, this period can be extended for another five-year period if the Administrator in consultation with the Administrator of USAID determines that: the country is making substantial progress towards adopting and implementing a REDD program measured against a national baseline; the GHG reductions are not resulting in significant leakage and are being appropriately discounted to account for any leakage that may be occurring.[145]

Standards

Developing countries must agree to meet the standards established for REDD activities.[146] The standards will appear in regulations made by the Administrator, in consultation with the Administrator of USAID, which will ensure that supplemental emissions reductions achieved through REDD programs are additional, measurable, verifiable, permanent and monitored, and account for leakage and uncertainty. The standards must:
- Require the establishment of a national deforestation baseline for each country which must: be national in scope; consistent with REDD NAMAs taking into account the average annual rates of historical deforestation during the previous five years, the applicable drivers of deforestation and other factors to ensure additionality; establish a trajectory that will result in zero net deforestation within 20 years from the date that the baseline is established; be adjusted over time due to changing national circumstances; and account for all significant sources of GHG emissions from deforestation in that country;[147]

- Require that the emissions reductions have actually been achieved and verified before any distribution of emissions allowances are made;[148]
- Apply a conservative discount factor to reflect the uncertainty regarding the levels of emissions achieved, where accounting for subnational deforestation reduction activities lack the standardised or precise measurement or monitoring techniques necessary for a full accounting of changes in emissions or baselines;[149]
- Ensure that activities carried out are managed: in accordance with the practices of environmentally sustainably forest management; in a way that promotes or restores native forest species and ecosystems and avoids the introduction of invasive species, and gives regard to the rights and interest of local communities, indigenous people, forest-dependent communities and vulnerable social groups; and consults with all of these stakeholders while ensuring the equitable distribution of profits.[150]

Recognised Activities and Their Implementation

A number of different activities are recognised including: national and subnational deforestation reduction activities and pilot activities that reduce GHG emissions even though they are subject to significant uncertainty; activities to measure, monitor and verify deforestation, avoided deforestation and deforestation rates; leakage prevention activities; MRV capacity building so as to facilitate the quantification of emissions reductions and the generation for sale of REDD offset credits; the development of REDD and illegal logging governance structures; the enforcement of REDD activities; policy reforms; and the monitoring and evaluation of the results of all REDD activities.[151] The Administrator, in consultation with the Administrator of USAID and on the advice of the Advisory Board,[152] may where appropriate expand the categories of activities to include reduced soil carbon-derived emissions associated with deforestation and degradation of forested wetlands and peat-lands.[153] A publicly accessible registry of supplemental emissions reductions achieved through REDD activities under the Act must be established by the Administrator.[154]

REDD activities will be selected for support and implementation primarily by the Administrator of USAID, in conjunction with the Administrator of the Act, and they must work together to develop and biennially update a strategic plan for meeting the objectives of the Act, described above.[155]

To implement the activities, the Administrator may distribute emissions allowances to an eligible country, to a private or public group (including international organisations), or to an international fund established by an international agreement to which the US is a party. The Administrator must get the concurrence of the Secretary of State before distributing allowances to the government of another country, or to an international organisation or international fund.[156]

Reports and Reviews

The Administrator and the Administrator of USAID must, no later than 1 January 2014, submit a report to: the Committee on Energy and Commerce and the Committee on Foreign Affairs of the House of Representatives, and the Committee on Environment and Public Works and the Committee on Foreign Relations of the Senate on the support provided through the allocation of emissions allowances. It must also be made publicly available. It must include: a statement of the quantity of supplemental emissions reductions for which compensation by way of emissions allowances was provided; and a description of the national and subnational deforestation reduction activities, and leakage prevention activities supported by the scheme, including a statement of the quantity of allowances distributed to each developing country for each activity during the prior fiscal year, as well as a description of what was accomplished by each of the activities.[157]

In addition, a review of the scheme must be undertaken by the Administrator and the Administrator of USAID no later than four years after the date that the legislation is enacted, and every five years thereafter. The advice of the Advisory Board must be taken into account. The review must include the effects of the activities on: total documented carbon stocks of each country that has received support either directly or indirectly, compared with the country's national deforestation baseline established as a requirement of receiving support; the number of countries with capacity to generate for sale REDD offset credits and the amount of such credits; forest governance in each country that has received support; the extent to which indigenous people and forest-dependent communities have been affected by the REDD activities; the protection of biodiversity and ecosystem services within forested areas associated with REDD activities; subnational and international leakage; and any REDD program or mechanism established under the *UNFCCC*.[158]

Key REDD Issues and Legal Implications

Given the extensive REDD provisions in the Negotiating Text and the *American Clean Energy and Security Act* of 2009 there can be no doubt that REDD has become the latest frontier in the international community's efforts to curb escalating GHG emissions. When one reviews the provisions of these two texts, however, it seems that they are rather too glib in making the assumptions they do about the capacity of tropical rainforest countries to implement REDD. As this section demonstrates, things are not quite as straightforward as they seem.

The Crucial Issue of Governance

It is axiomatic that good governance is essential to any tropical rainforest country government's capacity to effectively formulate and implement REDD policies and legislation. This is why the Negotiating Text and the US Act make frequent reference to the need for good governance. Yet, forest governance has long been a concern of the international community. The European Union developed its *Forest Law Enforcement Governance and Trade (FLEGT) Action Plan* in 2003, which sets out a range of measures to: tackle illegal logging by improving forest governance; strengthening peoples' tenure rights; developing a licensing scheme that assures that timber has been legally produced; and establishing a system to independently monitor implementation.

In spite of this initiative on the part of the EU, it seems that forestry governance remains a considerable barrier to REDD as indicated by World Resources Institute (WRI) in February 2009. The WRI released a review of 25 of the R-PINs submitted to the World

Bank's Forest Carbon Partnership Facility.[159] The WRI analysed the R-PINs with reference to 17 good governance criteria which it believes are vital for any country that wants to participate in REDD mechanisms. These were organised within six basic processes:

- law and policy development
- land tenure administration and enforcement
- forest management
- forest monitoring
- law enforcement
- forest revenue distribution and benefit sharing.[160]

The WRI found that although governance issues are generally well-recognised in R-PINs, none of the countries' submissions on this issue can be considered comprehensive. This is partly because the R-PINs are meant to be preliminary, but the WRI found that several critical issues were conspicuously missing in R-PINs. Consequently, the WRI notes that: law enforcement challenges require greater attention, particularly with regard to illegal logging and other forest crime; unclear tenure is a major challenge in most countries, and responding to this challenge will require much more effort (see below); measures to increase policy coherence between sectors, particularly with regards to land-use planning, need more attention; the adequacy of existing revenue distribution and benefit-sharing mechanisms should inform the development of a payment system under REDD; and transparency and accountability in forest monitoring systems for REDD need to be emphasised.[161]

Problems with inadequate governance around REDD credits were recently highlighted when the head of the Office of Climate Change in Papua New Guinea was suspended for entering into million dollar carbon trading agreements with foreign companies in the absence of any REDD policy and legislative frameworks. The PNG government has subsequently declared that carbon trading agreements will have no legal validity until such frameworks have been developed, which will be pursuant to the completion of an investigation into the Office of Climate Change. The government is also waiting for the outcomes on REDD at Copenhagen before establishing a domestic REDD framework.[162]

Establishing Baselines and National Reference Levels

As the Negotiating Text and the US Act make clear, the implementation of REDD policies requires accurate estimates of emissions reductions at the national level. As discussed above, these must be: additional in the sense that they would not have occurred but for the REDD policies); permanent; and not result in 'leakage'. Reliable data is essential for investor confidence especially where REDD is integrated into international carbon markets. Estimating emissions reductions has three essential elements. These are:

- Establishing the baseline (or 'business as usual' scenario);
- Establishing the crediting line (or reference scenario which enumerates the quantity of emissions reduction); and
- Monitoring actual rates of deforestation and degradation associated with emissions over time.[163]

The capacity of developing countries to adequately monitor rates of deforestation is elaborated on below. However, for lawyers the terms 'baseline' and 'reference scenarios' are highly significant as they are the standards against which a country's performance on REDD will be measured, against which compensation will be paid and against which carbon credits will be issued for sale on the international market. This is made clear both in the Negotiating Text and the *American Clean Energy and Security Act of 2009*. All of this assumes that certainty prevails, especially since, as discussed above, this is what the carbon trading market requires. Yet setting the baseline and reference scenarios for individual tropical rainforest countries is problematic.

Establishing a baseline is essentially a hypothetical assessment since it represents a counterfactual business-as-usual scenario where the question is: what would have happened had the REDD project not been implemented? Credits can only be issued by comparing the difference between emissions in the baseline scenario and emissions abatement resulting from the project. Since the baseline scenario will never occur, it is impossible to predict with certainty what the results of the scenario would have been. For this reason, baselines should be established conservatively so as to not overestimate the abatement outcomes of a REDD project.[164]

Meanwhile, Piraud has contended that 'reference levels' are also inherently problematic. There are two principal ways of determining reference levels: those designed on an historical basis by considering past trends, which may or may not be adjusted according to national contexts; and those based on modelled predictions, aiming to take into account a certain number of variables that are considered as determinants of the rate of deforestation. The trouble with reference levels designed on an historical basis is that this method does not take into account 'forest transition' phenomena whereby the level of economic development and resource scarcity might modify rates of deforestation from one period to another. As well, countries that have already started a massive deforestation are given a 'premium' in that their reference levels will be so high that such countries will be able to generate a disproportionately high number of credits, compared with countries that have low historical rates of deforestation.[165] On the other hand, many countries are reluctant to use 'predictive' scenarios given that the likely rate of deforestation will not only be influenced by relatively predictable factors, such as demography, road building, and annual economic growth rate, but also uncertain phenomena such as the price of various agricultural commodities on speculative and highly volatile markets.[166]

Clearly, like the issues of 'additionality', 'leakage' and 'permanence', the difficulties relating to baselines and reference scenarios will continue to plague the question of whether emissions reductions from REDD projects have been accurately measured. Questions will also continue to be asked about whether liable entities under emissions trading schemes should be allowed to rely on REDD offset credits rather than surrendering a carbon allowance.

Monitoring, Reporting and Verification: Its Link to Compliance and Enforcement

The enforcement of any international agreement and domestic legislation, which might be enacted by REDD countries, will rely on the ability of developing countries to monitor rates of deforestation for compliance with, and enforcement of, the law. Carbon credits, whether they are generated by a national government or on a project basis, will depend for their legitimacy and 'commerciability' on the proper monitoring of rates of deforestation against the baselines and reference scenarios. Without this, the credits are unlikely to satisfy any rules and modalities that might be developed by the COP.

A recent assessment of national forest monitoring capabilities in tropical rainforest developing countries has been undertaken for The Prince's Rainforests Project and the Government of Norway.[167] The report is rather salutary and seems to support the proposals put forward in the Negotiating Text for a REDD credits market to only be introduced in Phase 3 of the implementation process. Certainly, from a legal perspective, it has a number of important implications.

Essentially, the report assesses the capabilities of 99 tropical rainforest countries (REDD countries) to monitor their forest resources with respect to forest area change and associated carbon stock changes. It also makes recommendations for the implementation of operational forest monitoring within five years.[168] Although the details for a REDD mechanism are still under development, the authors of the report rely on the requirements for monitoring systems, carbon emissions estimation and reporting provided by the IPCC's international guidelines — Land-Use and Land-Use Change (LULUCF) and AFOLU — to draw their conclusions.[169] They rely also on the national reporting to the FAO Forest Resource Assessment,[170] and the submissions of national communications by REDD countries to the *UNFCCC*. For 25 of the countries they have also accessed their R-PINs, discussed above.

The report notes that although the development of the *UNFCCC* has stimulated REDD countries to establish national GHG inventories, less than 20% of countries are listed with a fully developed inventory. Of the 99 countries surveyed, 50 have inventories with a 50–100% range of completeness, while the rest have inventories which are less than 50% complete.[171] Consequently, the majority of countries have limitations in providing a complete and accurate estimate of GHG emissions and forest loss.[172] Furthermore, very few REDD countries report on soil carbon, even through emissions from deforested or degraded peat-lands may be significant. This will prove problematic if at Copenhagen it is agreed that the soil carbon pool is to be included in a country's strategy to receive REDD credits for reducing emissions from forest land.[173]

From a technological perspective there are many barriers to accurate monitoring of forest emissions. The most commonly used satellite dataset for forest monitoring at the national level is archived Landsat data. Other satellite sensor and optical remote sensing data can be used but access to this data depends on cost and availability. Other challenges associated with remote sensing technologies include persistent cloud cover, seasonality and topography which means that current Landsat 5 receiving stations fail to cover REDD countries in Central and West Africa and Central America. Also, although some remote sensing datasets are available free of charge, access to the internet is needed to access archived satellite data. Most of Africa experiences low available bandwidth which means that alternative means of data delivery need to be arranged, such as by mailing hard-discs or DVDs. Finally, remote sensing data needs to be pre-processed for interpretation which needs geometric and radiometric corrections and REDD countries may not have the capacity to undertake the corrections.[174]

The report makes recommendations for capacity building in REDD countries espe-cially on a regional level.[175] It also provides a very useful Appendix in which the indica-tors used to assess capacity are applied on a country-by-country basis.[176]

Australia's National Carbon Accounting System

It may be that Australia's world leading technology on measuring carbon from the land-based sector — the National Carbon Accounting System (NCAS) — will make an important contribution in the area of MRV. The NCAS uses satellite imagery of land cover change, land management data and climate and soils data and is designed to meet national and international reporting requirements, under the *UNFCCC*. Even more importantly for our purposes, a derivative of the NCAS — the National Carbon Accounting Toolbox (NCAT) — allows landholders to engage in carbon accounting from land based activities at the individual project level.

It has been announced recently that the technology is to be expanded and enhanced to support REDD. In a partnership between the Australian Government and the Clinton Climate Initiative, the NCAS will be extended into the international arena for global monitoring of carbon emissions. The partnership will facilitate access to technology that will help countries to make comprehensive legitimate forest assessments, and to monitor and manage their forests. It is intended that the system will be consistent with the requirements of the IPCC and the likely future needs for recognising REDD. The partnership will focus on large-scale projects that demonstrate the use of the NCAS in developing countries, and the development of a web based data delivery system that facilitates free and open access to wide-ranging data from satellites, aircraft and field measurements.[177]

So Who Really Owns the Carbon in Tropical Rainforest Countries: 'Land Tenure' or 'Resource Tenure'?

The Negotiating Text, the *American Clean Energy and Security Act of 2009* and the liter-ature are replete with references to 'land tenure' and the 'right and interests' of indige-nous people and local communities,[178] forest-dependent communities and vulnerable social groups. The reason for this is that ultimately, whether a public funding approach or a carbon credits approach is adopted for REDD, and in all likelihood it seems that it will be a combination of both, it is necessary to decide who 'owns' the carbon within existing tenure systems. Yet both the Text and the US legislation gloss over what this really means. It is as if these notions are so self-evident that their definition is not even required in the various sections that carefully provide definitions of a wide range of terms. Yet there are many reasons why the question of 'tenure' and who 'owns' the carbon is crucial.

First, there are concerns that indigenous peoples' and local communities' liveli-hoods, access to resources and other rights, such as cultural rights, will be disrupted where deforestation is substantially reduced or halted. Indigenous people face the risk that governments, companies and conservation NGOs will 'zone' forests thereby creating protected areas, biological corridors, forest reserves and sustainable forest man-agement zones in order to receive REDD payments while excluding or disadvantaging indigenous and traditional communities.[179] Second, there are concerns that insecure tenure may itself promote deforestation as resource users clear forests in order to show occupation where land claims are contested. Finally, tenure is the basis upon which all REDD benefit sharing arrangements will be determined since whoever has tenure over the forest 'owns' the carbon rights which emanate from that forest. Also, were forest carbon credits to be offered as 'eligible' credits in an emissions trading scheme, most likely in Phase 3 of implementing REDD, convention has it that credits are only legitimately generated and traded if they can be traced back to the 'owner' of the land. As Gray and Gray note, 'in order to be commerciable, rights must of course have a sufficient stability of definition to enable ready identification of the entitlement which is being traded.'[180]

Is it 'Land' Tenure or 'Resource' Tenure?

The issue of who 'owns' the carbon is an immensely complex legal question. First, it is not even clear what is envisaged when the Negotiating Text requires that issues of land 'tenure' be resolved. It is well-known that land tenure arrangements in many developing countries are diverse and not easily amenable of identification and resolution. Second, one cannot be absolutely sure whether successful REDD schemes require the identification of 'land' tenure or 'resource' tenure for their effectiveness and indeed legitimacy. The US legislation refers to 'the rights and interests' of various people without mentioning tenure at all, the Negotiating Text specifically mentions 'land' tenure and not 'resource' tenure, while the literature[181] uses the terms 'land' tenure and 'resource' tenure interchangeably without acknowledging the differences between them. Yet the difference between the two is significant, in a legal sense. It may be that even though 'land' tenure is not recognised, 'resource' tenure is, by way of either the explicit or implicit acknowledgement of rights of usufruct, for example, over the forests. If it is the case that the types of tenure which exist in tropical rainforest countries vary, and that in some countries land and forests are state-owned,[182] as discussed below, then the Negotiating Text should expand its definition of tenure beyond simply 'land' tenure to include 'resource' tenure, unless it really intends REDD to give the imprimatur for wholesale land reform in participating jurisdictions. Finally, it should not be automatically assumed that those who have either 'land' or 'resource' tenure will 'own' the forest carbon. As has occurred in Australia, carbon rights can be statutorily created as entirely separate instances of property from property in the forests.

'Tenure' and Proprietorial Rights

What cannot be avoided is the conclusion that references to 'land' tenure, 'resource' tenure or 'the rights and interests' of people reverts essentially to an analysis of some sort of proprietorial rights to forest carbon, and consequently to benefit-sharing in the proceeds from REDD.

It is not a simple task to identify what amounts to 'property' in forest carbon. This is so for a number of reasons not the least of which is that in trying to craft a definite legal answer to this issue, one realises that REDD schemes are multi-jurisdictional where either common law[183] or civil law[184] predominate, and where the constitutions of former colonies[185] may now include the protection of fundamental legal rights either by way of a Bill of Rights or Directives of State Principle.[186] We are then left to try to resolve the question of 'property' in carbon according to first principles about the notion of property.

Gray and Gray remind one that 'few concepts are quite so fragile, so elusive or so frequently misused as the notion of property',[187] and that one should not think of property as a 'thing but a power relationship'[188] which allows a person, or a community, 'to assert a significant degree of control over that resource'. The law of property meanwhile incorporates 'a series of critical value judgements, reflecting the cultural norms, the social ethics and the political economy prevalent in any given community'[189] while the 'limits of property are the "interfaces between accepted and unaccepted social claims".'[190] Consequently, 'every claim of "property" comprises the assertion of some quantum of socially approved power as exercisable in respect of some socially valued resource.'[191] Importantly, the common law has no concept of 'absolute title, property or ownership' and the quantum of property which anyone has in a resource can be measured on a sliding scale.[192] So claims of property can be graded across a 'spectrum of "propertiness"' and somewhere along that spectrum a right will be regarded as having sufficient 'gravamen to qualify for the appellation proprietary.'[193] The difficulty with this approach is that the identification of proprietary rights might be 'more instinctive than principled.'[194]

Tenure in Tropical Rainforest Countries

The International Institute for Environment and Development ('iied') report on tenure in REDD[195] states that effective local institutional capability, and good forestry skills and practices, are essential to make REDD work. According to iied, for this to be achievable, tenure should be the starting point and effective and equitable local property rights are needed.[196]

Land Tenure Arrangements

The iied report states that 77% of the world's forests are owned by government.[197] It also provides a typology of land tenure regimes in seven rainforest countries including: Brazil, Cameroon, Democratic Republic of Congo ('DRC'), Guyana, Indonesia, Malaysia and Papua New Guinea. State ownership is the predominant type of tenure in Cameroon (over 97% of land); DRC (all forested areas); Indonesia (most of the land); Malaysia (most forestland); PNG (less the 3% of the land). In Brazil most land and forests are privately owned, either individually or collectively as a result of national legislation. In some countries, the devolution and/or management of forestlands is being devolved to local government to try to promote accountability.

In some countries, customary tenure may be created by way of statute, such as has occurred recently in Mali, Mozambique, Tanzania and Uganda.[198] Even here, the enforcement of such rights may be weak. In Indonesia, the *Basic Agrarian Law of 1960* gives legal protection to customary land rights so long as customary systems still exist and their exercise is not inconsistent with the national interest and legislation. This makes holders of customary rights vulnerable to the exercise of administrative discretion while their rights may be taken for a public purpose which can include the business activities of private corporations.[199]

So, for example, in the Indonesian province of Kalimantan, to make an accurate assessment of land ownership one would need to determine the 'formal' legal position as well as the 'on the ground' customary practices (customary law, or 'adat'). Determining the formal legal position would require careful consideration of Indonesia's *Basic Agrarian Law of 1960* and the many government and ministerial regulations issued since then, many of which may be unclear and contradictory. As for customary practices, the relevant adat is likely to be highly localised and, therefore, there are likely to be many (dozens of) different land tenure arrangements throughout Kalimantan. Finding out what these are would require extensive field research as the vast majority of land in Kalimantan, like the rest of Indonesia, is not registered, despite Indonesia's decades-old registration system.[200]

Resource Tenure Arrangements

Customary systems of tenure, where resources are held by clans, families and other collective entities exist in much of rural Africa (including Cameroon and DRC) and Southeast Asia (Indonesia, Malaysia, and PNG) and areas inhabited by indigenous peoples in Latin America (including Brazil and Guyana). As iied reports, these are essentially rights to use the resources of forests but may have weak or no legal recognition. This exposes customary right holders to dispossession. Community forestry and co-management schemes may also exist in Cameroon, Brazil and Indonesia where the state transfers forest management rights and responsibilities to community based organisation which are the recipients of Payments for Ecosystem Services. These rights are vulnerable to termination.

'Ownership' of Carbon Based on Existing Tenure

If land tenure is the basis upon which forest carbon rights are to be recognised then in 77% of cases it is the state (whether federal or local) which 'owns' the carbon. According to the typology of tenure arrangements surveyed by iied, it is only the forests in Brazil and PNG which are privately-owned. Any existing customary tenure rights which are protected by statute in developing countries seems too fragile to be denoted as 'tenure' in land. Gray and Gray note that private ownership of land is legally recognised where certain irreducible features of property are present. They are: immunity from summary cancellation or extinguishment; presumptive entitlement to exclude others; and entitlement to prioritise resources values (to have decisional control).[201] It doubtful that all of these elements exist where communal tenure is statutorily recognised yet difficult to enforce.

Where private rights of land tenure do not exist, it seems that the rights might best be described as 'resource' tenure rights held by clans, families or other collective entities such as community-based organisations. Perhaps the closest one can come to a legal definition of this type of 'resource' tenure is the civil law notion of the right of usufruct. This is 'the right of reaping the fruits (fructus) of things belonging to others, without destroying or wasting the subject over which such right extended.[202] Yet this may be inappropriate given that in some cases forests are destroyed in the process of exercising the 'resource' tenure rights. Finally, can it be said that 'resource' tenure vests carbon property rights in those who exercise the rights? It is here that one reverts to the theory of property law that 'every claim of "property" comprises the assertion of some quantum of socially approved power as exercisable in respect of some socially valued resource.'[203] It may be that those exercising 'resource' tenure rights are able to assert that the power which they exercise over socially valued forestry resources is socially approved in which case they own the carbon rights in respect of the forests over which they have 'resource' tenure.

Relying on International Law to Assert Indigenous Peoples' Tenure Over Forests

There is a plethora of international law emanating from both international environmental law and international human rights law that seeks to protect the rights of indigenous peoples.

International Environmental Law

The international environmental law instruments which recognise the rights of indigenous peoples, and in some instances the rights of other local communities, include: the *Rio Declaration*,[204] *Agenda 21*,[205] the *UNFCCC*, the *United Nations Convention on the Conservation of Biological Diversity*[206] and the *Forest Principles*[207] all of which emerged from the United Nations Conference on Environment and Development in 1992 (the 'Rio Conference').

Principle 22 of the *Rio Declaration* acknowledges that

> Indigenous people and their communities, and other local communities, have a vital role in environmental management and development because of their knowledge and traditional practices. States should recognize and duly support their identity, culture and interests and enable their effective participation in the achievement of sustainable development.

Agenda 21 requires all countries to develop an effective strategy for tackling the problems of poverty, development and environment simultaneously which focus on resources, production and people and which considers the role of indigenous people and local communities in a democratic participatory process in association with improved governance.[208]

The *Convention on the Conservation of Biological Diversity* requires countries to respect, preserve and maintain the knowledge, innovations and practices of indigenous and local communities embodying traditional lifestyles relevant for the conservation and sustainable use of biological diversity.[209]

The *Forest Principles* meanwhile state that national forest policies should recognise and duly support the identity, culture and the rights of indigenous people, their communities and other communities and forest dwellers. Indigenous people and other communities should have an economic stake in forest use, perform economic activities, and achieve and maintain cultural identity and social organisation, as well as have adequate levels of livelihood and wellbeing, through, inter alia, land tenure arrangements which serve as incentives for the sustainable management of forests.[210]

International Human Rights Law

Part II of the *Convention (No 169) Concerning Indigenous and Tribal Peoples in Independent Countries*,[211] requires governments to respect the cultural and spiritual relationship between indigenous people and the lands which they occupy or otherwise use, and in particular the collective aspects of this relationship.[212] Their rights of ownership and possession over the lands that they traditionally occupy must be recognised as must their rights of access to lands which they do not occupy exclusively, for their subsistence and traditional activities. Adequate procedures must be established within the national legal system to resolve land claims by the peoples concerned.[213]

The convention recognises a few limited exceptions to the principle that indigenous peoples must not be removed from the lands that they occupy. Where the relocation of these peoples is considered necessary as an exceptional measure, the relocation must take place only with their free and informed consent. Where this cannot be obtained, the relocation must take place in accordance with appropriate procedures established by national laws and regulations, including public inquiries where appropriate, which provide the opportunity for effective representation of the peoples concerned. As soon as the grounds for relocation cease to exist, indigenous peoples must have the right to return to their traditional lands. If return is not possible, then they must be provided in all possible cases with lands of a quality and legal status at least equal to that of the lands previously occupied by them, which are suitable to provide for their present needs and future development. Where the peoples concerned prefer compensation in money or in kind, they must be so compensated under appropriate guarantees.[214] Indigenous people must be consulted whenever consideration is being given to transmit their rights outside their own community, and persons not belonging to these peoples must be prevented from taking advantage of their customs, or the lack of understanding of the laws on the part of their members, to secure the ownership, possession or use of land belonging to them.[215]

Relying on a Comparative Jurisprudence of the Courts to Assert Tenure Over Forests

Rights of tenure, whether of land or resource tenure, may be disrupted by REDD projects, while benefit-sharing regimes may not acknowledge the rights of indigenous peoples or local communities. In this case, a comparative jurisprudence which has emerged from a variety of judicial fora might be relied upon to make a claim. These courts have recognised that 'squatters' rights are integrally linked with rights to life and 'livelihood' and that communal indigenous rights to land might arise out of usufructury rights over that land.

The Common Law Jurisprudence of the Australian High Court

In the Australian case *Mabo v Queensland (No 2)*[216] a native title claim was brought by Mabo on behalf of the Meriam people who inhabit the island of Mer in the Murray Islands that lie in the Torres Strait. The Meriam people were in occupation of these islands before European contact and have continued in occupation to the present day. Anthropological records and research show that the Meriam people are descended from the people described in early European reports with whom they retain a strong sense of affiliation, including an affiliation with the society and culture of earlier times. One of the principal questions to be determined in *Mabo* was whether the Crown became the beneficial owner of all colonial land on first settlement. The High Court of Australia found that the Crown did not acquire beneficial ownership. Although this question might not arise in all instances where tenure is claimed for REDD purposes, it is instructive to refer to the reasoning applied by the High Court in recognising the rights of indigenous peoples.

In essence, the High Court found that, although native title is not an institution of the common law,[217] the common law of Australia recognises a form of native title which reflects the entitlement of indigenous inhabitants, in accordance with their laws or customs, to their traditional lands and that, subject to the effect of some particular Crown leases, the land entitlement of the Murray Islanders in accordance with their laws or customs is preserved, as native title, under the law of Queensland. This is so in cases where native title has not been extinguished. The court also held that the general principle that the common law will only recognise a customary title if it is consistent with the common law is subject to an exception in favour of traditional native title.[218]

Predicating its decision on Australia's ratification of the *Optional Protocol to the International Covenant on Civil and Political Rights*,[219] the High Court found that even if in earlier days there had been a refusal to recognise the rights and interests in land of the indigenous inhabitants of settled colonies, 'an unjust and discriminatory doctrine of that kind can no longer be accepted'.

As long ago as 1921, in *Amodu Tijani v Secretary, Southern Provinces*,[220] the Privy Council recognised not only usufructuary rights but also interests in land vested not in an individual or a number of identified individuals but in a community. In that case, Viscount Haldane observed:

> The title, such as it is, may not be that of the individual, as in this country it nearly always is in some form, but may be that of a community. Such a community may have the possessory title to the common enjoyment of a usufruct, with customs under which its individual members are admitted to enjoyment, and even to a right of transmitting the individual enjoyment as members by assignment inter vivos or by succession. To ascertain how far this latter development of right has progressed involves the study of the history of the particular community and its usages in each case. Abstract principles fashioned a priori are of but little assistance, and are as often as not misleading.[221]

Relying on this, the High Court of Australia concluded that if it is necessary to categorise an interest in land, arising out of individual rights of usufruct, as proprietary (in order that it survive a change in sovereignty), the interest possessed by a community that is in exclusive possession of land falls into that category. The court went on to say that 'it is not possible to admit traditional usufructuary rights without admitting a traditional proprietary community title'. While difficulties of proof of boundaries, or of membership of the community, or of representatives of the community, which was in exclusive possession might arise these should not be used as a reason to deny the existence of a proprietary community title capable of recognition by the common law.

Effectively, where a clan or group has continued to acknowledge the laws and (so far as practicable) to observe the customs based on the traditions of that clan or group, through which their traditional connection with the land has been substantially maintained, the traditional community title of that clan or group remains in existence.[222] As a result, indigenous rights, whether proprietary or personal and usufructuary in nature and whether possessed by a community, a group or an individual, may be protected by appropriate legal or equitable remedies.[223]

The relevance of this line of reasoning to indigenous peoples' claims of tenure over tropical rainforests is clear.

The Jurisprudence of the South African Constitutional Court

Customary indigenous rights have also been recognised by the South African Constitutional Court. The case *Alexkor Ltd v Richtersveld Community*[224] concerned a claim for restitution of land by the Richtersveld Community under the provisions of the South African *Restitution of Land Rights Act 1994* (the *Act*). The Richtersveld Community contended that, at the time that the *Natives Land Act (No 27) of 1913*[225] was enacted, it possessed a right of ownership and the right to exclusive beneficial occupation and use, or the right to use the subject land for certain specified purposes, including exploitation of natural resources. Both the interim and final South African Constitutions provided expressly for rights of restitution over land dispossessed as a result of the *Act*.[226]

According to the community, the rights of ownership arose under indigenous law and, after annexation, were protected under the common law of the Cape Colony or international law. Alternatively, the community contended that it held rights in the subject land under its own indigenous law which constituted a 'customary law interest', a right in land within the meaning of the *Act*, even if these rights were not recognised or protected.[227] The rights to land extended to rights of ownership over mineral and precious stones found in the land.

In this case, the South African Constitutional Court stated, in terms similar to those used by the Australian High Court in *Mabo*, that '[t]he Privy Council has held, and we agree, that a dispute between indigenous people as to the right to occupy a piece of land has to be determined according to indigenous law "without importing English conceptions of property law"'.[228]

The Constitutional Court held that in applying indigenous law, it is important to bear in mind that, unlike common law, indigenous law is not written but is a system of law, with its own values and norms, that was known to the community, practised and passed on from generation to generation. Throughout its history it has evolved and developed to meet the changing needs of the community.[229] With respect to indigenous Nama law, land was communally owned giving members of the community a right to occupy and use the land. These rights were not extended to non-members who had to obtain permission to use the land for which they sometimes had to pay.[230] The Court concluded that prior to annexation the Richtersveld Community had a right of ownership in the subject land under indigenous law and that accordingly the ownership of the minerals and precious stones vested in the community under indigenous law.[231]

The Constitutional Jurisprudence of the Indian Supreme Court

The Indian Supreme Court jurisprudence is helpful to those people who may be regarded as 'squatting' in tropical rainforests. The *Constitution of India* includes both fundamentally protected rights and Directive Principles of State Policy.[232] As noted earlier, these Directive Principles are not legally enforceable yet they serve to articulate the aspirations of the nation, and courts may have regard to them as an aid to statutory and constitutional interpretation. In *Tellis v Bombay Municipal Council*[233] the respondent sought to evict pavement and slum dwellers ('squatters') in the city of Bombay and in the process had demolished their dwellings. Relying on art 21 of the Constitution of India, which provides that 'No person shall be deprived of his life or personal liberty except according to the procedure established by law', the petitioners argued that their 'right to life' had been infringed. The court held that since the eviction of pavement and slum dwellers would lead, in a vicious cycle, to the deprivation of their employment this would infringe the right to life which included the 'right to livelihood'.[234]

The Indian Supreme Court relied on art 39(a) of the Constitution, which is a Directive Principle of State Policy. It provides that 'the State shall, in particular, direct its policy towards securing that the citizens, men and women equally, have the right to an adequate means of livelihood'. The Court also relied on art 41, another Directive Principle, which provides, inter alia, that 'the State shall, within the limits of its economic capacity and development, make effective provision for securing the right to work in cases of unemployment and of undeserved want'. Furthermore, art 37 of the Constitution provides that the Directive Principles, though not enforceable by any court, are nevertheless 'fundamental in the governance of the country'. Given this, the Court held that '[i]f there is an obligation upon the State to secure to the citizens an adequate means of livelihood and the right to work, it would be sheer pedantry to exclude the right to livelihood from the content of the right to life.'[235]

Should REDD Schemes Require the Vesting of Private Property Rights Over the World's Forests?

As mentioned above, one of the reasons for including references to land tenure in the Negotiating Text might be to give the imprimatur for greater recognition by governments in developing countries of private property rights in forests. The commercial reasons for doing so have been explained above. Heltberg[236] warns against the temptation for policy conclusions such as this where they are based upon simplistic analysis. For example, he argues that while it can rightly be argued that property rights in land help to internalise environmental externalities, the assertion that institutions are always optimal is absurd in the face of reality.[237]

Heltberg distinguishes between open access, common property, state property resources and private property. As does the iiea, he identifies many tropical rainforest countries as being state property which degenerate into *de facto* private or open access property regimes. He goes on to say that policymakers have used tenure reform in an effort to elicit improved natural resource management while paying insufficient attention to the local, institutional, cultural, technical and natural environment and the complex subtleties which underpin informal natural resource management. This has had adverse effects in some contexts and it needs to be acknowledged that resources under common property can serve vital economic functions that individual property cannot.[238] These include management, equity and insurance functions. Communal land rights may also be secure in that it is unlikely that someone would claim, encroach upon or expropriate usufruct land. Given this, Heltberg argues that is might be preferable to envision better regulation of the commons rather than assuming that individual, transferable property rights are the end goal. He also takes issue with the claim that private title is important for credit markets and trade since communal ownership may give sufficient security of tenure.[239] He concludes that there is a need for further research to be undertaken by anthropologists and field workers, who have detailed knowledge about existing institutions, and economists with insight into the functioning and efficiency of institutions such as private or communal property.[240]

Conclusions About Tenure

International law and the jurisprudence developed by courts in India, Australia, and South Africa, to mention a few, are replete with references to the right of indigenous people, other communities and even 'squatters'. Even if the analysis presented here is not definitive of who 'owns' the forest carbon in developing countries, it is indicative of the need for a more thorough legal analysis and exposition of who can rightly claim 'ownership' over forest carbon. At the very least, it seems that the analysis lends credence to the call for clear distinctions to be made, both at international and domestic law, between carbon rights based on 'land' tenure and 'resource' tenure.

Legal Arrangements for Benefit-Sharing: Payments for Environmental Services

As mentioned above, one of the primary reasons for needing to identify 'land' and/or 'resource' tenure is so that governments which participate in REDD programs, and derive the attendant financial incentives, can share the benefits with those who are managing the resources on the ground. These incentive payments have become known as Payments for Environmental Services ('PES'). Inevitably, developing countries which participate in REDD schemes will have to pass national legislation to determine the benefit sharing arrangements arising from REDD, except of course where REDD proceeds on a project basis.

Environmental services have been defined as 'the conditions and processors through which natural ecosystems, and species that make them up, sustain and fulfil human life'.[241] They include the provision of goods[242] as well as regenerating processes.[243] The services, which are regarded by economists as natural capital, supply 'for free' a stream of goods and other support services similar to those supplied by human and human-made capital. Heal et al[244] claim that there are two reasons that our ecosystem services are threatened. These include the unprecedented rate at which humanity is altering natural ecosystems and the processes that they control, and the failure of policy-makers to recognise the natural capital of these services. The result is that the ecosystem services underpinning the production of goods have no market value due to the fact that there is no market to capture and express their value. There are also no efficient price mechanisms to signal scarcity or degradation of the services. A further difficulty is that, as classic 'public goods', the ecosystem services cannot be exclusively controlled.

A single ecosystem may deliver carbon storage and sequestration services, biodiversity, salinity and water purification services. One of the keys, then, to PES is to recognise the many different goods and services delivered by a single ecosystem and the interaction between them.

There are now a multitude of PES systems in tropical rainforest countries such as Indonesia, Vietnam, Brazil, Mexico, DRC, Ecuador and Tanzania.[245] For the sake of explication, the Costa Rican Payment for Environmental Services Programme, which has received considerable international attention, will be assessed.[246] The framework for the PES programme has been established under discrete pieces of legislation including: the *Forestry Law No 7575 of 1996*, the *Public Services Regulatory Law*, the *General Law of the Environment*, the *Soil Conservation Law* and the *Biodiversity Law*. This suite of legislation indicates that a PES scheme needs to be implemented across all key environmental legislation relating to forestry ecosystem services. The scheme is implemented by the National Forestry Financing Fund (FONAFIFO) which was legally constituted in 1996.[247]

According to *Forestry Law No 7575*, the following forestry environmental services are recognised: greenhouse gas mitigation; the protection of water for urban, rural or hydroelectric purposes; the protection of biodiversity for conservation, sustainable, scientific and pharmaceutical uses; research and genetic improvement; and the protection of ecosystems and life forms, including natural scenic beauty for tourism and scientific purposes.[248]

Under the PES scheme, landholders are paid a flat rate for limiting their activities to specified land-uses. These include forest protection (five-year duration and USD210/ha dispersed over five years), sustainable forest management (15-year period and USD327/ha dispersed over five years) and reforestation (15- to 20-year duration and USD537/ha over five years).[249] In 2006, agroforestry projects were included in the scheme.[250] Payment for forest protection, the principal issue of concern for the present application, requires landholders to have a minimum of two hectares under protection. In return, carbon and environmental service rights are ceded to FONAFIFO for the length of the contract. Upon expiry of the contract, landholders can renegotiate prices or sell the rights to other parties. Obligations under PES contracts are registered in the public land register and are binding on future purchasers of the land.[251] Significantly, the regulatory framework for the PES scheme is different for indigenous territories given that land is often communally, rather than individually, owned. Consequently, FONAFIFO exempts indigenous territories from complying with land ownership regulations.[252]

There is some criticism of the way in which baselines, additionality and leakage are dealt with under the PES scheme although it seems that systems for monitoring, reporting and evaluation are comprehensive. These systems are supported by the use of Geographic Information Systems such as the Global Position System as well as visits by FONAFIFO staff to PES sites, and the auditing of FONAFIFO activities.[253]

The PES scheme is largely funded by a special fuel tax on the consumption of any crude-oil derivatives, as provided for under the *Forestry Act*. More recently, FONAFIFO has entered into PES agreements with foreign governments and private sector institutions in order to augment funding for the scheme.[254] It has also developed the issuing of Environmental Services Certificates to domestic companies and institutions that benefit from environmental services, to compensate forest owners for conserving them.[255]

Clearly, although the Costa Rican PES scheme provides a useful framework for understanding how benefit sharing might occur in a developing country, it will be necessary for each tropical rainforest country to devise a domestic regulatory framework for PES that is appropriate to its legal and institutional context.

Conclusions

The Negotiating Text discussed in this article provides some indication of the likely REDD outcomes at COP15 in December. However, given the number of options and alternatives for implementing REDD, to say nothing of the highly contingent nature of all of the provisions of the Text, one can only be cautiously optimistic that the details of a binding REDD agreement will emerge from COP15. It may be that a framework agreement emerges with the details to be determined at subsequent Conferences of the Parties. If this occurs, then the Prince's Rainforest Project proposal for an Emergency Package may gain increasing credibility and importance. Also, many jurisdictions, such as the European Union and Australia, are awaiting the outcomes of COP15 to determine their final emissions trajectories and the extent to which they will admit REDD offset credits into their emissions trading schemes.

The decision of the United States to accept REDD offset credits, and to rely on REDD programs to achieve supplemental emissions reductions for the US, is interesting. It seems that the US has accepted that many of the issues raised in this article will not be easily resolved in all tropical rainforest countries. Hence its decision to issue offset credits on a national basis to 'Category A' countries and only on a project level to 'Category B' countries. However, the decision to adopt this approach to REDD upon enactment of the *American Clean Energy and Security Act of 2009* is curious given that the Negotiating Text seems to represent the consensus of the international community that REDD be phased in, with a credits/project based approach emerging only in Phase 3. The project based approach, together with the decision to distribute credits on a subnational basis to provinces, seems to disregard the serious concerns about carbon leakage which have prevented REDD offset credits from entering the emissions trading market to date. The US seems to deal with these issues by requiring that project-based and subnational crediting programs be halted within five years after the enactment of the legislation. This indicates that these programs should be regarded only as interim strategies until such time as more tropical rainforest countries satisfy the 'Category A' country criteria. It will be interesting to see how the US legislation is reconciled with the outcomes of Copenhagen if it secures Senate support later in 2009.

What is clear, is that writing the international rules for REDD at Copenhagen, and beyond, will be an extremely complex and challenging undertaking. Yet, for all the reasons enumerated in this article, this represents only the first step in implementing REDD schemes. A major prerequisite for the successful implementation of REDD in tropical rainforest countries will be the flow of financing from developed countries. Following this, individual developing countries will need to develop domestic policy and legal frameworks which are consistent with international legal frameworks. In addition, major institutional reforms will be needed with respect to the recognition of land/resource tenure, appropriate governance arrangements, alternative livelihood and compensation plans for those displaced by REDD, to say nothing of all of the scientific capacity that will have to be developed for monitoring, reporting and verification purposes.

It is the author's contention that the complexity of the REDD project cannot be glossed over or, worse still, ignored in a rush to profit from REDD offset credits. Albeit that the financial incentives from REDD for developing countries, investors and liable entities under emissions trading schemes are undeniable, REDD schemes must ultimately satisfy the imperatives of ecologically sustainable development. This requires a thorough integration of all of the economic, social and environmental facets to REDD.

Endnotes

1   AP Sari, M Maulidya, RN Bhutarbar, RE Sari, Wisnu Rusmantoro, *Executive Summary: Indonesia and climate change* (DFID, PEACE: 2007) 3.

2   On the need for a new definition of forests and forest degradation see N Sasaki and F Putz 'Critical Need for New Definitions of "Forest" and "Forest Degradation" in Global Climate Change Agreements' (2009) 2 *Conservation Letters* 226.

3   *An Emergency Package for Tropical Forests* (The Prince's Rainforests Project: March 2009) 9.

4   United Nations Intergovernmental Panel on Climate Change (IPCC), *Climate Change 2007 — The Physical Science Basis: Working Group I Contribution to the Fourth Assessment Report* (2007).

5   Sir Nicholas Stern, *Stern Review on the Economics of Climate Change* (2006) 538 (*Stern Review*).

6   Ibid 547.

7   K Karousakis 'Initial Review of Policies and Incentives to reduce GHG emissions from Deforestation' (Organisation for Economic Co-operation and Development (OECD): October 2006) xxx; see also R Lyster '(De)regulating the rural environment' (2002) *Environmental Planning and Law Journal* 34; G Heal, GC Daily, PR Ehrlich, J Salzman, C Boggs, J Hellmann, J Hughes, C Kremen, T Ricketts, 'Protecting Natural Capital Through Ecosystem Service Districts' (2001) 30 *Stanford Environmental Law Journal* 333, 336. See also K Karousakis and J Corfee-Morlot 'Financing Mechanisms to Reduce Emissions from Deforestation: Issues in design and implementation' (OECD and IEA: December 2007).

8   L Peskett et al, 'Can payments for avoided deforestation to tackle climate change also benefit the poor?' (Overseas Development Institute: November 2006) 1.

9   *United Nations Framework Convention on Climate Change*, opened for signature 4 June 1992, 1771 UNTS 107 (entered into force 21 March 1994) (*UNFCCC*).

10   *Kyoto Protocol to the United Nations Framework Convention on Climate Change*, opened for signature 16 March 1998, 2303 UNTS 148 (entered into force 16 February 2005) (*Kyoto Protocol*).

11   13th Conference of the Parties to the *United Nations Framework Convention on Climate Change* (COP 13), *Bali Action Plan*, -/CP.13, art 1(b)(iii).

12   As part of this process, a number of workshops have been organised under the auspices of the Subsidiary Body of Scientific and Technical Advice ('SBSTA') of the UNFCCC including in Italy in September 2006, in Australia in March 2007, in Bonn in May 2007 and in Bali in December 2007; see, eg UNFCCC, *Reducing Emissions from Deforestation in Developing Countries*, SBSTA, 26th sess, FCCC/SBSTA/2007/L.10 (17 May 2007).

13   These are developed countries with emissions reduction targets under the *Kyoto Protocol*.

14   Afforestation is the artificial establishment of forests by planting or seeding in an area of non-forest land.

15   Reforestation is the restocking of existing forests and woodlands which have been depleted, with native tree stock.

16   Although reliance on this is limited in accordance with the Marrakesh Accords negotiated at COP7 in 2001; see UNFCCC, *Land use, land-use change and forestry*, Decision11/CP.7, 8th plen mtg, FCCC/CP/2001/13/Add.1 (10 November 2001).

17   The Clean Development Mechanism (CDM), *Kyoto Protocol* art 12, allows developed countries to invest in emission reducing projects in developing countries, and to obtain certified emission reductions (CER) towards meeting their assigned amounts.

18   'Leakage' refers to greenhouse gas emissions which occur outside the project boundary but which are nevertheless attributable to its activities.

19   'Permanence' refers to the possibility that carbon is released into the atmosphere as a result of fire, illegal logging or a change in government.

20   Credits can only be generated for emissions below the 'baseline'; ie, GHG emissions reduction that would have occurred even in the absence of a CDM project.

21   It must be demonstrated that the carbon sequestration would not have occurred without the incentives provided by the project.

22   See, for example, IPCC, *Guidelines for National Greenhouse Gas Inventories — Agriculture, Forestry and Other Land Use* (vol 4, 2006) and GOFG-GOLD, *Reducing Greenhouse Gas Emissions From Deforestation and Degradation in Developing Countries: A sourcebook of methods and procedures 47 for monitoring, measuring and reporting* (2009) ('*REDD Sourcebook*') <http://www.gofc-gold.uni-jena.de/redd/ at 31 January 2008, which uses remote sensing to monitor and measure greenhouse gas emissions from forests.

23   See UNFCCC, *Approved A/R Methodologies* <http://cdm.unfccc.int/methodologies/ARmethodologies/approved_ar.html> at 8 February 2008; see also C Streck et al, 'The Role of Forests in Global Climate Change: Whence we come and where we go' (2006) 82(5) *International Affairs* 868.

24   An extensive search of Westlaw, LexisNexis, LegalTrac, Legal Journal Index, AGIS, APA-FT, Kluwer Law Online, HeinOnline in early 2008 failed to produce any articles in this area of research although since then a body of work in beginning to emerge. See I Fry, 'Reducing Emissions from Deforestation and Forest Degradation: Opportunities and pitfalls in developing a new legal regime' (2008) 17(2) *Review of European Community and International Environmental Law* 166 and the (2008) 3 issue of the *Carbon and Climate Law Review.*

25   UNFCCC, *Ad Hoc Working Group on Long-Term Cooperative Action Under the Convention — Negotiating Text*, 6th sess, 1–12 June 2009, FCCC/AWGLCA/2009/8 (19 May 2009) ('Negotiating Text' or 'Text').

26   UNFCCC, *Ad Hoc Working Group on Long-Term Cooperative Action Under the Convention — Revised Negotiating Text*, 6th sess, 1–12 June 2009, FCCC/AWGLCA/2009/INF.1 (22 June 2009).

27   Negotiating Text, above n 25, 3.

28   This chapter proposes the long-term global goal for emissions reductions with the stablisatiojn of GHG concentrations in the atmosphere being set at the following possible levels: Option 1 — 400 ppm; 450 or lower ppm; not more than 450 ppm; or 450 ppm and a temperature increase limited to 2°C above pre-industrial levels with a commitment collectively by the parties to reduce global emissions by 50% by 2050; Option 2 — stabilization well below 350 ppm and a temperature increase limited to 1.5°C above pre-industrial levels whereby the parties agree to collectively reduce global emissions by either 81–71% or more than 85% from 1990 levels by 2050; Option 3 — limiting global temperature increase to 2°C above pre-industrial levels; Option 4 — a reduction in global average GHG emissions per capita to about 2 t CO_2 ; Option 5 — on the basis of historical responsibility, or emissions debte, or per capital accumulative emissions convergence, or an equitable allocation of the global atmospheric resources. Various reductions by 2020 and 2050 are proposed as are various options for developing countries to reduce their emissions; at 8–10.

29   Above n 25, 5.

30   Above n 26, [70] 86. Note that five alternatives to [70] are offered in the Text.

31   Above n 25, 23.

32   Above n 26, [70.1] 87.

33   Ibid [73] 89.

34   Ibid [74] 90–3.

35   Ibid [75]–[87] 93–9.

36   Ibid [x.1], [x.2], [x.3] 100.

37   Ibid [91]–[93] 102–3.

38   Ibid [101], [x.1]–[x.22] 105–8.

39   Ibid [x.1], [x.2] preceding [106] 110.

40   Ibid [106] 111. Note that four alternatives to this paragraph are proposed.

41   Ibid [106.1] 111.

42   Ibid [106.3] 111–12.

43   Ibid [107] 112. Various alternatives are proposed to this paragraph.

44   Ibid [108.1] 113.

45   Ibid [109] 113. There are two alternatives to this paragraph.

46   Ibid [108] 113. Four alternatives are proposed for this paragraph.

47   Ibid [110] 114–15. Three alternatives are proposed for this paragraph.

48   Ibid [111] 116. There are four alternative proposals to this paragraph.

49   Ibid [112.1] 117.

50   Ibid [112.2] 117.

51   Ibid [112.3] 117.

52   Ibid [x.1], [x.2] 117.

53   Ibid [115] 122.

54   Ibid alternative to [115] 122. See also [120], [121] 123–4.

55   Ibid [117] 122.

56   Ibid [117.1] 122.

57   Ibid [117.2] 123.

58   Ibid [118] 123.

59   Ibid [119] 123.

60   Project Catalyst, *Scaling up Climate Finance: Finance Briefing Paper September 2009* <http://www.project-catalyst.info/images/publications/climate_finance.pdf> at 21 September 2009.

61   *Reducing Emissions From Deforestation and Forest Degradation (REDD): An Options Assessment Report* (Meridian Institute: March 2009); available at http://www.redd-oar.org/ (viewed 22 July 2009).

62   Ibid 43.

63   Karousakis, above n 7.

64   T Griffiths, *Seeing 'RED'? 'Avoided' deforestation and the rights of Indigenous Peoples and local communities* (Forest Peoples Programme: 2007).

65   Above n 26 [IV(A)(x.16)] 148.

66   Ibid [C(2)] 117.

67   Ibid [175] 162.

68   Ibid [113] 118–19.

69   See Karousakis, above n 7; Griffiths, above n 64; R Haverfield '*Hak Ulayat* and the State: Land reform in Indonesia' and D Fitzpatrick 'Beyond Dualism: Land acquisition and law in Indonesia' in T Lindsey (ed), *Indonesia: Land and society* (1999).

70   Above n 26, [114] 119–20.

71   Forest Carbon Partnership Facility <http://www.forestcarbonpartnership.org/fcp/>.

72   International Bank for Reconstruction and Development, *Charter Establishing The Forest Carbon Partnership Facility* (June 2008).

73   Ibid.

74   World Bank to launch two new CDM funds, *Point Carbon* (10 September 2007).

75   *About the FCPF*, Forest Carbon Partnership Facility <http://www.forestcarbonpartnership. org/fcp/node/12> at 22 July 2007.

76   The Prince's Rainforest Project, *An Emergency Package for Tropical Forests* (2009) 16.

77 Ibid 20–1.

78 Ibid 35–6.

79 Ibid.

80 M Ogonowski et al, 'Reducing Emissions from Deforestation and Degradation: The dual markets approach' (Centre for Clean Air Policy: August 2007); B Schlamadinger et al, 'Should we include avoidance of deforestation in the international response to climate change? *Tropical Deforestation and Climate Change*' (IPAM, Istituto de Perquisa Ambiental de Amazônia; Belém, Pará (Brazil)); PM Fearnside 'Mitigation of Climatic Change in the Amazon' in WF Laurance and CA Peres (eds) *Emerging Threats to Tropical Forests* (2006) 353; L Pedroni et al, 'Mobilizing Public and Private Resources for the Protection of Tropical Rainforests' (CATIE Tropical Agricultural Research and Higher Education Center); K Karousakis 'Incentives to Reduce GHG Emissions from Deforestation: Lessons learned from Costa Rica and Mexico' (Organisation for Economic Co-operation and Development ('OECD'): 2007).

81 See Karousakis, above n 7, 30–2.

82 Ogonowski et al, above n 80, i.

83 See World Bank, *State of Carbon Market 2006* (2006); see also Meridian Institute, above n 61, 59.

84 Ibid.

85 R Piraud *The fight against deforestation (REDD): Economic implications of market-based funding* (Insitutue du developpement durable et des relations internationals (iDDRi): No 28/2008.

86 Such as shifting cultivators, small farmers, soya or palm oil growers, livestock farmers, spontaneous or government-induced migrants and indigenous peoples,

87 Ibid 8.

88 Fearnside, above n 80, 3.

89 Ibid.

90 Streck, above n 23, 868.

91 D Mollicone et al, 'Elements for the expected mechanisms on "reduced emissions from deforestation and degradation, REDD" under UNFCCC' (2007) 2 *Environmental Research Letters* (IOP Publishing) 5.

92 American Carbon Registry, *Forest Carbon Project Standard*, (version 1, 2009).

93 Climate, Community and Biodiversity Alliance, *Climate, Community and Biodiversity Project Design Standards* (2008). Note that the CCB Standards cannot be used on their own to generate carbon credits. Rather, the Standards can be used to enhance other Standards since they deal specifically with incorporating socio-economic and environmental co-benefits in addition to consideration of carbon abatement.

94 Plan Vivo, *Plan Vivo Standards* (2008).

95 Social Carbon, *Social Carbon Guidelines* (version 3, 2009).

96 VCS, *Voluntary Carbon Standard 2007.1* (2008).

97 UNFCCC 'Methodologies for Afforestation and Reforestation CDM Project Activities' <http://cdm.unfccc.int/methodologies/ARmethodologies/index.html> at 17 August 2009.

98 IPCC, *2006 IPCC Guidelines for National Greenhouse Gas Inventories*, 'Volume 4: Agriculture, Forestry and Other Land Use' (2006) <http://www.ipcc-nggip.iges.or.jp/public/2006gl/vol4.html> at 17 August 2009.

99 See A Kollmuss, H Zink, C Polycarp, *Making Sense of the Voluntary Carbon Market: A Comparison of Carbon Offset Standards* (2008) WWF <http://assets.panda.org/downloads/vcm_report_final.pdf> at 17 August 2009.

100 Ibid 14.

101 Ibid 33.

102 To understand the way in which such risk is typically distributed under CDM offset projects see the Emissions Reduction Purchase Agreement ('ERPA'): International Emissions Trading Association, 'CDM Emissions Reduction Purchase Agreement' version 2.0 (2004) <http://www.ieta.org/ieta/www/pages/download.php?docID=450> at 17 August 2009.

103 Ibid.

104 VCS, *Voluntary Carbon Standard: Tool for AFOLU methodological issues* (2008) 4 <http://www.v-c-s.org/docs/Tool%20for%20AFOLU%20Methodological%20Issues.pdf> at 17 August 2009. Note that AFOLU stands for 'Agriculture, Forestry and Other Land Use Projects'.

105 Baker & McKenzie, *CDM Rulebook*, 'Establishing Additionality — The A/R additionality tool' <http://cdmrulebook.org/658> at 17 August 2009.

106 See Eduard Merger, *Forestry Carbon Standards 2008: A comparison of leading standards in the voluntary carbon market* (2008) 20 <http://www.carbonpositive.net/fetchfile.aspx?fileID=133> at 17 August 2009. See also Kollmuss et al, above n 99, 15.

107 See J Speckman 'REDD under way' (2008) *Environmental Finance* 56.

108 See Rosemary Lyster 'Chasing Down the Climate Change Footprint of the Public and Private Sectors: Forces converge — Part II' (2007) 24 Environmental Planning and Law Journal 450.

109 Karan Capoor and Philippe Ambrosi, State and Trends of the Carbon Market 2009 (2009) World Bank <http://wbcarbonfinance.org/docs/State_Trends_of_the_Carbon_Market_2009FINAL _26_May09.pdf> at 16 August 2009.

110 Ibid 1.

111 Ibid 2.

112 Ibid 4.

113 *American Clean Energy and Security Act of 2009* 111 HR 2998 ('the Act' or 'US Act').

114 'Costa Rica Sells Avoided Deforestation Credits to US Foundation', *Point Carbon* (19 September 2007).

115 Speckman, above n 107, 58.

116 Macquarie Group and Fauna & Flora International, 'Macquarie Group and Fauna & Flora International form Carbon Forest Task Force' (Press Release, 24 June 2008).

117 The Australian Government, for example, has launched the International Forest Partnership to establish agreements between itself and the governments of Indonesia and Papua New Guinea. One such agreement is the *Indonesia-Australia Forest Carbon Partnership* under which the Australian government has committed AUD10 million to build capacity in Indonesia in anticipation of a REDD carbon market; see *Indonesia-Australia Forest Carbon Partnership*, signed 13 June 2008. The Australian government has also committed AUD30 million to establish the *Kalimantan Forests and Climate Partnership*. Under this partnership, Australia and Indonesia are implementing a demonstration activity in the carbon rich peat-land forests of Central Kalimantan; *Kalimantan Forests and Climate Partnership*, Australia—Indonesia, signed 9 September 2007. The Government of Norway's International and Forest Initiative is also a significant funder of REDD capacity building projects in developing countries; see <http://www.regjeringen.no/en/dep/md/Selected-topics/klima/the-government-of-norways-international-/why-a-climate-and-forest-initiative.html?id=547202> at 17 August 2009.

118 *American Clean Energy and Security Act of 2009* 111 HR 2998.

119 Beginning in 2012, the following sectors are covered: all electricity generators; any facility or entity that produces or imports petroleum or coal-based liquids, petroleum coke, or natural gas liquids if combustion of the fuel results in the emission of more than 25,000 mt CO_{2-e} per year; emissions from sites that geologically sequester CO_2. Under a separate cap, beginning in 2012, producers and importers of HFCs, and importers of products containing HFCs, would be required to submit allowances for the carbon dioxide-equivalent tons of HFC they produce or import. Beginning in 2014, industrial facilities that manufacture a wide variety of products are covered if their activities result in more than 25,000 $mtCO_{2-e}$ of emissions. Beginning in 2016, natural gas distributors that deliver at least 460 million cubic feet of natural gas to customers would need to submit allowances for the GHG emissions that would result from the combustion of the gas delivered to those customers; ibid s. 700(13).

120 *American Clean Energy and Security Act of 2009* 111 HR 2998 / 722(c)(1)(A).

121 *American Clean Energy and Security Act of 2009* 111 HR 2998 / 722(c)(1)(B). The percentage is determined by dividing 2 billion by the sum of 2 billion plus the number of emission allowances established under / 721(a) for the previous year multiplied by 100. For 2012, for example, the number of allowances is 4,770 million allowances.

122 *American Clean Energy and Security Act of 2009* 111 HR 2998 / 743(e).

123 Ibid 743(e)(3).

124 Ibid 743(e)(2).
125 Ibid 743(e)(6)(A).
126 Ibid 743(e)(5)(A)(iii).
127 Ibid 743(e)(1)(B).
128 Ibid 743(e)(4).
129 Ibid 743(e)(1)(C).
130 Ibid 743(e)(1)(D).
131 Ibid 743(e)(1)(E).
132 Ibid 743(e)(6)(B).
133 Ibid 743(e)(6)(C).
134 Ibid 743(e)(6)(D).
135 Ibid 743(f).
136 Ibid 743(e)(5)((C)(iii).
137 Ibid 743(e)(5)(D).
138 Ibid 743(g).
139 Ibid 754(j).
140 Ibid 752.
141 Ibid 753(a).
142 Ibid 754(b).
143 Ibid 781.
144 Ibid 754(h).
145 Ibid 754(g).
146 Ibid 754(a).
147 Ibid 754(d).
148 Ibid 754(d)(3).
149 Ibid 754(d)(4).
150 Ibid 754(d)(5), (6).
151 Ibid 754(b)(1).
152 This is the 'Offsets Integrity Advisory Board' that is established under the Act: *American Clean Energy and Security Act of 2009* 111 HR 2998 / 731.
153 *American Clean Energy and Security Act of 2009* 111 HR 2998 / 754(e)(2).
154 Ibid 754(f).
155 Ibid 754(b).
156 Ibid 754(c).
157 Ibid 755(a).
158 Ibid 755(b).
159 See Crystal Davis, Florence Daviet, Smita Nakhooda, and Alice Thuault, 'A Review of 25 Readiness Plan Idea Notes from the World Bank Forest Carbon Partnership Facility' (Working Paper, World Resources Institute, 2009).
160 Ibid 2.
161 Ibid 2–3.
162 Chris Lang, 'PNG government does not support Voluntary Carbon Agreements', *REDD-Monitor* (2 September 2009).
163 See I Bond et al, *Incentives to Sustain Forest Ecosystem Services: A Review and Lessons for REDD* (International Institute for Environment and Development: 2009) 26.
164 Kollmuss et al, above n 99, 18.
165 Ibid 5.
166 Ibid 6.
167 M Herold, *An Assessment of National Forest Monitoring Capabilities in Tropical non-Annex 1 Countries: Recommendations for capacity building* (GOFC-GOLD, Friedrich Schiller University Jena: 2009).
168 Ibid 7.
169 Ibid 8.

170 Every 5–10 years the FAO undertakes a global forest resources assessment requiring developing countries to submit detailed information on their national forest resources.

171 Herold et al, above n 167, 13.

172 Ibid 14.

173 Ibid 20.

174 Ibid 22–4.

175 Ibid 24–5.

176 Ibid Annex A.

177 Available at <http://www.greenhouse.gov.au/ncas/factsheets/fs-gcms.html> (viewed 2 August 2008). For academic commentary on difficulties associated with monitoring carbon emissions from forests see F Achard et al 'Pan-tropical monitoring of deforestation' (2007) 2(4) *Environmental Research Letters* 045022; Holly K Gibbs et al 'Monitoring and estimating tropical forest carbon stocks: making REDD a reality' (2007) 2(4) *Environmental Research Letters* 045023.

178 The term 'communities' is defined in the Climate, Community as Biodiversity Project Design Standards as 'all groups of people including Indigenous Peoples, mobile peoples and other local communities, who live within or adjacent to the project area as well as any groups that regularly visit the area and derive income, livelihood or cultural values from the area. This may include one or more groups that possess characteristics of a community, such as shared history, shared culture, shared livelihood systems, shared relationships with one or more natural resources (forests, water, rangeland, wildlife etc), and shared customary institutions and rules governing the use of resources.'; see CCBA, *Climate, Community and Biodiversity — Project Design Standards* (2nd ed, 2008).

179 Griffiths above n 64, 10. Griffiths also provides evidence of the impact of carbon forestry on indigenous peoples and peasant communities in the Ecuadorian Andes ibid 11.

180 K Gray and S Gray, *Elements of Land Law* (5th ed, 2009) 98.

181 L Cotula and J Mayers, *Tenure in REDD: Start-point or afterthought* (iied: 2009); see also F Daviet, C Davis, L Goers, S Nakooda, *Ready or Not: A review of the World Bank Forest Carbon Partnership R-Plans and the UN-REDD Joint Program documents* (June 2009) and G Barnes and S Quail, Property rights to carbon in the context of climate change (Paper presented at the Conference on Land Governance in Support of MDGs: Responding to New Challenges, 9 March 2009).

182 Ibid.

183 All former British colonies, for example.

184 All former Dutch, French and other continental colonies.

185 For example, India, Pakistan, and the Philippines.

186 These are aspirational goals which may be relied upon by Supreme Courts to interpret other rights such as the fundamentally protected 'right to life'.

187 Gray, above n 180, 86.

188 Ibid 87.

189 Ibid 88.

190 Ibid.

191 Ibid 90.

192 Ibid.

193 Ibid 95.

194 Ibid.

195 Above n 162.

196 Ibid vi.

197 Ibid 11, citing A White and A Martin, *Who Owns the World's Forests: Forest tenure and public forests in transition* (Forest Trends: 2002).

198 Above n 162, 16.

199 Ibid.

200 Interview with Dr Simon Butt, Sydney Law School, University of Sydney.

201 Gray, above n 179, 103.

202 J Burke, *Jowitt's Dictionary of English Law* (2nd ed, 1977) 1844.

203 Gray, above n 179, 90.

204 Report of the United Nations Conference on Environment and Development, Annex I, Un Doc A/CONF.151/26 (Vol I) (12 August 1992) ('*Rio Declaration*').

205 UNCED, *Agenda 21: Earth Summit — The United Nations Programme of Action from Rio*, UN Doc A/CONF.151/26 (14 June 1992) ('*Agenda 21*').

206 *United Nations Convention on Biological Diversity*, opened for signature 5 June 1992, 1760 UNTS 79 (entered into force 29 December 1993) ('*Biological Diversity Convention*').

207 *Report of the United Nations Conference on Environment and Development*, Annex III, UN Doc A/CONF.151/26 (Vol III) (14 August 1992) ('*Forest Principles*').

208 *Agenda 21* art 3.2.

209 *Biological Diversity Convention* art 5.

210 Principle 5(a).

211 *Convention Concerning Indigenous and Tribal Peoples in Independent Countries*, No 169, ILO, 76th sess (entered into force 5 September 1991).

212 *Convention Concerning Indigenous and Tribal Peoples in Independent Countries* art 13.

213 Ibid art 14.

214 Ibid art 15.

215 Ibid art 16.

216 (1992) 175 CLR 1('*Mabo*').

217 *Mabo* [65].

218 *Mabo* [65].

219 *Optional Protocol to the International Covenant on Civil and Political Rights*, opened for signature 19 December 1966, 999 UNTS 171 (entered into force 23 March 1976).

220 [1921] 2 AC 399 ('*Amodu Tijani*').

221 *Amodu Tijani* 403–4; cited *Mabo* [52] (Brennan J).

222 *Mabo* [66].

223 *Mabo* [68].

224 *Alexkor Ltd v The Richtersveld Community* (Constitutional Court of South Africa, CCT 19/03, 14 October 2003) ('*Alexkor*').

225 The *Native Land Act* deprived black South Africans of the right to own land and rights in land in the vast majority of the South African land mass.

226 Above n 218 [36].

227 *Alexkor* [47].

228 *Alexkor* [50].

229 *Alexkor* [53].

230 *Alexkor* [58].

231 *Alexkor* [62].

232 *Constitution of India* pt IV.

233 [1985] 2 Supp SCR 51 (India); (1987) LRD (Const) 351 (Supreme Court of India).

234 *Tellis* [21].

235 *Tellis* [33].

236 R Heltberg, 'Property Rights in Natural Resource Management in Developing Countries' (2002) 16(2) *Journal of Economic Surveys* 189.

237 Ibid 191.

238 Ibid 197.

239 Ibid 205.

240 Ibid 210.

241 G Heal, G C Daily, P R Ehrlich, J Salzman, C Boggs, J Hellmann, J Hughes, C Kremen, T Ricketts, 'Protecting Natural Capital Through Ecosystem Service Districts' (2001) 30 *Stanford Environmental Law Journal* 333, 336.

242 Such goods include food, pharmaceuticals, durable materials, energy, industrial products and bio-diversity: ibid 337.

243 These include cycling and filtration processes, translocation processes, stabilising processes, and life-fulfilling functions: ibid.

244 Ibid 340–1.

245 See Bond et al, above n 162; see also S Engel and C Palmer 'Payments for environmental services as an alternative to logging under weak property rights: The case of Indonesia' (2008) 65 *Ecological Economics* 799 .

246 See Karousakis, above n 12, 16–23.

247 FONAFINO <http://www.fonafifo.com/english.html> at 5 February 2008.

248 FONAFIFO, *Environmental Services* <http://www.fonafifo.com/paginas_english/environmental_services/servicios_ambientales.htm> at 5 February 2008.

248 R Sierra and E Russman, 'On the Efficiency of Environment Service Payments: A forest conservation assessment in the Osa Peninsula, Costa Rica' (2006) 59(1) *Ecological Economics* 131, cited by Karousakis, above n 12, 19.

250 T Wunscher, S Engel, S Wunder, 'Payments for environmental services in Costa Rica: increasing efficiency through spatial differentiation' (2006) 45(4) *Quarterly Journal of International Agriculture* 317, cited by Karousakis, above n 12, 19.

251 Karousakis, above n 12, 19.

252 Ibid 17.

253 Ibid 21.

254 FONAFINO, *Invest in Forest — Other Investment Options* <http://www.fonafifo.com/paginas_english/invest_forest/i_ib_convenios.htm> at 5 February 2008.

255 FONAFINO, *Invest in Forest — ESC* <http://www.fonafifo.com/paginas_english/invest_forest/i_ib_que_es_csa.htm> at 5 February 2008.

Contractual Perspective
of Climate Change Issues

Elisabeth Peden

The interrelationship between contract law and climate change may not be readily apparent. However, considering that the approach adopted by the *Kyoto Protocol* to reduce carbon emissions has been to develop carbon credit trading, contract law has been given a significant role in the future of climate change. The trading of carbon credits is achieved through contracts to buy and sell, just as other commodities are traded. Contract law can also operate in other contexts that have an impact on the environment, such as conservation agreements between governments and landowners. With the developing approach to include forests in carbon credit regimes, contract law will also play a role in agreements to maintain forests, and grow new ones. Insurance will no doubt have a role to play, and will be effected by contracts of insurance. The list could go on.

In considering the role of contract law, it is important to remember that the creation of carbon credit trading is a result of governments creating markets for the commodity of carbon credits through the introduction of emission targets and corresponding penalties. Where there is such government control or regulation in relation to a policy-driven goal (here, reduced emissions), then contract law may provide a useful tool in reaching that goal. However, in contexts where there is no regulation, contract law will not necessarily deliver the protection that might be expected.

It must be remembered that contract law is concerned with the formalisation of promises and the allocation of risks between parties. The parties are taken to have agreed either to perform their primary obligation, that is their promise, or their secondary obligation, that is to pay for the value, in money terms, of the performance they have failed to provide. Generally, the only possible remedy from the perspective of contract law is to provide financial compensation; there will be no orders, for example, to pollute less or to replant trees. So the limitations of contract law (without regulation) to guarantee grand environmental goals must be considered.

If contract law is now to be relied on as a mechanism to assist the fight against climate change and to assist in the preservation of the earth's longevity, it is important to consider questions such as 'What if it all goes wrong?' and 'What are the possible remedies?'

This chapter first seeks to provide a perspective of how contract law has fared when given the role of protecting rights relating to environmental issues through the case example of *Tito v Waddell*.[1] This case highlights that often contract law cannot provide a remedy that is ideal in situations that do not concern purely commercial issues, because contract law is designed to allocate risks in markets rather than ensure important concerns, such as slowing climate change or protecting forests or flora, are addressed.

With this background, this chapter then considers the operation contract law may have when the Australian Government introduces a carbon trading scheme. Some observations are also offered relating to the standard term contract that was introduced in 2008 in Europe, the home of the largest carbon credit trading market. This analysis focuses on remedies, and provides some insights into issues that might be considered in Australia should a similar standard form contract be adopted.

Tito v Waddell

To understand the case of *Tito v Waddell* it helps to understand the history of the Banaba island. Banaba, named 'Ocean Island' by English explorers, is a small island of 595 hectares, almost on the equator, due north of New Zealand. It is about 180 kilometres from its nearest neighbour Nauru and forms part of Kiribati. It is almost round in shape with a diameter of about 2.5 kilometres.

In the late 1800s, phosphates were discovered to provide an effective fertiliser, and phosphate deposits were sought-after and valuable. Guano, or accumulated bird deposits, is rich in phosphates, and when mixed with poor soil leads to improved harvests. Banaba had a large concentration of guano from migrating birds that stopped over on the island.

This fact was discovered in 1900 by a New Zealander, Albert Ellis, who was working for the Pacific Island Company. The company had been mining small Pacific islands for guano, but their sources were almost exhausted. Ellis tested a rock that had been brought back from Nauru, and reasoned that as Nauru and Banaba were so close together it was more than probable that both islands would have similar guano deposits. The advantage of finding guano on Banaba, rather than Nauru, was that Banaba was annexed to Great Britain, whereas Nauru was German territory.

According to one writer:

> The chiefs of the island were gathered together and invited aboard the Pacific Island's Company steamer, the Archer, where they were given food, drinks and presents. A paper was obtained from some of these guests, giving the company rights to raise and export phosphate from the island for a period of 999 years for the payment of £50 a year, or trade goods to that amount from the company store. The document was signed on behalf of the company by Albert Ellis.

Ellis soon discovered that this signed paper meant nothing to the natives. Those who had signed, had no rights or authority to lease the whole island on behalf of others, as each family independently owned its own land. He would have to obtain a contract or lease from each and every landowner to work the land. By the 10 May, 1900 he had obtained contracts with waterfront access, to several small plots of land, each allowing the removal of phosphate rock from the site for a period of one year, provided that the gardens and coconut trees were not interfered with. Each contract provided allowances for the building of houses to store the guano and the laying of tramlines.

Local native labour was offered the sum of 8 shilling a ton for the phosphate rock, collected, bagged and delivered into the ships on arrival. Meanwhile the phosphate was to be stored in 'houses' near the shore, within easy access of the landing.

...

By the year 1909, over two hundred and forty acres of land on Ocean Island had been stripped by mining the phosphate, including the food trees the company had pledged to leave untouched. Some three hundred acres of land was nothing more than coral pinnacles, leaving just nine-hundred and sixty acres of land to support the food for the population of the Banaban people remaining on the island.[2]

In July 1920, the Pacific Phosphate Company was wound up and bought by the British Phosphate Commission, which continued mining until about 1940. The plight of the Banaban people became even worse during the Second World War, when the island was captured by the Imperial Japanese Army, and many were killed or sent to war camps. By the time the island was recaptured by the Australians in 1945, there were only 280 Banaban people left and the island was completely uninhabitable. The commission used some money that was held on trust for the Banaban people to buy from the Government of Fiji an island called Rambi for their resettlement.

Various agreements were made over the years in relation to the mining of the phosphate and later sand and shingle from the beaches for cement-making.

The Banaban people decided they wanted compensation for the loss of their island. Two elders, Tito and Tebuke, first went to Australia and New Zealand seeking compensation, but without success. They then went to London, found lawyers, and in 1971 commenced an action against the commission. There were two actions brought — numbered 1 and 2. The case numbered '2' was heard first. It concerned a claim that the Crown, in the form of the British Phosphate commission, owed the Banaban people a fidiciary duty when deciding the royalties paid to the people, which was breached in agreements made in 1931 and 1947 as the rates were too low. In the case numbered '1', various Banaban plaintiffs sued mainly for specific performance of the contractual promises to replant the island, or in the alternative, for damages for the failure to replant. It was the longest case that had ever been heard in England, lasting 221 days and costing the Banaban people about £750,000.

The legal action turned on the various agreements that had been entered into between the company and commission on the one hand and the Banaban people on the other. The land on the island was divided up into a large number of small plots (most of them being less than one acre in extent) owned by individual Banabans or groups of Banabans. Under King's Regulations made by the High Commissioner under the Pacific Order in Council 1893, there were severe restrictions on the purchase and lease of land from native landowners, and the transactions that were permitted required the approval of the Resident Commissioner. The company sought to avoid those restrictions by evolving 'P and T deeds', under which the company merely bought the right to remove phosphate and trees from the land for 5 or 10 years.

By 1909, the legality of the P and T deeds was being questioned, and the company was finding it difficult to obtain further land for mining from the Banabans. Prolonged negotiations took place between the company on the one hand and the Colonial Office in London and the High Commissioner and the Resident Commissioner on the other hand. Finally, the terms to be put before the Banabans for the acquisition of further land were agreed.

In November 1913, an agreement based on those terms ('the 1913 agreement') was made between the Banaban landowners and the company, with the Resident Commissioner as witness to the signatures or marks. The 1913 agreement provided, inter alia, for the acquisitions to be made only in three specified areas of the island. In addition to agreeing to pay certain sums to each landowner who granted mining rights to the company, the company agreed to pay the government an additional royalty of 6d per ton. The first year's additional royalty (apart from £300) was to be expended for the benefit of the existing Banaban community. Subject to that, the £300 and the interest on those royalties were to be distributed as annuities to all Banabans who thereafter leased mining land to the company. The agreement also provided that the company should return all worked-out lands to the original owners, and should 'replant such lands — whenever possible — with coconuts and other food-bearing trees, both in the lands already worked out and in those to be worked out'.[3]

In accordance with that agreement, many Banaban landowners entered into deeds granting the company the right to remove phosphate and trees from their lands until 1999. Two new forms of deed were used, the 'A' deeds to replace a P and T deed, and 'C' deeds for new company acquisitions. All the deeds included a promise by the company that when it ceased to use the land the company:

> ... shall replant the said land as nearly as possible to the extent to which it was planted at the date of the commencement of the company's operations under Clause I (i) hereof with such indigenous trees and shrubs or either of them as shall be prescribed by the Resident Commissioner for the time being in Ocean Island.[4]

Further, it was agreed that the land was to revest in the landowner when in the Resident Commissioner's opinion that might be without prejudice to the company's operations.

The contract claim is of interest to the issue of environmental protection. While there was no concern in the early 1900s about global warming and deforestation, the Banabans clearly and naturally wanted to keep their environment the way they had known it. Furthermore, there was no protection provided by mining legislation, such as exists in Australia where mining companies are required to preserve or restore the environment if they wish to mine.

The Banabans main concern may have been to protect their food sources, and for that reason the deeds mentioned replanting with coconuts and food-bearing trees. Nevertheless, the case is interesting when considering the remedies available and whether contract law can provide sufficient protection to environmental interests in the 21st century.

The cases were heard by Megarry J, and are still considered good law. Admittedly, the judgments run into hundreds of pages and concern important legal issues beyond contract law, such as trust and assignment of benefits and burdens. So, what was the decision on the contract claims? Megarry J held that the Banabans should not be entitled to specific performance in the form of replanting, nor substantial damages.

Specific Performance

The claim for specific performance failed because of the wording of the deeds. In particular, they had required replanting 'where possible'. It was held that the intention at the time of the agreements was that trees could be replanted in the few feet of phosphate after mining an area that had been contemplated. There was no agreement that there would be extensive levelling and then importation of soil.[5] Megarry J refers[6] to the fact that in the first 12 years of the company's work on the island, they had replanted coconut trees in the mined out areas and that they had grown well. The company was prepared to continue replanting this way. However, most coconut trees planted did not bear fruit, because of the very low rainfall on the island that was prone to droughts.[7] Megarry J makes a very important statement:

> ... in deciding legal liability it is plainly impossible to take a bargain struck on a basis of no reinstatement but limited replanting and then say that because environmental ideas are changing for the better, the legal burdens accepted by one party to the bargain ought to be correspondingly increased. However potent such arguments may be in political or social fields, they cannot affect the law of contract.[8]

The agreements required was what was reasonably practicable, and not what might be achieved with vast amounts of time, effort and money.[9] Just because it might be 'possible' to restore the island did not mean that a contract requiring replanting 'where possible' required anything more than what was commercially possible. In considering this, Megarry J referred to evidence that it might cost AUD50 million before one coconut was planted, and restoring the soil might take 100 years.[10] Megarry J was concerned that even if soil of two to six feet was imported, it would still not ensure success in the replanting. After the mining operations, what was left was dolomotised limestone and tree roots would have no chance of penetrating this rock.[11]

Megarry J stated: 'The complexities of specific performance are weighty and discouraging, but by themselves I do not think that they suffice to induce the court to refuse specific performance.'[12] However, specific performance was rejected on a number of grounds including:

- for specific performance to be granted, every owner of the property in question would have to be before the court, which was not the case;
- the plaintiffs in the case were just a small number of Banabans, whose land was about 11 1/4 acres (out of over 1500 acres on the island), and on that basis no order for specific performance for replanting over the whole island was appropriate;[13]
- the order for specific performance would be futile, considering that the decision had been made that the obligation to replant just meant planting some coconut trees, and there was little evidence that the trees would survive in any event.[14]

Damages

As specific performance was not going to be ordered, Megarry J had to consider damages for breach of the promise to replant. The plaintiffs were required to show a loss and that might be either the diminution in the value of the land, because of the failure to replant or the cost of replanting.[15] As it was not clear that the Banabans, as a whole, would actually replant, should the plaintiffs be awarded damages to do so?[16] Megarry J held that the appropriate sum for diminution in value of the land in its unplanted form, compared to it being planted with coconuts, was merely AUD75 per acre.[17]

What Can Be Learned From This?

Several points may be made. First, the framing of an action and the ownership of land is very important to a successful claim for restoration of land. Second, it is not wise to enter into long-term contracts, because contracts will be construed at the time of formation with any views of environmental protection that exist at the time and construction of contractual promises cannot 'move with the times' to take account developments that may occur. Third, specific performance of an obligation, such as one to replant, is possible, but will depend on a construction of the contract and whether damages are an appropriate alternative. Pausing there, it might be said that damages will never be an appropriate alternative to situations where the environment has been damaged. However, if a court feels that such an order is 'futile' as there is little chance of success, then there is no help. What if the reason why the order may be futile is because of the behaviour of the defendant? Fourth, damages will be ordered on the basis of the diminution in the value of the property. So if there is no commercial value in native flora and fauna and they are damaged by a contracting party in breach of a promise, then the damages could be zero, as no loss could be proved. Finally, in Australia, mining is regulated by legislation that requires that environmental concerns be taken into account, and requires restoration following mining. However, not all nations have laws that protect the environment in this way.

As a postscript to the situation on Banaba, it might be said that contract law provided no assistance to the Banaban people, who had been mistreated, and their environment so badly scarred. However, in May 1977, after consultation between the Commonwealth countries involved, the following statement was issued in the British House of Commons:

> The three Governments [Britain, Australia and New Zealand] are prepared to make available, on an ex gratia basis and without admitting any liability, a sum of 10 million Australian dollars. The money would be used to establish a fund which will be preserved for the benefit of the Banaban community as a whole, the annual income being paid to the Rambi Council of Leaders for development and community purposes ...[18]

Also, Ocean Island is no longer a barren waste; nature has covered the effects of mining with new growth and vegetation.

Carbon Trading Agreements: Contract Law in the Context of Government Regulation

There has been much written about carbon trading following the *Kyoto Protocol*.[19] The protocol requires emission reduction targets. Specifically, developed countries[20] ('Annex 1' nations) must reduce their greenhouse gas emissions by 5.2% over 1990 levels by 2012. Some flexibility was included in the *Kyoto Protocol* for countries to reach their required targets, and so there are three ways in which governments or corporations can comply by:

- reducing direct emission, for example, turning off lights at night
- investing in emission offset projects
- suffering financial penalties if emissions go beyond the legal limit, or purchasing carbon credits to cover emissions over the limit.

Carbon credits are generated by those who emit less than the legal limit or those whose activities remove carbon from the atmosphere. These can be sold to those whose emissions exceed the legal limit. Obviously, the market value of a carbon credit must remain below the cost of the penalty that would otherwise be imposed.

Carbon trading can be done by one party preparing a contract that describes the activity being undertaken to reduce or offset emissions. This contractual promise is then sold to another party that wants the carbon reduction or offset. These contracts are usually traded 'over the counter' (OTC), that is, without a market. These OTC contracts are usually single trades and often the terms are at least in part confidential. The alternative is the carbon trading market, which operates like a share market. There carbon credits are proved with carbon sequestration certificates for those activities which meet the rules defining carbon sequestration.

In 2006, the World Bank estimated the size of the global carbon market was about USD30 billion. By 2007, it was estimated to be worth USD70 billion. There is a suggestion that by 2020 it will be worth USD700 billion.[21] So it is now true that money grows on trees! Since the *Kyoto Protocol* in 2005, the market for carbon credits has grown 24 fold. Europe leads the carbon market, where futures and options contracts over carbon credits are traded on dedicated 'climate exchanges'.[22] NSW has the world's second largest carbon credit market and the Federal Government is working on a national scheme. There is also a trading market in the United States;[23] however, participation is voluntary, unlike in Europe. Having said that, at least 22 US States are currently exploring cap and trade systems and several bills are before Congress.[24] Furthermore, at a climate summit in Bali in December 2007, the United States pledged to work out a new arrangement by the end of 2009.[25]

Position in Australia

In 2009, the Australian Government is in the process of creating an emissions trading scheme (ETS). Before the release of the 'Green Paper' in mid-2008, it had outlined five requirements for an ETS which have been explained in more detail in the Green Paper:

A cap and trade scheme to be internationally consistent. The *Kyoto Protocol* and European Union emissions trading schemes are based on a cap and trade scheme. In this scheme, total emissions are 'capped', permits allocated up to the cap and trading allowed to let the market find the cheapest way to meet any necessary emission reductions.

Designed to effectively reduce emissions. The plan is to reduce emissions by 60% by 2050.

Economically responsible. The government is concerned that a domestic scheme does not undermine Australia's competitiveness and ensures that Australian firms that are energy-intensive are not disadvantaged.

Fair. The government would like the scheme to allow both costs and benefits to be shared across the community, so additional government policies will be introduced to ensure households reduce energy consumption as well.

Recognise the need to act now. The scheme is to be introduced by 2010.

EU Agreements

The Australian Government plans for the ETS to be internationally consistent. Therefore, in considering how contracts for the Australian trading scheme will operate, it is useful to consider the operation of the EU scheme and in particular the standard contracts that are used.

In February 2008, the International Emissions Trading Association (IETA) released version 3 of its *Emissions Trading Master Agreement* for the EU Scheme ('*Master Agreement*'). It explains in the recitals that the agreement relates to the scheme established by the EU and the Member States under which participants may buy and sell allowances for greenhouse gas emissions. The agreement is for parties who have or intend to enter into one or more transactions for transferring allowances or credits.

This contract is governed by English law (cl 14.7), which makes it particularly relevant to Australia, as we have a very strong tradition of Anglo-Australian law, and while English approaches are not binding in Australia, they are persuasive. Furthermore, England is recognised as an international trading centre with excellent commercial courts.

In these contracts, there are express provisions relating to remedies, which are interesting for this discussion. Of particular interest are the clauses dealing with transfer failure and liquidated damages, which are discussed below.

Consequences of 'Transfer Failure'

Clause 6 of the *Master Agreement* deals with 'transfer failure' and provides that if there is a failure to transfer either the allowance or the money payable, then the other party may by notice require the party in breach to remedy the breach within one day after receipt of the notice. If this notice is not complied with, then the innocent party may terminate the contract.[26]

If parties to a trade relied on the common law, rather than a contractual right, they would also be allowed to terminate the contract. However, at common law there is no obligation to give a party in default the option to rectify the breach. The agreement gives the innocent party the option of allowing rectification, but does not require it.

Under the contract, interest is payable for late delivery, and replacement costs can be recovered.[27] If the receiving party is charged an Excess Emission Penalty (EEP) because of the failure to deliver, then the delivering party will be liable to compensate the receiving party for this penalty.[28] The common law would arguably reach a similar result. Damages are payable in contract law for any loss that can be proved to be caused by a breach, so long as it is not too remote to be recoverable. Loss of the use of money may be recoverable at common law, but does cause problems.[29] The usual approach of parties aware of their rights in contracts generally is to expressly state that interest is recoverable, as in the *Master Agreement*. As for the requirement to compensate for EEPs, arguably such a fee would be recoverable as common law damages as well. It would either be considered a 'usual loss' or within the contemplation of the parties that if there is delay or failure to transfer allowances that the purchasing party might be fined, as the whole reason for the contract is for the purchasing party to avoid being fined. Alternatively, it might be said that there was actual knowledge by the trading party that the purchasing party would be liable to be fined if it did not receive the allowances in time, and therefore the loss is not too remote under the second limb of *Hadley v Baxendale*.[30]

Under the contract, if a party terminates the contract for breach of the other party, 'termination payments' are due.[31] The EU contract provides mechanisms for the non-defaulting party to determine the amount that must be paid by a party in default, rather than requiring the parties to go to court to prove their loss. However, it is conceivable that if the defaulting party disagreed with the amount calculated by the non-defaulting party, litigation would be possible.

Liquidated Damages

Clause 13.4 is particularly interesting. It provides that:

> Reasonable Pre-estimate and Maximum Liability. Each Party acknowledges that the payment obligations in clause ... 6 ... are reasonable in the light of the anticipated harm and the difficulty of estimation or calculation of actual damages. Each Party waives the right to contest those payments as an unreasonable penalty. Each Party further acknowledges that the payment obligation in clause 12 (Termination) shall constitute the maximum liability in the event of termination of this Agreement.[32]

This clause purports to be a liquidated damages clause and a limitation of liability clause in one. Liquidated damages clauses are included when parties have decided to predeter-mine the amount of money that should be payable in the event of a breach of contract. The advantage of including such a clause is that if a breach occurs, there is no need to prove loss, but rather the nominated sum is payable like a debt. However, the effect of a liquidated damages clause is to oust the jurisdiction of the court to assess damages. Courts will only allow this if the amount that is chosen can be seen to be a genuine pre-estimate of loss and not a penalty. The law concerning the distinction comes from a 1915 decision of the Privy Council in *Dunlop Pneumatic Tyre Co Ltd v New Garage and Motor Co Ltd*,[33] which was recently approved as the still relevant law by the High Court in 2005 in *Ringrow Pty Ltd v BP Australia Pty Ltd*.[34]

In *Dunlop*, Lord Dunedin set out the 'tests' for whether a clause is a penalty or a liqui-dated damages clause.[35] He stated that the label that the parties use will not be conclusive. Instead, the court construes the clause to decide if it was a genuine pre-estimate of loss or whether instead it was 'extravagant or unconscionable' in amount. The High Court has said that in circumstances where it is difficult to estimate the loss that might flow from a breach with certainty, then the amount chosen must not be 'out of all proportion' with the greatest loss that might flow.

In clause 13.4 the parties are trying to 'have their cake and eat it'. They are labelling the clause as a 'reasonable pre-estimate of loss' and waive the right to challenge the amounts as penalties. They are purporting to prevent a court from inquiring into whether the clause is a penalty or an enforceable liquidated damages clause, and this would not be upheld by a court.

Arguably this is meant to be an incentive to perform, however, if it is unenforceable, which I suggest, then the innocent party would have to rely on the common law for an assessment of its damages. The disadvantage of not having a liquidated damages clause to rely upon include:

- the parties need to ask a court to assess damages, rather than just requiring payment of a known sum;
- the court requires proof of the breach, the quantum of the loss, and must also be convinced that the loss is not 'too remote';
- it is likely a party could recover the difference in the market value of the carbon credits and the contract price, on the assumption that the innocent party has gone into the market and bought replacement credits;
- if there is no available market, then assessment of damages is more difficult and requires proof of the loss;
- an innocent party is likely to recover any fine or penalty payable if it has exceeded the quota of carbon emissions that have not been offset by carbon credits;
- an innocent party could recover transaction costs.

Conclusion

Will economic incentives, such as carbon emission targets coupled with carbon trading schemes, be sufficient to achieve the desired result of reduced carbon emissions and thereby reduce the impact of climate change? In developed countries based on market economies, the answer appears to be 'yes'. Contract law can provide the mechanism for the sale of carbon credits in Australia, and other jurisdictions with similar principles, providing it has government regulation behind the scheme, which will control penalties and ensure that a market for carbon credits exists. The idea behind trading carbon credits is that polluting will be less economically viable, as it will have to be paid for, by buying credits from those who do not pollute. So carbon credit contracts are just to facilitate this process.

However, in developing countries, *Tito v Waddell* shows that there is a real possibility that the environment and local people can suffer if contract law is their only protection and if the resource available for exploitation is valuable. Contract law will not be able to ensure compliance with promises. Furthermore, even if financial compensation is available for a non-defaulting party, that compensation may never be sufficient to restore the innocent party back to their original position.

Consider for example a developing country with forests that are able to produce tradable carbon credits. What is the remedy for the local people if a mining corporation damages that land and destroys or damages the forest? The international world will need to consider the ways in which these issues are regulated. Contract law will not be enough. Regulation at an international and national level will be essential.

Endnotes

1   *Tito v Waddell* (No 2) [1977] Ch 106 ('*Tito*').
2   Anthony Flude, Ocean Island: *A rocky land of hidden treasure* (2002) <http://homepages.ihug.co.nz/~tonyf/ocean/ocean.html>.
3   *Tito* [1977] Ch 106
4   *Tito* [1977] Ch 106
5   *Tito* [1977] Ch 106, 279.
6   *Tito* [1977] Ch 106, 279.
7   *Tito* [1977] Ch 106, 280.
8   *Tito* [1977] Ch 106, 281.
9   *Tito* [1977] Ch 106, 282.
10 *Tito* [1977] Ch 106, 282.
11 *Tito* [1977] Ch 106, 283.
12 *Tito* [1977] Ch 106, 323.
13 *Tito* [1977] Ch 106, 325–6.
14 *Tito* [1977] Ch 106, 326–7.
15 *Tito* [1977] Ch 106, 335.
16 *Tito* [1977] Ch 106, 336.
17 *Tito* [1977] Ch 106, 341.
18 Flude, above n 2.
19 *Kyoto Protocol to the United Nations Framework Convention on Climate Change,* opened for signature 16 March 1998, 2303 UNTS 148 (entered into force 16 February 2005) ('*Kyoto Protocol*').
20 Developing countries are not required to do so.
21 Barclays research report quoted in Dow Jones & Co, As the World Heats Up, So Does Carbon Trading (2008) *Business Spectator*. Available at www.businessspectator.com.au
22 See generally <www.ieta.org>. There are currently two frameworks in Europe. First, there is the European Union Allowances ('EUA'), which allocate companies their emissions targets and allow trade of carbon. Second, there are Certified Emission Reduction ('CER') credits, which are generated when companies in the developed world invest in clean-technology projects that trim emissions in the developing nations.
23 See <www.chicagoclimatex.com>.
24 Dow Jones & Co, above n 17.
25 Ibid.
26 International Emissions Trading Association (IETA), Emissions Trading Master Agreement for the EU Scheme, Version 3, cll 6.1.1(y), 6.1.2.1(y) (2008) <http://www.ieta.org/ieta/www/pages/download. php?docID=3021>.
27 Ibid cl 6.1.2.1(y).
28 Ibid cl 6.1.2.2.
29 See, eg, *Hungerfords v Walker* (1989) 171 CLR 125.
30 *Hadley v Baxendale* (1854) 9 Ex 341; 156 ER 145.
31 IETA, above n 22 Clause 12.5(a) provides: 'On, or as soon as reasonably practicable after, the Early Termination Date, the Non-Defaulting Party shall in good faith, calculate the termination payment … which is the Loss for all Transactions unless the Market Amount is specified as the termination payment method in Schedule 2 (in which case it is the Market Amount).'
32 IETA, above n 26, cl 13.4.
33 *Dunlop Pneumatic Tyre Co Ltd v New Garage and Motor Co Ltd* [1915] AC 79 ('*Dunlop*').

34 *Ringrow Pty Ltd v BP Australia Pty Ltd* (2005) 224 CLR 656. For a discussion see JW Carter and Elisabeth Peden, 'A Good Faith Perspective of Liquidated Damages' (2007) 23 *Journal of Contract Law* 157.
35 *Dunlop* [1915] AC 79, 87–8.

Climate Change and Tax Law:
Tax Policy and Emissions Trading

Celeste M. Black

Touted as the 'most significant economic and structural reform undertaken in Australia since the trade liberalisation of the 1980s',[1] the Australian Government's response to the challenge of climate change is the proposed establishment of an emissions trading scheme, to be operational by 2010. The Carbon Pollution Reduction Scheme ('CPRS'), as described in the discussion paper (the Green Paper),[2] is a permit trading scheme based on the 'cap and trade' model with additional measures to assist households and business to transition to a low carbon economy. The details of the proposal are extensive. After a period of consultation, it is proposed that draft legislation will be released for public comment in December 2008. This chapter provides an overview and analysis of the income tax law implications of the emissions trading scheme as currently proposed by the government in the Green Paper. Given the relatively recent release of the proposal, the comments provided here are of a preliminary, though considered, nature and, of course, as government policy is refined and further details of the scheme are released with the draft legislation in December, additional taxation issues may later arise that are not currently apparent.

The tax treatment of any emissions trading scheme is of importance as the taxation system has the potential to either distort or support the scheme. As pointed out in the Green Paper, the primary objective in the design of the tax treatment of emissions trading is to support the objectives of the scheme, of which reducing emissions in the most cost-effective manner is paramount.[3] Other objectives include tax neutrality over time; that is, a matching of deductions for costs with the income or benefits derived, and simplicity.[4] With these objectives in mind, the various proposals put forward in the Green Paper, especially those contained in Chapter 11 (titled 'Tax and accounting issues'), will be considered and the income tax implications of other features of the proposed scheme will also be analysed. The discussion will proceed by incorporating a description of the various design features of the scheme within the analysis of the tax implications. The proposed scheme also raises important additional taxation issues, such as the treatment of the trading of permits under the Goods and Services Tax (GST) and international trade and tax issues, but these issues are beyond the scope of the current paper.

The paper will begin with a description of the discrete income tax regime that is proposed to apply to trading in carbon pollution permits (here referred to simply as 'permits') and an evaluation of the regime against the stated tax policy objectives. In this context, the interaction between the proposed tax regime and the compliance timeline will be illustrated and the issues raised by this will be discussed. The paper will also comment on taxation issues raised by each of the following proposed scheme features: liability under the scheme; assistance to emissions-intensive trade-exposed ('EITE') industries, including the treatment of free permits for tax purposes; assistance to strongly affected industries; the price cap; foreign ownership of permits; and the development of forward markets.

Tax Treatment of Permits

The income tax treatment of the acquisition, trading, holding and acquittal of carbon pollution permits (to be referred to as 'Australian emissions units' in the legislation)[5] will turn on the legal nature of the permit and the characterisation of the various transactions for tax purposes. In several areas of tax policy, there has been a move by government to adopt taxation laws which provide treatment that is similar to that prescribed by international accounting standards.[6] However, in this area, no standards have yet been proposed by the International Accounting Standards Board, although an exposure draft of standards may be available in late 2009.[7]

Tax Policy Objectives

In its Green Paper, the government adopts three tax policy objectives that should guide the development of taxation rules specific to permits. Foremost is cost-effectiveness, that is, the tax system should support the main objective of the trading scheme of reducing emissions in the most cost-effective manner.[8] This concept, closely linked with the common policy objective of economic efficiency, requires that the tax system minimises distortions to decision-making with respect to the purchase, sale or acquittal of permits or engagement in pollution abatement activities. This is closely linked to the second policy objective: tax neutrality.[9]

In addition to neutrality as between responses to the scheme in a given year, the Green Paper also establishes, as a goal, tax neutrality over time.[10] This objective is soon to be realised under the current income tax system through the link between the nature of the benefits obtained by virtue of an outgoing and the treatment of the expense for tax purposes.[11] This could be interpreted as requiring a correspondence between the costs of compliance under the scheme and the income generated by the activity creating the compliance obligation (that is, the pollution generating activity).

A final policy objective is simplicity, which is another touchstone of tax policy generally. A more simple system will keep compliance and administrative costs to a minimum and will support efficient decision-making on the part of taxpayers.[12] It remains to be seen how well the proposed treatment of permits under the tax system meets these policy objectives.

Discrete Regime for Permits

As noted above, the treatment of permits for tax purposes will turn in part on their legal character. Under the proposals of the Green Paper, a carbon pollution permit will be defined as personal property which incorporates the right to surrender the permit to meet obligations under the scheme as well as the right to transfer the permit by assignment.[13] Permits will be represented by entry on an electronic national registry (no physical certificates) and each permit will be given both a unique identification number and a vintage.[14] Under the scheme it is proposed that there be limited short-term borrowing[15] and unlimited banking.[16]

Under the scheme as currently described, permits will be considered property and therefore assets rather than a cost of doing business. Such permits may be held by taxpayers for a variety of reasons such as to meet current compliance obligations, in anticipation of future obligations, for trading or speculative purposes, for marketing purposes, as an investment, or a combination of these.[17] Under the Australian income tax system, the purpose for which an asset is held will often dictate tax consequences such as the character of the gain or loss (income or capital) and the timing of the realisation of that gain or loss. For example, a permit could be characterised as 'trading stock' where it is held for the purpose of sale in the ordinary course of business (such as where a speculator 'trades' in permits).[18] For tax purposes, div 70 of the *Income Tax Assessment Act 1997* (Cth) would be triggered, which provides a specific regime for the taxation of trading stock. A permit could also be considered a 'CGT asset'[19] for capital gains tax purposes where another taxation regime would be triggered.[20] Such an outcome is considered by the government to be undesirable, given the objective of neutrality and simplicity, and instead, the Green Paper proposes a discrete taxation regime for permits[21] that is based on the current trading stock regime.

The preferred position, as outlined in the Green Paper, is to create a rolling balance method with the following features:

- acquisition costs would be deductible;
- proceeds on sale would be assessable; and
- account will be taken of the difference between the opening and closing values of permits held, with any increase in value being included in assessable income and any decrease being allowable as a deduction.[22]

For those not familiar with trading stock accounting, some further elaboration may be helpful. Through the use of the rolling balance method (measuring the change in value of permits held), the mechanism will effectively defer the deduction for the cost of a permit that is retained and provide for a deduction for the value of the permit in the year in which it is sold or acquitted.[23] More specifically, where permits are acquired in the year, a deduction is provided for the cost, but if the permits are still held at the

Table 8.1

	2010/11	Tax effect
Permits purchased: 10 @ $10 → deduction		(100)
Permits sold: none → income		0
Opening balance	nil	
Closing balance	100	
Increase in balance → income	100	100
Net effect		nil

year end, those additional permits will affect an increase in the value of permits held which will give rise to assessable income which will, as a result, cancel out the earlier deduction. Table 8.1 shows a very simple illustration of this process. It assumes that the taxpayer has purchased permits of one vintage in the 2010/11 year and holds them at year end to meet compliance obligations which will come due in the 2011/12 year.

Then, assume that, in the 2011/12 year, the taxpayer surrenders eight permits to meet compliance obligations and sells the additional two permits that are not required, at a price of $12. Assuming that there are no other relevant transactions in the year, the rolling balance method would give the following results shown in Table 8.2.

The two transactions could be described as follows. On the sale of the two permits for $12 each, the taxpayer has realised a profit of $2 per permit, or a total of $4. In addition, the taxpayer has surrendered eight permits that have an original cost of $10 each. On surrender or acquittal, the cost of $80 total is now incurred. The net of the two transactions is an expense of $76 (the profit of $4 net against the expense/loss of $80).

Application of the Method

The proposed rolling balance method is intended to apply to all taxpayers who carry on a business or other activity that produces assessable income, regardless of the purpose for the holding of permits, thereby supporting the policy objectives of efficiency and simplicity.[24] Expenditure in relation to permits that is of a private or domestic nature will not be deductible.[25] This could apply where an individual wishes to voluntarily acquire permits to offset his/her 'carbon footprint' and is in accordance with general tax policy to deny deductions for private or domestic expenditure.[26]

Table 8.2

	2011/12	Tax effect
Permits purchased: none		0
Permits sold: 2 @ $12 → income		24
Opening balance	100	
Closing balance	0	
Decrease in balance (sell 2, surrender 8) → deduction	(100)	(100)
Net effect		(76)

It appears that the rolling balance method is intended to operate to the exclusion of other tax accounting regimes by virtue of the comment that the capital gains tax (CGT) provisions will not apply to transactions involving permits.[27] A similar reconciliation currently exists with respect to the CGT regime and the treatment of trading stock and depreciating assets, whereby any gain or loss which might otherwise be recognised by the CGT rules with respect to these asset classes is disregarded.[28] The same result can be achieved for permits with a similarly drafted exclusion.

It may also be necessary to determine whether it is desirable to exclude permits from the operation of the proposed Taxation of Financial Arrangements (TOFA) regime.[29] Under the most recent version of the draft regime, the TOFA rules, which are generally given priority over all other tax regimes,[30] apply to a 'financial arrangement' which is, broadly, a cash settlable legal or equitable right or obligation to receive or provide a financial benefit.[31] The key element to test for in this type of case is whether the right/obligation is 'cash settlable' as defined.[32] This generally requires that the benefit is money or a money equivalent or that you intend or it is your practice to settle the right/obligation with money. Often, this would not be the case as the permit entitles the holder to have its scheme obligations discharged on surrender, which is not a benefit in money. However, there is an inclusion for taxpayers who deal in the rights/obligations in order to generate short term profits from price fluctuations and/or a dealer's margin[33] where this could apply to taxpayers trading in permits. It may be preferable, in order to maintain consistency and simplicity, for permits to be specifically excluded from the operation of the TOFA regime, as have a number of instruments listed in Subdivision H.[34] Whether this would be appropriate for permit derivatives should also be considered.

Valuation Issues

A critical element of the rolling balance method is the valuation of permits held at the tax year end, as any increase in the value would be assessable and any decrease in the value held will be deductible. As canvassed in the Green Paper, there would appear to be two options for valuation: historical cost and market value.[35] The government, at this stage, has not expressed a preference for either method.[36] Another option, that was not specifically mentioned in the Green Paper, would be to give taxpayers the option of choosing their preferred method, rather than prescribing one or the other. The trading stock regime, upon which the rolling balance method is based, allows taxpayers to elect to value each item of trading stock on hand either at its cost, market selling value, or replacement value.[37] Under these rules, a taxpayer need not use the same method for all items of stock and can change valuation methods from year to year, the only prescription being that the opening value for the current year is the closing value from the year before.[38]

The use of either the historical cost or market value option will have implications for taxpayers. That said, historical cost would most likely be the most simple method to apply and it would not give rise to the cash flow implications of the market value method that is described below. However, financial accounting standards may require mark-to-market of permits as there will be an easily identifiable market and market price, which will then require adjustment back to use of historical cost for tax purposes. As is apparent in the TOFA rules, there has been a policy objective to, where appropriate, align tax and financial accounting methods. In addition, the use of historical cost may affect the selection of permits to acquit in a given year, which may impact on efficiency. This last point is addressed below with the other acquittal issues. The requirement to use historical cost will also necessitate additional rules regarding the treatment of free allocations, also described below, which will impact on simplicity.

The most serious objection to a requirement that taxpayers use market value is the cash flow implications that this method can create.[39] Also, implicit in this method is reliance on spot prices. Given the likely volatility in the permit market, it may not be appropriate to fix one's tax liabilities on a price on a particular day. As an illustration of the impact of using the market value method, one can compare the results shown below to those described above, which relied on historical cost. If we assume the same facts as above, with the additional fact that the market value of a permit has risen to $11 as at the tax year end, the tax consequences would be as shown in Table 8.3.

The result under this method is to pick up the unrealised increase in the value of the permits held as an amount of assessable income. However, this method would similarly also allow a deduction where the market price dips below the original purchase price. Again, we can assume that, in the 2011/12 year, the taxpayer surrenders eight permits to meet compliance obligations and sells the additional two permits which are not required, at a price that has now risen to $12. Assuming that there are no other relevant transactions in the year, the market value method would give the result shown on Table 8.4.

Table 8.3

	2010/11	Tax effect
Permits purchased: 10 @ $10 → deduction		(100)
Permits sold: none income		0
Opening balance	nil	
Closing balance: MV=$11	110	
Increase in balance → income	110	110
Net effect		10

Table 8.4

	2011/12	Tax effect
Permits purchased: none		0
Permits sold: 2 @ $12 → income		24
Opening balance	110	
Closing balance	0	
Decrease in balance (sell 2, surrender 8) → deduction	(110)	(110)
Net effect		(86)

The two transactions could be described as follows. On the sale of the two permits for $12 each, the taxpayer has realised a profit of $1 per permit in addition to the accrued gain of $1 from the previous year, or a total of $2. In addition, the taxpayer has surrendered eight permits that have a carry-over value of $11 each. On surrender or acquittal, this value of $88 total is now an incurred cost (this picks up both the original cost of $10 each and the $1 gain that was accrued in the prior year but was not in the end realised). The net of the two transactions is an expense of $86 (the profit of $2 net against the expense/loss of $88). It should be noted that, across the two years the net result (expense of $76) is the same as under the historic cost method but the timing differences are apparent.

Disposal of Permits

Permits may be surrendered or acquitted to meet compliance obligations or may be sold in the trading market or in an over-the-counter transaction. As the previous examples have shown, upon surrender or sale of a permit, it will no longer be on hand as at the tax year end, its carrying value will no longer be included in the year end closing balance and the carrying value (cost or market value) will effectively be allowed as a deduction. Under the proposed scheme, participants will be allowed unlimited banking (carrying forward) of permits and limited borrowing.[40] In addition, each permit will have a vintage year and will be given a unique identification number.[41] The combination of these features with links to purchase price/market values may produce incentives for taxpayers to select specific permits for surrender or sale, both across vintages and within vintages, which are tax related rather than dictated by the scheme. For policy reasons, it may be preferable to prescribe the basis upon which permits are selected. This issue has been addressed for income tax purposes in the context of shares but it was not addressed in the Green Paper.

The Commissioner of Taxation has issued a number of public rulings that address the identification of shares sold.[42] For both the purposes of CGT and the trading stock rules, a taxpayer may identify which shares have been sold, out of a pool of otherwise identical holdings, for the purposes of determining the acquisition date and relevant cost of the shares. Where shares can be identified by individual numbers, this provides sufficient identification. Where this is not the case, appropriate accounting records may be maintained to identify the relevant shares. Otherwise, a first-in first-out (FIFO) basis is considered acceptable and the last-in first-out (LIFO) method is rejected.

Given the designation of permit ID numbers, it would be straightforward to identify particular permits surrendered or sold at a given time. Depending on the circumstances of a taxpayer, there may be an incentive to select for surrender or sale those permits with the highest cost or carrying value so as to maximise the deduction on surrender or minimise the profit realised on sale. Given the position taken by the Commissioner of Taxation with respect to shares, it is submitted that the same approach should also apply to permits.

Timing Issues

By overlaying the compliance timeline with the income tax year, several timing issues raised by the scheme as proposed are highlighted. The timelines may be represented as indicated in Figure 8.1.

The Green Paper proposes that the compliance period be aligned with the financial year (ending 30 June)[43] with the relevant reporting date being 31 October under the National Greenhouse and Energy Reporting System.[44] The final surrender date to meet compliance obligations for that period would be 15 December.[45] Once the scheme is up and running, it is proposed that there be eight auctions for each vintage, three in years prior to the relevant compliance period, four during the period and one between the reporting date and the final surrender date (approximately on 15 November).[46] This would be combined with a tax year that ordinarily is the financial year.

Although one of the stated policy objectives is tax neutrality over time, the combined effect of these timelines would see the income derived from the pollution-generating activity being assessed with respect to the X1 year but the expenses for the permits required as a result of that activity being incurred on the surrender date in the following (X2) year. Simply put, the expenses are not matched with the relevant income. This mismatch could provide an incentive to surrender early, before the final surrender date, perhaps before 30 June so as to generate the deduction earlier. The Green Paper states that early surrenders will be accepted.[47] However, this could have a distortionary effect on permit prices as the final instalment of the vintage will not be on the market until the final auction date of 15 November. These issues could be further complicated where a taxpayer has a substituted accounting period for tax purposes.

Assistance to Industry: Free Permits

Not unlike emissions trading schemes in other jurisdictions, the Australian scheme will provide assistance to EITE industries where the preferred form of this assistance is by way of free permits.[48] These permits may make up 20% of all permits issued.[49]

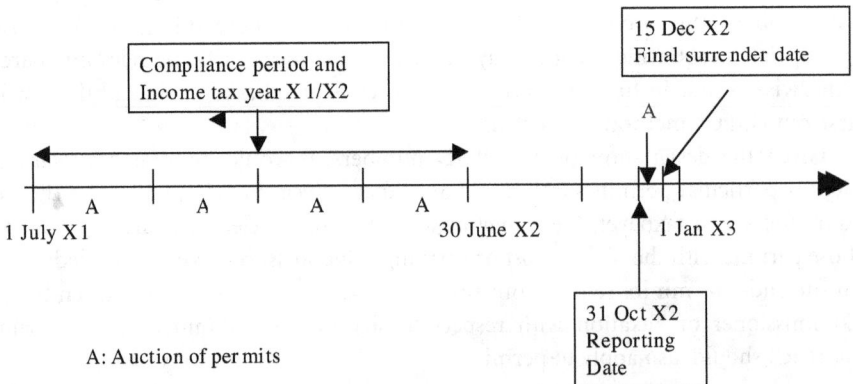

Figure 8.1

An activity must qualify as both energy-intensive[50] and trade-exposed.[51] The amount of assistance will be linked to permit liability (direct emissions) and increased electricity costs (indirect up-stream emissions)[52] and will be by way of annual assistance of free permits allocated at the beginning of the compliance period.[53] This assistance raises tax issues in the treatment of the allocations as well as the treatment of the permits under the rolling balance system. It should be noted up front that many of the complications (from a tax perspective) raised by the allocation of free permits would disappear if the assistance were by way of cash grants. Cash assistance is mentioned as an option in the Green Paper[54] but it is not clear how favourably this option is viewed.

Free Permits as Income

Under Australian income tax law, a government payment to a taxpayer to assist that taxpayer in the carrying on of a business will often be considered 'business income' and therefore assessable as ordinary income.[55] In a recent public ruling issued by the Commissioner of Taxation, the view was taken that this type of assessable payment, often called a subsidy or bounty, would include, inter alia, situations where the payment assists with meeting operating costs or providing income support.[56] Specifically, '[a] GPI [government payment to industry] to assist with business operating costs or liabilities is ordinary income in the hands of the recipient and is assessable under section 6–5 in the income year in which it is derived.'[57] It is submitted that an allocation of free permits is of this character as it is designed to assist an activity that has increased obligations under the emissions trading scheme. Alternatively, if not considered ordinary income, a bounty or subsidy will be assessable by virtue of a specific statutory provision.[58] The treatment of free permits as income also achieves matching as the additional expenses incurred under the scheme will generally be deductible outgoings.

In the case of free permits, the subsidy is given in kind, rather than in cash. Under the income tax regime, if non-cash receipt is convertible to money, that 'money value' will be taken to be the amount of income or outgoing.[59] Alternatively, the 'non-cash business benefit' rule could be triggered, which would also serve to include the market value of the permits in income.[60] The preferred position in the Green Paper is to apply these general rules to free permits so as to treat the value of the permits as income derived when the permits are received, even if they cannot be used until a future year.[61] Any valuation mechanism would need to rely on the spot prices generated by the permit markets (auction and/or secondary) as these would provide a 'market value' for the permits. However, difficulties may be experienced if the free permits are significantly future-dated (the example given in the Green Paper is the issue in 2010 of free permits dated for 2020)[62] such that a liquid market may not have yet developed. In such a case, an alternative valuation method will need to be explored.

A significant issue for taxpayers from this approach is cash flow. Under the preferred approach, a recipient of free permits (in kind) will be assessable on their value, where the tax, of course, must be paid in cash. This could place pressure on the recipient's business cash flow, although there are several options available to free-up some of the value in the permits. The most straightforward option would be to simply sell some of the permits on the secondary market or through the double-sided auction process described in the Green Paper.[63] No doubt, other financial instruments will also be developed to allow the value in the permits to be realised without necessitating sale given that the proprietary nature of the permits will allow for the taking of security over permits and creating equitable interest in them.[64]

Surrender and Sale of Free Permits

Where a free permit is received and surrendered in the same income year, it is proposed that there be no resulting net tax impact.[65] If the value of the permit is included in income at the time of receipt so to achieve this result, a special deduction provision will be required to allow a deduction for that same amount. As the permit will come on hand during the year but no longer be on hand by the year end, it will not affect the balance of permits held.

An additional issue arises when a free permit is received in the first year and surrendered in the following year in respect of emissions from the first year. This could occur often given that the reporting date and final surrender date for a given year will always fall in the second year. Under the basic rules outlined above, the value of the free permit will be assessable but it will still be on hand as at the year end. However, if it is surrendered in the following year, matching would only be achieved if the deduction for the value is allowed in the first year. A policy decision would need to be made on this point.

It is envisioned that free permits may be sold, either through the double-sided auctions in the early phase of the scheme[66] or on the secondary markets. Under the ordinary rules described above, the proceeds on sale of a purchased permit are included in assessable income but these proceeds are effectively net of the purchase price/cost through the rolling balance method. Where a free permit is acquired and sold in the same year, the initial value is included in income and any additional profit or loss on sale should also be included in income or allowed as a deduction.[67] This result could only be achieved if another special rule is included in the tax act for this category of permits. The treatment of free permits will therefore require a departure from the simplicity objective in order to achieve the correct tax results.

The following examples are given to illustrate how the treatment of free permits will operate as part of the rolling balance method. The following facts are assumed in all cases:

- There is no opening balance of permits held.
- The taxpayer receives 10 free permits in year 1 with a current value of $10 each.
- Of those 10, five are surrendered to meet scheme obligations.
- In year 2, two of the permits are sold for $12 each.
- In year 2, three of the permits are surrendered to meet scheme obligations, leaving a nil balance of permits held.

Free Permits and Historical Cost

Some comment should be made regarding this analysis (Table 8.5). The deduction for the free permits surrendered that were acquired in the same year must be based on the amount previously picked up as income, rather than any spot price that might be available. To obtain the closing balance of permits held, the cost must be deemed to be the

Table 8.5

	2010/11	Tax effect
Free permits issued: 10 @ $10 → income*		100
Permits surrendered: 5 @ $10 → deduction to offset income*		(50)
Opening balance	nil	
Closing balance (5 permits): deem cost of $10*	50	
Increase in balance → income but must reduce for value already assessed as income on issue*	50 less 50	0
Net effect (net income)		50

Note:* denotes a step requiring a special rule

amount previously assessed with respect to those five free permits still held. This closing balance figure is needed as it will become the opening balance for year 2 and only any additional profit or loss on those permits should be recognised on disposal. However, the increase in rolling balance must be reduced for the value of those five that have already been assessed as income.

This scenario in Table 8.6 assumes that only a total of eight permits are required to meet scheme obligations with respect to year 1 so the extra two permits are sold for a spot price of $12 each. If it is considered desirable to match the free permits with the compliance year, the 'cost' of the three permits surrendered in year 2 would need to be carried back to year 1. The net result for the year would then only show the profit of $4 realised on the two permits sold.

Free permits and mark-to-market

The same scenario under a mark-to-market system would produce the results shown in Table 8.7.

By relying on a mark-to-market basis, there is no requirement to have a 'deemed cost' rule that is otherwise required under the historic cost basis (see Table 8.8). The balance for the year therefore picks up the value of the five permits received for free and still held (both the original value and the increase in value in the year).

An alternative approach

It is suggested that the following alternative approach to the treatment of free permits, based on the mark-to-market valuation rule, has the advantage of simplicity as it dis-

Table 8.6

	2011/12	Tax effect
Permits purchased: none		
Permits sold: 2 @ $12 → income		24
Opening balance	50	
Closing balance (sell 2, surrender 3)	0	
Decrease in balance → deduction	(50)	(50)
Net effect (net loss)		(26)

Table 8.7

	2010/11	Tax effect
Free permits issued: 10 @ $10 → income*		100
Permits surrendered: 5 @ $10 → deduction to offset income*		(50)
Opening balance	nil	
Closing balance (5 permits): MV=$11	55	
Increase in balance → income but must reduce for value already assessed as income on issue*	55 less 50	5
Net effect (net income)		55

penses of the need to construct special rules for this category of permits. As a starting point, it is not necessary to include the value of free permits in income on receipt. Rather, the value of those permits will be picked up as income through the increase in the value of permits on hand as at the year end if they are still held. If free permits are surrendered in the same year as issue, no proceeds are received and, as the desired effect is no tax effect, this result can stand. If permits are sold in the year, the mechanism can simply pick up the proceeds as income. The same scenario under this approach would appear as shown in Table 8.9.

The only special rule required would seem to be one which says that the value of the free permits should not be included in income (see Table 8.10).

Other Income Tax Issues

Many additional issues are raised from the proposed design of the emissions trading scheme and no doubt more issues will become apparent as the additional detail of the scheme is finalised and released by the Australian Government. This section merely makes note of some of the income tax issues so far identified by the author.

Liability under the Scheme

Obligations for compliance under the scheme are placed on the controlling corporation of a company group where a member of the group has operational control over the emitting activity.[68] It is this entity which ultimately must report emissions and acquit the requisite permits. The entity with operational control may not, however, be

Table 8.8

	2011/12	Tax effect
Permits purchased: none		
Permits sold: 2 @ $12 → income		24
Opening balance	55	
Closing balance (sell 2, surrender 3)	0	
Decrease in balance → deduction	(55)	(55)
Net effect (net loss)		(31)

Table 8.9

	2010/11	Tax effect
Free permits issued: 10 @ $10 → no income		0
Permits surrendered: 5 @ $10 → no deduction		0
Opening balance	nil	
Closing balance (5 permits): MV=$11	55	
Increase in balance → income	55	55
Net effect (net income)		55

Table 8.10

	2011/12	Tax effect
Permits purchased: none		
Permits sold: 2 @ $12 → income		24
Opening balance	55	
Closing balance (sell 2, surrender 3)	0	
Decrease in balance → deduction	(55)	(55)
Net effect (net loss)		(31)

the same as the entity with financial control and ownership of the business, which would be the relevant taxpayer. It may be that the operator will be deemed to be acting as an agent on behalf of the taxpayer/owner such that the expenses are directly incurred by the taxpayer. It is unclear from the Green Paper how joint operators or owners will be treated under the scheme.

Another issue linked to liability is the power of the regulator to review emissions reports for a period of four years unless in the case of fraud, where the review period would be unlimited, where these amendment periods match those applicable to business income tax returns.[69] Upon review, it may be determined that a 'refund' or additional liability arises. Although this could take the form of permits (requiring additional acquittal or getting new permits for free), the spot price for the then current year may be quite different to the spot price of the vintage relevant to the compliance year. It must then be considered whether any adjustments should be made to take this differential into account. The tax consequences would be quite straightforward if adjustments were in cash as one would simply derive income or incur an expense for the adjustment.

Adjustments to Free Allocations

Under the mechanism to provide assistance to EITE industries, the level of assistance, or allocation of free permits, is initially based on projected output.[70] The mechanism then provides for a 'true-up' once the actual output is known. It has been suggested that, where the initial assistance has been overly generous, the extra permits could be required to be handed back or future allocations could be adjusted to reflect the initial over-allocation.[71] Hand-back would only be required where the entity has ceased the

EITE activity, as adjustments to future allocations would obviously not be an option.[72] From a tax perspective, the correct result would seem to be reached where the rolling balance method registers a reduction in permits on hand due to the hand back with no proceeds on sale — this effectively produces a deduction for the cost or carrying value of the permits handed back.

Strongly Affected Industries

As proposed, the CPRS will also include a mechanism to compensate strongly affected industries for the negative effect of the scheme on asset values.[73] Given the criteria set out in the Green Paper, coal-fired electricity generators are the initial sector which may be eligible for assistance.[74] Aside from any adjustments which may be triggered under the current tax rules,[75] it is proposed that assistance be provided to the financial owner of the affected assets[76] by way of direct payments or free permits.[77] As the financial owner may not be the operator, and may therefore not have direct compliance obligations under the scheme, it is suggested that it may be more appropriate for the assistance to be in cash.

Like the case of EITE industries discussed above, this government assistance will be assessable income to the recipient, whether it is in cash or by way of permits (where the market value will be the amount of income derived). It is suggested in the Green Paper that the total amount of assistance to an owner may be determined up-front[78] but may be payable over time.[79] For tax purposes, this could raise the issue of whether the value of the assistance is derived when the right to a total amount of assistance is determined, where the taxpayer/owner is returning income on an accruals basis.[80] Another way of expressing this accruals basis is that income is derived when it is earned.[81] More details of the mechanism for assistance are be required for this issue to be fully considered.

From a capital gains tax perspective, the taxpayer/owner would acquire a CGT asset, being the right to assistance under the scheme, when that right is fixed. However, as the value in this right would generally be assessed as ordinary income, the CGT mechanism would have no residual effect.[82] The CGT rules may have operation if the taxpayer deals with the right, such as if the right were assigned to a third party.[83]

Price Cap and/or Penalty

Another feature of the scheme proposed by the Green Paper is the setting of a maximum cost of compliance, at least in the early years of the scheme, as a 'safety valve'.[84] Two mechanisms are proposed that could operate as this price cap. An administrative penalty (without a make-good provision) could operate to set the ceiling on the cost of compliance where, if permit prices exceeded the price cap, an operator would pay the penalty rather than purchase permits.[85] The other option canvassed would be for additional permits to be released to keep the price down to the specified level.[86]

The characterisation of the price cap as a 'penalty' in this discussion is misleading. Perhaps better referred to as the 'scheme fee', if this mechanism is designed to cap the cost of compliance, the scheme fee should be fully deductible in the same way as the cost of permits would otherwise have been (thereby achieving neutrality once permit price levels reach the cap level). To use the term 'penalty' is to confuse the price cap mechanism with a true penalty for failure to comply. The administrative penalty for non-compliance would serve a different policy purpose and would likely always remain a feature of the scheme, in contrast with the price cap that is designed to only be temporary. For policy reasons, penalties are not deductible outgoings[87] and this rule would apply to a penalty for non-compliance. It would therefore be important to determine whether both a price cap and a penalty be included in the scheme, where these would serve different purposes and should be treated differently for tax.

Foreign Owners/Traders

To encourage liquidity in the market, the Green Paper proposes that foreign entities be allowed to own permits, which could be acquired through the auctions or by way of trading on the secondary market.[88] The tax treatment of such trading by foreign entities should be considered and two issues arise initially. The first is that foreign residents are now subject to capital gains tax with respect to only those limited asset categories specified in the legislation.[89] These rules have recently been relaxed with respect to shares in companies so as to encourage foreign investment in shares.[90] As permits would not generally fit in one of the listed categories,[91] this exclusion from CGT would appear to be appropriate as it compares to shares.

Where a foreign resident is engaged in trading in permits, it is likely that any profits or losses will be considered foreign-sourced and therefore not subject to taxation in Australia. The source of profits derived from the sale of property, whether trading stock or not, is generally the place of contract.[92] As the contracts for sale of permits are most likely to be formed overseas, the source of the profits will be foreign and therefore not generally subject to Australian tax in the hands of non-resident traders.

Other Markets

As has been evidenced in other environmentally-based markets, the efficiency of the permit market will depend on forward price discovery through the development of forward markets for permits, such as through options and future contracts. It is expected that these markets will develop (they may even precede the secondary market) and that the volumes will be significant. The likely existence and tax treatment of such markets is not addressed in the Green Paper so it is presumed that current tax rules would apply to these instruments. The tax treatment may depend on whether the instruments are deliverable derivatives. The application of the proposed Taxation of Financial Arrangements regime should also be considered.

It is also suggested in the Green Paper that scheme obligations may, subject to time restrictions, be met through the surrender of *Kyoto* units being: emission reduction units (ERU), removal units (RMU) and certified emission reductions (CER).[93] It may therefore be appropriate to apply the rolling balance method to these units as well, if held for compliance purposes.

Conclusions

A firm's decision whether to engage in pollution abatement activities or to purchase scheme permits will depend on, in large part, not simply the cost of the alternatives but rather the after-tax cost, making the design of the taxation rules triggered by a scheme such as the CPRS critical. This paper has sought to identify and analyse some of the income tax issues that may arise from the CPRS as currently proposed. There are clear opportunities to establish an efficient tax regime up-front and it is encouraging that the Australian Government has identified the taxation issues as being of enough significance to devote an entire chapter of the Green Paper to them. As the details of the CPRS are refined, more developed tax rules will likely follow, providing the opportunity for more in-depth study.

Endnotes

1   Senator Penny Wong, Minister for Climate Change and Water, 'Government Announces Detailed Timetable on Emissions Trading' (press release PW 35/08, 17 March 2008). Retrieved July 15, 2008, from http://www.environment.gov.au/minister/wong/2008/pubs/mr20080317.pdf

2   Department of Climate Change, Carbon Pollution Reduction Scheme: Green Paper (July 2008) ('Green Paper') <http://www.climatechange.gov.au/emissionstrading/index.html> at 28 July 2008.

3   Ibid 395.

4   Ibid 396.

5   Ibid 38.

6   For example, see the proposed Taxation of Financial Arrangements regime contained in Tax Laws Amendment (Taxation of Financial Arrangements) Bill 2008.

7   Green Paper, above n 2, 419. The previous guidance from the IASB, IFRIC 3 'Emissions Rights', was withdrawn in June 2005 and work continues on the development of a new standard. International Accounting Standard Board, Project Report: Emission trading schemes (November 2008) <http://www.iasb.org/NR/rdonlyres/D0D0B44A-254A-4112-9FCE34178B236D07/0/0811 ProjectEmissionTrading.pdf > at 16 February 2009.

8   Green Paper, above n 2, 394.

9   Ibid 395.

10 Ibid.

11 Ibid. The example given is the treatment of depreciating assets where the cost of the asset is deductible over the asset's effective life. See *Income Tax Assessment Act 1997* (Cth) ('ITAA 1997') div 40.

12 Green Paper, above n 2, 395. Another recent example of the Federal Government's emphasis on the need for simplicity in the tax system can be seen in the objectives of the current review of the income tax system (the Henry Review) and the Review's first report, *Architecture of Australia's Tax and Transfer System*. Retrieved 10 September 2008 from http://taxreview.treasury.gov.au/content/ Content.aspx?doc=html/home.htm

13 Green Paper, above n 2, 149–50.

14 Ibid. The 'vintage' of a permit will determine the first year in which the permit may be surrendered.

15 Ibid 158. The term 'borrowing' is used to describe a situation where an entity with a compliance obligation is allowed to use a future-dated permit to meet the current liability.

16 Ibid 155. The term 'banking' is used to describe a situation where an entity holding permits can choose to carry them forward to meet future compliance obligations.

17 Ibid 398–9.

18 *ITAA* 1997 s 70-10.

19 *ITAA* 1997 s 108-5.

20 *ITAA* 1997 divs 100-152.

21 Green Paper, above n 2, 401.

22 Ibid 403.

23 The current trading stock rules provide this same result through the valuation of trading stock 'on hand' at the beginning and the end of the income year. See ITAA 1997 s 70-35.

24 Green Paper, above n 2, 401.

25 Ibid.

26 *ITAA* 1997 s 8-1(2)(b).

27 Green Paper, above n 2, 401.

28 See *ITAA* 1997 ss 118-24, 118-25. A permit issued under the scheme would meet the definition of 'CGT asset' for the purposes of the regime, ITAA 1997 s 108-5.

29 The TOFA regime was re-introduced to the Australian Parliament in December 2008 by way of Tax Laws Amendment (Taxation of Financial Arrangements) Bill 2008 (the 'TOFA Bill 2008').

30 See TOFA Bill 2008 s 230-20.

31 TOFA Bill 2008 s 230-45, where the term 'financial benefit' is defined by reference to s 974-160 as anything of economic value and includes property and services.

32 TOFA Bill 2008 s 250-45(2).

33 TOFA Bill 2008 s 250-45(2)(e).

34 See, eg, the exclusion for specified leasing arrangements and insurance policies at s 250-460.

35 Green Paper, above n 2, 404.

36 Ibid 410.

37 *ITAA* 1997 s 70-45. The replacement value option would not be relevant in the case of permits.

38 *ITAA* 1997 s 70-40.

39 One issue encountered in the use of market selling value in the trading stock context is the identification of the appropriate market and therefore market price (i.e., a wholesale versus retail market). See, eg, *Benwerrin Developments Pty Ltd v Federal Commissioner of Taxation* (1981) 12 ATR 335. In the case of permits, this should not be problematic as the permit market will be established under the scheme.

40 Green Paper, above n 2, 155, 158.

41 Ibid 150.

42 Commissioner of Taxation, Taxation Ruling TR 96/4 'Income Tax: Valuing shares acquired as revenue assets' and Commissioner of Taxation, Taxation Determination TD 33 'Capital Gains: How do you identify individual shares within a holding of identical shares?'.

43 Green Paper, above n 2, 212.

44 Ibid 207.

45 Ibid 214.

46 Ibid 264, 269.

47 Ibid 213.

48 Ibid 302.

49 Ibid 316. Although the total of free permits which would be allocated to EITE industries under the model is 30%, due to the proposed exclusion of agriculture from the scheme at least in the early year (and agriculture's relatively emissions levels) the total estimate of free permits would only be 20% (presumably the other 10% would have been allocated to agriculture had it been included).

50 The meaning of energy intensive activity on one that falls into one of two qualifying groups: those activities with emissions intensity above 2,000t $CO2$ –e/$ million revenue (given assistance at 90 per cent of industry average emissions per unit of output) and those with emissions intensity between 1,500 and 2,000t CO_2 –e/$ million revenue (60% assistance level). Green Paper, above n 2, 321.

51 Ibid 308–09.

52 Ibid 309.

53 Ibid 302.

54 Ibid 414.

55 *ITAA* 1997 s 6-5.

56 Commissioner of Taxation, Taxation Ruling TR 2006/3 'Income tax: government payments to industry to assist entities (including individuals) to continue, commence or cease business'.

57 Ibid [12].

58 *ITAA* 1997 s 15-10.

59 *ITAA* 1936 s 21.

60 *ITAA* 1936 s 21A. The Commissioner canvasses both options in the subsidy ruling: TR 2006/3, above n 56, [125]–[127].

61 Green Paper, above n 2, 414.

62 Ibid 413.

63 The Green Paper acknowledges that recipients of free permits will be allowed to offer those permits for auction along with new issue permits through the 'double sided' feature of the auctions.

64 Green Paper, above n 2, 150.

65 Ibid 412.

66 Ibid 273.

67 In the Green Paper, it is stated that the proceeds would be assessable but it is considered that this is not likely to be the intention. See Ibid 412.

68 Ibid 196.

69 Ibid 209.

70 Ibid 328–9.

71 Ibid.

72 Ibid 329–30.

73 Ibid ch 10.

74 Ibid 353.

75 Such as the mechanism to re-calculate the effective life of a depreciating asset after a change in circumstances, *ITAA* 1997 s 40-110.

76 Green Paper, above n 2, 377.

77 Ibid 385.

78 Ibid 389.

79 Ibid 389–90.

80 That is, where income is considered to be derived as it accrued rather than when it is received (cash basis).

81 *J Rowe & Son Pty Ltd v Federal Commissioner of Taxation* (1971) 124 CLR 421, 448 (Menzies J).

82 The *Tax Act* provides that a capital gain is reduced by the amount already included in assessable income, *ITAA* 1997 s 118-20.

83 This could give rise to a CGT event A1 with respect to the right, whereby any capital gain or loss would be calculated by comparing the relevant cost base to the capital proceeds: ITAA 1997 s 104-10.

84 Green Paper, above n 2, 165.

85 Ibid 165–66.

86 Ibid 165.

87 *ITAA* 1997 s 26-5.

88 Green Paper, above n 2, 151.

89 *ITAA* 1997 div 855.

90 The new rules apply as from 12 December 2006. Prior to this change, assets were taxable if they had the 'necessary connection' with Australia, which included, inter alia, shares in private companies and non-portfolio interests (10% or more interest) in public companies. See the former *ITAA* 1997 s 136-25 repealed by *Tax Laws Amendment (2006 Measures No 4) Act 2006* (Cth).

91 The only potential inclusion which could be triggered would be if the permit is used in carrying on a business through a permanent establishment. *ITAA* 1997 s 885-15.

92 *Commissioner of Taxation (WA) v D&W Murray Ltd* (1929) 42 CLR 332.

93 Green Paper, above n 2, 236–8.

PART IV
Litigation and Business Liability

Climate Governance and Corporations:
Changing the Way 'Business does Business'?

Susan Shearing

There is a common understanding that business leaders can no longer get away with rhetoric. Climate change affects their business. They will have to think through the consequences of climate change in every aspect of their operations and plan for the risks that will have a significant impact.

Sir Nicholas Stern[1]

[D]o boards talk about climate change? Most certainly they do … And in making decisions and spending money to deal with climate change, for example, I think there are still two clear questions that boards need to be able to answer and activities that they need to engage in. First, they need to ensure that the shareholders are aware … that the company is spending money on particular activities and why … And secondly, boards need to be able to answer the question that the money is being spent on a legitimate corporate purpose and that the money isn't being spent on the pet fad of the CEO or the chairman or whatever.

Ron McNeilly, Deputy Chairman, Bluescope Steel and Chairman
Worley Parsons Ltd[2]

In recent years, the demand for an enhanced regime of corporate environmental and social accountability has increased dramatically, reflecting intense scrutiny of corporations and their actions by a variety of stakeholders. In a number of jurisdictions, such calls have underpinned a broader review by policy-makers and governments of the fundamental nature of corporations, in particular, their role, the extent of legal responsibility for the environmental and social impacts of corporate activities, public disclosure obligations and the nature of duties owed by directors. However, as the impacts of global climate change become more widely understood and quantifiable,[3] the role of business and, in particular, corporations, in contributing to industrial greenhouse gas ('GHG') emissions is becoming a key focus of regulators and the corporate sector itself. In many jurisdictions, including Australia,[4] efforts are being made to comply with international commitments under the Kyoto Protocol to the *United Nations Framework Convention on Climate Change*[5] to reduce GHG emission levels through the development of a range of measures, including emissions trading schemes ('ETS') and other market mechanisms that will have significant effects on the way in which business 'does business'.

In addition to the direct regulation of heavy GHG emitting corporations, many corporations are, to varying degrees, finding themselves subject to demands from consumers, financiers, insurers, supply and production chains, potential investors and shareholders to minimise their GHG emissions and to consider the risks and opportunities posed by climate change in business planning and management. Viewed in this context, climate change initiatives may be seen as more than voluntary actions undertaken by a corporation and as part of its wider corporate social responsibility agenda to achieve a reputation for responsible corporate citizenship.

This chapter considers a number of mechanisms under Australian corporations law that may require corporate managers to consider climate change within the existing corporate legal regulatory framework. It will be argued that in some cases, directors of companies that fail to consider climate change risks and opportunities as an integral part of decision-making practices risk breaching their obligations as directors under corporations law. In other situations, a failure to recognise the impacts of climate change may carry other risks to business that warrant careful consideration by directors and a proactive approach to management.

Business Risks of Climate Change

Identifying the Risks

What are the key risks faced by business from climate change? Most discussion of this issue breaks the range of climate change related risks into four broad categories, namely, regulatory, physical, reputation and litigation ('climate risks').[6] For every corporation, the risk profile will vary depending on the business and economic environment within which it operates. Almost every sector is exposed to one or more of these risks and there are markedly differing levels of preparedness across sectors.[7] Even where a corporation may not currently face climate risks in Australia, it may be affected by developments in overseas jurisdictions in which it operates; or from which it is reliant on inputs from businesses affected directly by one of the risk factors, such as increases in prices of energy and raw materials.

Physical

The most direct threat is that posed by the physical impacts of climate change on the operations and assets of a corporation.[8] These impacts may result from severe weather related events such as increased incidence of severe storms and flooding or droughts and forest fires. Physical assets such as buildings and equipment may be directly affected by climate change impacts with numerous consequences, including implications for the availability and cost of insurance.

Clearly, the scope and nature of the physical risks posed will vary depending upon the sector and location within which a company operates. For some sectors, the physical impacts of climate change are already forcing companies to adapt — one example is tourism. Businesses that rely on cold temperatures and the availability of snow are looking to diversify the range of activities offered to tourists to maintain visitor numbers.[9] Other businesses need to consider how climate change impacts on the availability of resources for production of goods.

For many companies, understanding whether and how the physical impacts of climate change may impact upon decision-making and planning remains an area of uncertainty.

Regulatory

The second area of risk for companies is regulatory risk. What is meant by 'regulatory' in the climate change context? Regulatory risk generally focuses on law that regulates GHG emissions into the atmosphere. Such regulatory measures are being developed and implemented at regional, national and local levels.[10] Lyster notes that the range of 'climate law' that is emerging globally generally comprises legislation that regulates GHG emissions of those corporations that are the heaviest emitters (for example, through regulating emissions from the stationary energy sector and mandating reductions of emissions from transport and agricultural sectors), and requires corporations to adopt energy efficiency measures or mandates the achievement of renewable energy targets.[11]

Climate law regulatory risk is of primary concern to business.[12] This is not surprising. As a practical matter, corporations need to identify their responsibilities under applicable climate laws and adapt their operations to comply with the requirements of the regulatory environment created by governments. Where such measures have not yet been developed or are in the policy planning stage, there is a strong need for certainty and direction from regulators to enable long term planning and adaptation measures to be put in place. In this regard, there has been a convergence of business and environmental groups calling on governments for clarity in regulation (for example, the Australian Business Roundtable on Climate Change and the United States Climate Action Partnership).[13]

Reputation

Corporations face increasing pressure from a variety of sources to ensure that their green credentials are in order.[14] The approach adopted by a company to climate change can affect employee retention rates and the capacity of an organisation to attract quality staff.

For consumers, the issue of climate change is increasingly becoming a mainstream concern — the way in which a company responds to the climate change impacts of its operations can be a critical factor driving consumer choices.[15] This is particularly (although not exclusively) the case in those sectors that have high GHG emission levels.

Consumers are also concerned more generally with the climate change profile of businesses with which they deal. This presents significant opportunities for a business to differentiate itself from competitors. However, a company needs to ensure that it is not only seen to be responding to climate change risks but that claims made regarding carbon offsets by providers or purchasers of carbon offsets can be substantiated.[16]

On the production side, companies can face reputational risks through demands made within their supply chains for information as to carbon management strategies that they have developed and implemented to enable consumers to determine the total emissions in the supply chain for products.[17] The Carbon Disclosure Project has undertaken a project that aims at developing a consistent approach for reporting by suppliers of GHG emission reduction measures being undertaken along product supply chains.[18]

Litigation

In a number of jurisdictions, climate change litigation has emerged as a significant risk for corporations. Lyster has comprehensively outlined the trend toward climate change litigation in the United States and Australia.[19] Clearly, companies that are subject to climate laws that mandate reductions in GHG emissions need to ensure that they comply with applicable regulatory requirements. However, in Australia and the United States, avenues for climate change litigation have focused largely on public law actions challenging the decisions of public agencies on the basis of a failure to consider climate change in relation to a variety of actions (such as the granting of a licence or development approval to a corporation). In such cases, the outcome of such litigation can have significant consequences for the operations of corporations whose activities are affected by the decisions of such agencies.[20]

The other key area of litigation risk for corporations is in the area of tort law — the possibility of an action for negligence or creating a nuisance. In relation to negligence, there is the issue of establishing that a duty of care exists with respect to climate change damage due to the requirement to establish that the harm was reasonably foreseeable. There is also the need to establish that the negligence caused or materially contributed to the climate change damage.[21]

Notwithstanding the growth in climate change litigation over the past 10–15 years, it is not clear that climate litigation risk is currently viewed as a key concern for businesses.[22] This is probably due to a number of factors that will determine whether a corporation is likely to be the subject of scrutiny and potential litigation, such as the scope of the regulatory framework, and the sector and jurisdiction in which a company operates. However, as the climate related impacts of actions taken by corporations become more widely understood, this area of climate risk is likely to become an increasingly prominent concern for the corporate sector.

Understanding the Risks

Internationally, directors and chief executive officers of companies are beginning to regard climate change as a key business issue. A survey of 390 chief executive officers whose companies have endorsed the United Nations Global Compact (2000) (UNGC)[23] found that:

- 61% believed that companies should 'have the board, as part of its risk management and fiduciary responsibilities, discuss and act on' environmental and social governance issues; and
- 38% identified responding to climate change as 'critical' to addressing the future success of their business.[24]

The most recent report of the Carbon Disclosure Project, 'CDP5', showed that 79% of the FT500 companies that responded to the CDP survey consider that climate change poses a 'commercial risk', while 82% consider that it presents commercial opportunities for new and existing products.[25]

To what extent do Australian business leaders regard climate change as a risk to business? A recent survey[26] of the chief executive officers or chief financial officers of 303 Australian corporations across a broad range of sectors revealed the following:

- the majority (53%) are unsure whether climate change poses any risk to their businesses;
- 8% were certain that climate change poses no such risk, 15% believe it will be a risk in 12 months, and 29% believe it will be a risk in 2012;
- the majority of respondents have only taken minimal action to respond to climate change risks — 78% have not taken any action and 98% are yet to implement a strategic response;
- The resources sector was an exception: all leaders surveyed consider that climate change represents a risk to business — a result that reflects the higher degree of exposure of this sector to regulatory risk and the fluctuations of international markets.

Uncertainty within the business sector generally is understandable, given the range of projections available to business concerning short, medium and long term climate change impacts. The survey indicated that leaders of the companies interviewed did not understand their company's climate change reporting obligations, and had a poor level of confidence in their GHG emissions data, with most respondents either having no data or being unsure of the data they do have. In addition to the regulatory compliance implications of such findings, there are also consequences for the ability of such corporations to recognise the climate change opportunities and risks.

Australian Corporations Law and Climate Change

The various climate risks discussed above may be faced by all forms of business structure, including corporations, sole traders, partnerships, trusts and joint ventures. However, the legal regulation of corporations differs from that of other unincorporated business structures in that specific duties are imposed upon those individuals who manage the business of companies. The control and management of a company, its property and operations is usually vested by the constitution of the company in its board of directors.[27] To ensure that this power is exercised appropriately having regard to the interests of shareholders, the directors of a company are subject to a range of duties under both general law and statute. Corporations are also subject to statutory disclosure obligations that do not apply more broadly to other forms of business entity.

What is the impact of climate risks and opportunities on directors' duties? At the outset, it is clear that these will vary significantly across sectors. In the Australian context, the companies that are most 'at risk' in terms of regulation under the proposed ETS and other climate laws are those in the stationary energy, transport, fugitive emissions, industrial processes, waste and forestry sectors.[28] For the majority of companies, the issues of whether a director is under a duty to address physical or regulatory climate risks or whether the company should publicly disclose such risks are unlikely to arise. However even for such companies, issues of reputation may be relevant — whether consumer or supply/production chain driven.

Looking beyond climate laws, what impact might corporate law have in shaping the response of directors to climate risks and opportunities? Most discussion of whether (and to what extent) corporate managers may consider the environmental impacts of a company's activities falls squarely within the corporate social responsibility debate (see discussion below). Indeed, directors' concerns are often cast in terms of whether it is lawful to consider such issues[29] rather than the possibility of directors being subject to an existing duty to consider them.

In this paper it is argued that corporate law, and specifically the law in relation to directors' duties, may require managers to consider, identify and address climate change risks. Viewed in this way, corporate law is not a vehicle or 'device' for securing environmental regulatory goals. Rather, climate change provides an impetus for reviewing how corporations are governed by potentially extending the standards by which directors' conduct is measured.

Directors' Duties

The fiduciary relationship between a director and a company gives rise to a number of positive duties, namely, a duty of loyalty and a duty of care to the company. The common law also recognises a duty of care that arises from the proximity between a company and its directors.[30]

In addition to general law duties, the *Corporations Act 2001* (Cth) imposes a number of statutory duties on directors, the most relevant being the duty to act with care and diligence (s 180(1)) and the duty to act in good faith (s 181(1)).

Climate Change and the Duty to Act With Care and Diligence

Section 180(1) is extremely broad in scope and is one of the most frequently prosecuted provisions dealing with directors' duties.[31] At common law, this duty is known as the duty to act with 'care, skill and diligence'. The duties owed under s 180(1) are essentially the same as the duties of a director under common law.[32]

Directors or other officers must exercise their powers and discharge their duties with the degree of care and diligence that a reasonable person would exercise if he or she were a director (or officer) of a company in the company's circumstances and occupied the office held by, and had the same responsibilities within the company as the director or officer.

The leading cases in this area establish that all company directors are required to take the necessary steps to enable them to 'effectively' guide and monitor the management of the company. This requires a director to:

- become familiar with the fundamentals of the company business;
- keep informed about the company's activities on a continuing basis;
- regularly attend board meetings so as to partake in directorial management (requiring a general monitoring of company affairs and policies); and
- maintain familiarity with the financial status of the company through a regular review of financial statements.[33]

There is no standard of skill referred to under s 180(1). The standard applied is objective, requiring the court to ask:

> what an ordinary person, with the knowledge and experience of the defendant might be expected to have done in the circumstances if he or she was acting on their own behalf.[34]

This means that the court looks to the circumstances of the company (its nature, size and complexity), the particular officer's position and responsibilities in the company. The 'reasonable person' requirement in ascertaining the standard of care has been construed as similar to that imposed under the law of negligence.[35]

In determining what a reasonable person would do in responding to a risk, it is relevant to consider:

> the magnitude of risk and the degree of probability of its occurrence, along with the expense, difficulty and inconvenience of taking alleviating action and any other conflicting responsibilities the defendant may have.[36]

In considering the magnitude of the risk, the court must consider the gravity, frequency and imminence of the danger to be guarded against.[37] The statutory standard requires an inquiry into what degree of care and diligence a reasonable person 'would exercise', not what such a person might do.[38]

Does the duty to act with care and diligence require a director to consider and address climate risks faced by a corporation? As far as the author is aware, there is no case law in Australia or elsewhere that has imposed an obligation to consider climate risks and opportunities as part of the common law or statutory duties of directors.

However, given the increasing profile of climate change issues in the business community, and in particular the proposed introduction of the Carbon Pollution Reduction Scheme (CPRS) to be implemented from 1 July 2011 by the Australian Government,[39] directors who fail to identify and address climate risks may be vulnerable to assertions that they have breached their s 180(1) duty.

For example, a breach of duty may occur where a director fails to prevent a breach by the company of a climate law that regulates GHG emissions by the corporation.[40] In addition to the impact of any pecuniary penalty imposed, the adverse effects on a company's reputation arising from breaching environmental laws can be significant.

Directors of corporations operating in carbon intensive industries will need to carefully assess their exposure to regulatory risk under the CPRS. Such corporations will be required (among other responsibilities) to ensure that appropriate systems are in place to monitor GHG emission levels and prepare an annual report of their annual emissions.[41] Companies will also be required to maintain adequate records to enable assurance of those reports. In the case of large emitters (those with obligations under the scheme of 125 000 tonnes of carbon dioxide equivalent or more), the Australian

Government's White Paper proposes that assurance will be carried out by an independent accredited third party.[42]

It is noted that the *Carbon Pollution Reduction Scheme Bill 2009 (Cth)* provides that where a corporation contravenes a civil penalty provision and an executive officer knew that (or was reckless or negligent as to whether) the contravention would occur, the officer will be subject to a civil penalty if he or she was in a position to influence the conduct of the body corporate, in relation to the contravention, but failed to take all reasonable steps to prevent it.[43] Civil penalty provisions under the Bill include obligations relating to the provision of information required by the scheme regulator and record-keeping requirements. In considering whether an executive officer failed to take all reasonable steps, the court may have regard to all relevant matters matters including what action (if any) the officer took directed towards ensuring that the body corporate arranges regular professional assessments of its compliance with civil penalty provisions.[44]

Further, it might be argued that in some circumstances, the duty to act with reasonable care and diligence will require directors to consider the climate change implications of their decisions independently of any climate specific regulatory requirements. This may be particularly relevant with regard to companies that are exposed to physical climate risks, including companies that:

- are required to make long term capital investment decisions (such as infrastructure and equipment);
- operate globally and rely on resource supply from countries/areas that are subject to adverse climate impacts;
- operate in sectors in which climate is integral to production/function (such as agriculture, tourism and construction);
- rely on infrastructure (such as roads) in their supply chains; or
- deal with reflected risks (such as insurance and banking).[45]

As noted above, the duty to act with care and diligence requires a court to consider whether a director has exercised their powers and discharged their duties with the degree of care and diligence that a reasonable person would exercise if he or she were a director (or officer) of a company in the company's circumstances and occupied the office held by, and had the same responsibilities within the company as, the director or officer. Depending on the sector within which a company operates, the scope of its activities and the position and responsibilities of the director concerned, a director who fails to identify and make recommendations as to how a company might adapt to climate risks may be vulnerable to assertions of a failure to have acted with the care and diligence that a 'reasonable' person would exercise.

Internationally, such concerns have led to suggestions that directors who fail to consider climate change issues may find themselves liable for breach of duty of care as the standards by which their conduct is judged shift to encompass obligations to inquire, to be adequately informed and to employ adequate internal monitoring systems.[46] In the United States, for example, it has been suggested that failure by a board to undertake a thorough enquiry as to the climate change implications of the board's decision may enable shareholders to attack, alleging that the impending business risks and opportunities (for example, through evolving product markets) presented by climate change would have been evident to the board had it adopted a thorough approach to researching and evaluating information relevant to this issue.[47]

In a 2002 US case *Salsitz v Nasser*,[48] shareholders alleged that director oversight had damaged the financial results of the Ford Motor Company and possibly contributed to deaths and injuries globally. The case was based upon alleged mismanagement of an internal investigation of technical matters concerning defective ignition switches and an alleged 'sustained or systematic failure of the board to exercise oversight', through management reliance on under-qualified employees, to implement a complex metals trading strategy leading to significant write-offs. The shareholders were unsuccessful, the court relying upon the strong presumption afforded by traditional protections for directors in carrying out their responsibilities as managers of business. However, some commentators have suggested that it is not inconceivable that directors might be faced with similar challenges in relation to the oversight of climate change decision-making.[49]

Indeed, one US attorney has noted that an area of potential lawsuit that companies need to consider may be based on shareholder claims that senior executives have been guilty of gross neglect of their fiduciary duties in relation to climate change matters.[50]

The Business Judgment Rule

At common law, courts have been reluctant to intervene to review business decisions made by a board of directors:

> There is no appeal on merits from management decisions to courts of law: nor will courts of law assume to act as a kind of supervisory board over decisions within the powers of management honestly arrived at.[51]

> However, where the decision of a board is one that no reasonable board could consider to be in the interests of the company, or one that no reasonable board could think to be substantially for a purpose for which a power was granted, a decision might be found to be in breach of the common law duty to act in good faith and in the interests of the company or to make a decision for a proper purpose.

Under s 180(2), directors and other officers of a corporation who make a business judgment are taken to have met their duty of care and diligence under s 180(1) and their equivalent common law and equitable duties if they:

- make the business judgment in good faith and for a proper purpose; and
- do not have a material personal interest in the subject matter of the judgment; and
- inform themselves about the subject of the judgment 'to the extent they reasonably believe to be appropriate'; and
- rationally believe that the judgment is in the best interests of the corporation.

A business judgment is defined as 'any decision to take or not take action in respect of a matter relevant to the business operations of the corporation'.[52] A belief is rational 'unless the belief is one that no reasonable person in their position would hold'.[53]

In relation to compliance with climate laws, the s 180(2) defence may not be available where directors have failed to ensure that adequate emissions monitoring and recording mechanisms are in place to enable necessary reporting and assurance requirements to be met.

However, looking beyond regulatory risk, it may be difficult for directors to rely on s 180(2) if it can be shown that there has been a failure to take steps to ascertain and incorporate data on climate risk into management decisions. A key requirement for the application of the business judgment rule is that directors or officers have informed themselves about the subject matter of the judgment to the extent they reasonably consider being appropriate. For example, where the board is required to make long term capital investment decisions (such as infrastructure and equipment) and climate risks are not identified and addressed, it may be argued that the directors do not hold a 'reasonable' belief as to the extent to which it is appropriate that they inform themselves in relation to that business judgment.

Nonetheless, it is important to recall that there remains a clear reluctance on the part of courts to intervene in relation to directors' decisions. In the recent case of *Ingot Capital Investments v Macquarie Equity Markets* (*No. 6*),[54] McDougall J noted that the mere foreseeability of harm does not of itself dictate that directors must be found to have failed to exercise due care and diligence for the purposes of s 180(1). Rather:

> [i]t is necessary to balance risk and reward, or, more accurately, to be satisfied that the directors, acting reasonably and in the best interests of [the company] and employing their individual knowledge and skills and taking account of relevant circumstances, did so.[55]

This may be so where directors have considered climate risks and chosen to act in a particular way, taking the circumstances of the company into account. However, the position may differ where there has been a complete failure to consider the climate risks attendant on management decisions.

Climate Change and the Duty to Act in Good Faith

Section 181(1) of the *Corporations Act* requires directors and other officers to exercise and discharge their duties in good faith in the best interests of the corporation. This duty also arises in general law. The test as to whether directors have acted in the best interests of the company is objective.[56]

A key issue here is the meaning of the phrase 'interests of the company'. The 'company as a whole', does not mean 'the company as a commercial entity, distinct from the corporators as a general body'.[57] Nor does it refer to individual shareholders although, in some circumstances, a duty may be found to exist between an individual shareholder and a director if there is a relationship of confidence or reliance on information or advice provided by the director.[58]

While directors are required to consider the interests of the existing shareholders, to what extent does this incorporate the long term interests of shareholders? Arguments supporting the consideration of long term corporate viability have been made in the context of compliance with environmental laws generally[59] and board responses to hostile takeover bids.[60]

However, the 'short term versus long term interests' debate is pivotal in relation to climate governance. Directors may be concerned that decisions that focus on long term planning to deal with climate change risks and opportunities at the expense of short term gains may leave them exposed to shareholder action based on a failure to act in the best interests of the corporation.

There has been much analysis of the ability of directors to consider stakeholders other than shareholders. Such analysis proceeds on the basis that directors may be subject to allegations of breach of duty if such interests are considered at the expense of maximising profits (the traditional 'shareholder primacy' view). These arguments are particularly relevant in the context of corporate philanthropy cases where donations are made for purely altruistic purposes without shareholder approval and cannot be justified as being made for the benefit of the company.[61]

However, if directors can demonstrate that in acting to identify and deal with climate risks they are motivated to promote the benefit of the company and that a reasonable director would think that the action is in the interests of the company, the action should be valid.

In some cases, a failure to consider climate risks and opportunities may expose directors to shareholder claims that they have breached their duty to act in the best interests of the company. Examples may include permitting the company to breach climate laws or failing to take action to reduce the GHG emissions of the company and avoid ensuing damage to reputation.

Further, a failure by directors to consider climate change risks may constitute a breach of duty under s 181(1) and corresponding common law and equitable duties if a corporation faces material climate related risks and potential consequential losses that affect the value of shareholdings. For example, action by directors is particularly critical where there are clear and immediate physical threats based on current climate conditions; for example, where decisions are being made to situate buildings in areas susceptible to sea level rises. As the availability of information enabling the assessment and measurement of climate risks grows, the scope for arguing that directors need to consider and mitigate such risks increases commensurately.

Further, the imposition of a carbon price will have a significant impact on the cost of emissions intensive goods and services and will require a careful review by directors of production methods and inputs, supply chain requirements and consumer demand to ensure the long term viability of the company.

In the United Kingdom, the government has recently introduced a shadow carbon pricing policy that is to be incorporated into public sector policy and investment appraisals in the same way as any other cost–benefit analysis to support decision-making. This approach is to apply when choosing among alternative carbon abatement measures and 'to all other option appraisals which have a significant carbon impact'.[62] UK companies are already adopting a shadow carbon price when making infrastructure decisions following the introduction of this policy.[63]

Increasingly, the strategic investment decisions made by Australian corporations are being affected by climate change. In the areas of mergers and acquisitions of companies and major asset sales, the viability of transactions are already being measured by reference to financial models that include a carbon price.[64] In the short term, uncertainty concerning the timing and effect of the proposed CPRS may result in debt refinancing risks for corporations operating in heavily exposed sectors such as power.[65]

Climate change poses important issues for corporate managers in protecting the long term viability of the company. A failure to identify and consider climate risks may jeopardise the company's short term and long term commercial interests and leave directors vulnerable to shareholder action.

Shareholder Action and Climate Change

What action can be taken by shareholders concerning climate risks and opportunities? Directors' duties are owed to the company and can only be enforced by the company or the regulator, the Australian Securities and Investments Commission. If the directors are unwilling or unable to bring an action based on an alleged breach of director's duty, shareholders would need to rely on the statutory derivative action provided under ss 236 and 237 of the *Corporations Act*. These provisions enable shareholders to institute action in the name of the company with the leave of the court where the court is satisfied as to the matters listed in s 237. This section requires the court to be satisfied that the company, is unlikely to bring the action, that the applicant is acting in good faith, that it is in the best interests of the company that the applicant be granted leave, that there is a serious question to be tried and that notice of the application has been given to the company, or the Court is of the view that it is appropriate that leave be granted notwithstanding that notice has not been given.[66] Action might also be taken under these provisions if the constitution of a company requires directors to consider climate and other environmental risks and there is a failure to do so.

Turning from the issue of directors' duties, minority shareholders may act to ensure that climate change is recognised and addressed by those managing the company. Under s 249D of the *Corporations Act*, shareholders may require the board of directors to call an extraordinary general meeting at which resolutions may be proposed for the consideration and vote of members.[67] This obligation arises only where the request is made by members with at least 5% of the votes that may be cast at the general meeting or at least 100 members entitled to vote at the meeting.

While this process can be an effective means of influencing the environmental policies of some large public companies, such resolutions are rarely passed.[68] Nonetheless, boards are likely to face increasing shareholder pressure to address climate change issues as part of an effective framework of corporate management. For example, shareholders might propose low carbon resolutions that include calls for specific goals to be set limiting GHG emissions and requiring the company to engage in renewable energy research.[69] In the United States, the largest single category of shareholder resolutions filed in 2008 was climate change proposals.[70]

Corporate Social Responsibility and Climate Change

It has been argued that in some cases, directors may be required to consider and address climate risks as part of their duties under corporations law. However, the growing recognition that business needs to identify and deal with climate risks and opportunities can also be viewed within the context of a broader debate concerning the corporate social responsibility[71] of corporations. That is, whether, and if so how, directors should have a duty to consider interests beyond those of members, or whether they should be required to 'have regard' to such other interests (including environmental and social) in carrying out their duty to 'the company'.

The 'enlightened self-interest' or 'business approach' to corporate social responsibility is that 'beyond the obligations of companies to comply with environmental and other societal laws … it is likely to be in the company's own commercial interests, in terms of long term value creation and risk reduction, to take into account the environmental and social context in which it operates.'[72] Thus, this approach incorporates both corporate risk management and enhancing corporate values and opportunities - issues of key importance in the climate change debate.[73]

Options for extending those matters to which directors may have regard in carrying out their duties to the company are numerous. For example, the constitution of a company might require directors to consider particular interests in managing the company. Given the contractual status of corporate constitutions between members and officers under the *Corporations Act*,[74] directors must comply with the terms of such provisions.

Some commentators have suggested self regulation through the adoption of a replaceable rule in the *Corporations Act* that in the absence of any alternative provision in the constitution, directors and other officers may 'have regard to' a set of objectives extending beyond short term profitability.[75] Alternatively, legislation in other areas of the law may prescribe factors that directors are required to consider. Again, a failure to comply with such laws will result in a director acting in breach of their duty where the company is found liable.

However, in some jurisdictions, directors' duties are being recast to impose an obligation to consider a range of external stakeholders in managing the company, including the environment. For example, recent amendments to the *Companies Act 2006* (UK) impose a new duty to 'promote the success' of a company in place of the general law duty of good faith in the best interests of the company. Specifically, s 172 of the *Companies Act 2006* provides that:

(1) A director of a company must act in the way he considers, in good faith, would be most likely to promote the success of the company for the benefit of its members as a whole, and in doing so have regard (amongst other matters) to-

(a) the likely consequences of any decision in the long term,

(b) the interests of the company's employees,

(c) the need to foster the company's business relationships with suppliers, customers and others,

(d) the impact of the company's operations on the community and the environment,

(e) the desirability of the company maintaining a reputation for high standards of business conduct, and

(f) the need to act fairly as between members of the company.[76]

This duty embodies the so-called 'enlightened shareholder value' principle that views the purpose of a company as being to create value for the benefit of its shareholders by taking a long term view — including having regard to those matters that might impact on long term outcomes. Under s 172, the requirement to 'have regard to' the non-exhaustive list of factors must be viewed in the context of the primary obligation to act in a way that is 'most likely to promote the success of the company for the benefit of its members as a whole'.[77]

The climate change impacts of a company's activities may well fall within the scope of s 172(1)(d) depending on the sector within which the company operates. However, the climate risks and opportunities discussed in the first part of this chapter may also warrant the attention of directors in having regard to the matters listed in s 172(1)(a)–(c) and (e).

Section 172 may be criticised on a number of bases. For example, there is no guidance as to how directors are expected to balance the different factors they are required to consider or what weight they should attach to them. There is also a lack of clarity as to the meaning of 'success' for the benefit of the company's members as a whole. Currently, the law requires directors to act in good faith in the best interests of the company. According to the Explanatory Notes to the Act, determining 'success' is a matter for the directors' good faith and judgment.[78] Further, in having regard to the factors listed in s 172, the duty to exercise reasonable care, skill and diligence[79] is to apply:

> It will not be sufficient to pay lip service to the factors, and, in many cases the directors will need to take action to comply with this aspect of the duty.[80]

This suggests that, provided that directors make a decision in good faith, and have exercised reasonable care, skill and diligence in reaching that decision, it should not be open to challenge in the courts. This position accords with the courts' traditional reluctance to overturn directors' commercial decisions provided they have been made in good faith.

The UK is not the only jurisdiction that has undertaken a sweeping review of directors' duties. In Hong Kong a consultation paper has recently been released dealing with proposals for the proposed rewrite of the *Companies Ordinance (Cap 32)* (HK). Among the issues referred to in this paper is the introduction of a statement of directors' duties similar to that enacted in s 172 of the *Companies Act 2006* (UK).[81]

In 2006 the Australian Corporations and Markets Advisory Committee (CAMAC) reviewed Australia's corporations legislation to consider the question of how corporate social responsibility should be regulated and in particular, whether the *Corporations Act 2001* should be revised:

- To clarify the extent to which directors may take into account the interests of specific classes of stakeholders or the broader community when making corporate decisions.

- To require directors to take into account the interests of specific classes of stakeholders or the broader community when making corporate decisions.[82]

CAMAC concluded that it was not necessary to revise the *Corporations Act 2001* in this regard. The committee's view was that environmental and social matters raised in the debate concerning corporate social responsibility are matters that 'directors should already be taking into account in determining what is in the best interests of the corporation in its particular circumstances'.[83] Further, the committee concluded that the existing common law and statutory duties of directors and officers under ss 180 and 181 of the *Corporations Act 2001* are:

> sufficiently broad to enable corporate decision-makers to take into account the environmental and other social impacts of their decisions, including changes in societal expectations about the role of companies and how they should conduct their affairs. [84]

In this regard, a recent survey of Australian company directors provides some interesting insights into the understanding of directors themselves concerning their legal obligations. In particular, the survey considered the extent to which shareholders were perceived to be the most important among a range of key stakeholders comprising shareholders, the company, employees, customers, suppliers, lenders/creditors, the community, the environment and the country.[85] The survey covered both public (listed and unlisted) and proprietary companies of varying sizes and across a mix of industries.[86] Significantly, it was found that:

- A majority of directors (55%) understood that their obligation to act in the 'best interests of the company' meant that they are to balance the interests of all stakeholders.

- A significant minority (38.2%) believed that that in order to act in the interests of all stakeholders, they must ensure the long term interests of shareholders.

- No directors believed that they are required to act only in the short term interests of shareholders.

- In examining the directors' understanding of their responsibilities to the different stakeholders, priority rankings among stakeholders indicated that shareholders were most commonly ranked number one (78.2%) followed by the company (71.1%).

- A significant majority of directors (94.3%) believe that the current law is broad enough to allow them to take into account the interests of stakeholders other than shareholders.[87]

These findings support the CAMAC report's conclusions concerning the scope of existing directors' duties under Australian law and, in particular, the ability of directors to adopt an expansive approach to discharging their duty to act in the best interests of the company.

A Proactive Approach to Climate Governance

In the current climate of enhanced shareholder scrutiny of company managers, good corporate governance suggests that directors would be prudent to adopt a proactive approach in considering the nature of climate risks affecting their activities and the most effective ways of addressing them. Appropriate steps include undertaking necessary inquiries as to the climate impacts of a decision and implementing adequate risk management and monitoring systems as part of the adaptation process to minimise adverse business consequences.

Referring to the physical risks of climate change, the Pew Centre on Global Climate Change has suggested that businesses need to act promptly to identify and adapt to such risks by implementing a climate risk screening framework under which companies ask the following questions:

- Is climate change important to business?
- What is the nature of the threat? Is it an immediate risk (for example, a physical risk posed by changing climactic conditions) or are long term assets, investments or decisions being locked into place?
- Is a high value at stake if a wrong decision is made?[88]

There is no doubt that a number of companies are voluntarily adopting a proactive approach in dealing with climate change impacts upon business.[89]

However, a key step is the integration of climate change into the decision-making process at board level. While in some areas, such as the resources sector, climate change risks have already been elevated to board level, awareness and action at senior management levels in other sectors of the Australian business community has been extremely limited to date. This is consistent with international trends.[90]

Corporate governance initiatives that may be implemented to effectively enhance the integration of governance and climate change include:

- ensuring that climate change is a governance priority for board members and chief executive officers;
- linking executive remuneration to GHG emission reduction targets and the implementation of other climate change related measures;
- establishing corporate energy efficiency targets and mandating energy efficiency evaluations for major capital investments.[91]

Conclusion

Corporations are operating in a business environment that is characterised by increasing climate regulation, stakeholder demands for climate change action and accountability, and climate law litigation. At the same time, there is growing recognition of the potential business opportunities provided by climate change in relation to adaptive measures. In such an environment, corporate managers cannot afford to ignore or downplay the significance of climate related risks and opportunities. Indeed, it may be argued that climate change provides an impetus for reviewing how corporations are governed by potentially extending the standards by which directors' conduct is measured. As such, it calls for a response on the part of corporate managers that fundamentally changes the ways in which business 'does business'.

Endnotes

1   KPMG International, *Climate Changes Your Business* (2008) 17 <http://www.kpmg.com/Global/IssuesAndInsights/ArticlesAndPublications/Pages/Climatechangesyourbusiness.aspx> at 4 August 2008.

2   Ron McNeilly, Climate Change: A new challenge for directors? (Panel discussion at ASIC Summer School, Melbourne, 18 February 2008) <http://www.watchdog.asic.gov.au/asic/pdflib.nsf/

LookupByFileName/Corporate_governance_and_directors_duties.pdf/$file/Corporate_governance_and_directors_duties.pdf> at 15 December 2008.

3   Sir Nicholas Stern, *Stern Review on the Economics of Climate Change* (2006); Ross Garnaut, *Garnaut Climate Change Review — Final report* (2008); Intergovernmental Panel on Climate Change, *IPCC Fourth Assessment Report (AR4)* (2007).

4   In December 2008 the Australian Government outlined its approach to the design of a carbon pollution reduction scheme. Commonwealth, *Carbon Pollution Reduction Scheme: Australia's low pollution future*, White Paper (2008) ('White Paper') <http://www.climatechange.gov.au/whitepaper/report/index.html> at 17 December 2008.

5   *United Nations Framework Convention on Climate Change*, opened for signature 4 June 1992, 1771 UNTS 107 (entered into force 21 March 1994) ('*UNFCCC*'); *Kyoto Protocol to the United Nations Framework Convention on Climate Change*, opened for signature 16 March 1998, 2303 UNTS 148 (entered into force 16 February 2005) ('*Kyoto Protocol*').

6   Above n 1. This report analysed 50 reports from numerous industry sectors addressing the business risks and economic impacts of climate change across sectors. Half of the reports reviewed raised the physical risks of climate change as a matter of concern.

7   Ibid 11–12.

8   Ibid 36.

9   Frances G Sussman and J Randall Freed, *Adapting to Climate Change: A business approach* (Pew Center on Global Climate Change: 2008).

10   The International Energy Agency ('IEA') online database of IEA member countries' climate change policies and measures (the database includes information on policies in Brazil, China, the European Union, India, Mexico, Russia and South Africa): <http://www.iea.org/textbase/pm/index_clim.html> at 4 August 2008.

11   Rosemary Lyster, 'Chasing Down the Climate Change Footprint of the Public and Private Sectors: Forces converge' (2007) 24 *Environmental and Planning Law Journal* 281.

12   Above n 1, 7. Regulatory risk was the most commonly identified risk for businesses in the reports surveyed in the KPMG Report (raised by 72% of respondents).

13   See Australian Business Roundtable on Climate Change <http://www.businessroundtable.com.au/> and the United States Climate Action Partnership <http://www.us-cap.org/>.

14   The KPMG Report noted that 28% of reports raised reputation as an issue of concern: above n 1, 29.

15   PriceWaterhouse Coopers, *Carbon Countdown — A survey of executive opinion on climate change in the countdown to a carbon economy* (2008).

16   See Australian Competition and Consumer Commission, *Carbon Claims and the Trade Practices Act* (2008) <http://www.accc.gov.au/content/index.phtml/itemId/833279> at 4 August 2008.

17   See Carbon Disclosure Project, *Supply Chain Leadership Collaboration (SCLC) Pilot Results and Findings Report* (2008) <http://www.cdproject.net/sc_documents.asp> at 4 August 2008.

18   Ibid.

19   Lyster above n 11, 301–9.

20   Ibid.

21   Ibid.

22   The KPMG Report notes that only 14% of the reports analysed dealt with litigation as a climate related concern: above n 1, 29.

23   Under the UNGC, companies voluntarily commit themselves to 10 principles to guide their conduct in areas including human rights, labour standards and practices, the environment and anti-corruption. UNGC, *Overview of the UN Global Compact* <http://www.unglobalcompact.org/AboutTheGC/index.html> at 31 August 2009.

24   Debby Bielak, Sheila MJ Bonini and Jeremy M Oppenheim, 'CEOs on Strategy and Social Issues' (2007) (4) *The McKinsey Quarterly* 8.

25   Carbon Disclosure Project, *Carbon Disclosure Project Report 2007 Global FT500* (2007) 8 <http://www.cdproject.net/historic-reports.asp> at 31 August 2009. It is noted that at the World Economic Forum in January 2007, climate change was chosen as the shift most likely to affect the world in the future. World Economic Forum, *The Shifting Power Equation: Exploring the implications* <http://www.weforum.org/en/knowledge/Themes/Enviroment/ClimateChange/KN_SESS_

SUMM_20026?url=/en/knowledge/Themes/Enviroment/ClimateChange/KN_SESS_SUMM_2002
6> at 4 August 2008.

26 PriceWaterhouse Coopers, above n 15. The survey covered corporations with an annual turnover in excess of AU$150M.

27 *Corporations Act 2001* (Cth) s 198.

28 White Paper, above n 4, ch 6 'Coverage'.

29 ASIC Summer School 2008, above n 2.

30 RP Austin, HAJ Ford and IR Ramsay, *Company Directors: Principles of law and corporate governance* (2005) 211.

31 Examples of breaches of the s 180(1) duty are numerous and include the following: failure to take part in the active supervision of the management of the company (*Daniels v Anderson* (1995) 37 NSWLR 438); failure to supervise the accuracy of the company's financial accounts (*Sheahan v Verco* (2001) 79 SASR 109); allowing a company to trade in an unreasonably risky manner, contrary to industry practice (*Circle Petroleum (Qld) Pty Ltd v Greenslade* (1998) 16 ACLC 1577); failure to observe management, to ascertain the financial position of the company and to inform the board of developments that might adversely affect the company (*ASIC v Rich* [2003] NSWSC 85; *Re One.Tel Ltd (in liq)* [2003] NSWSC 186).

32 *ASIC v Adler* [2002] NSWSC 171 [372].

33 *Daniels v Anderson* (1995) 37 NSWLR 438; *ASIC v Adler* [2002] NSWSC 171.

34 *ASIC v Adler* [2002] NSWSC 171 [372].

35 *Vines v ASIC* [2007] NSWCA 75.

36 *Wyong Shire Council v Shirt* (1980) 146 CLR 40, 47–8 (Mason J); *ASIC v Vines* [2005] NSWSC 738 [1071] (Austin J).

37 *Mercer v Commissioner for Road Transport and Tramways (NSW)* (1936) 56 CLR 580 (Dixon J).

38 *ASIC v Vines* [2005] NSWSC 738 [1075] (Austin J).

39 Penny Wong, Minister for Climate Change and Water, 'Carbon Pollution Reduction Scheme: Support in managing the impact of the global recession' (Media Release, 4 May 2009) <http://www.environment.gov.au/minister/wong/2009/mr20090504a.html> at 15 May 2009.

40 The extent to which directors fail to implement appropriate controls to prevent breaches of law have been found to be relevant to the s 180 duty in a number of contexts: see *ASIC v Adler* [2002] NSWSC 171. In this case, a failure by a director to implement safeguards to prevent unlawful investments being made was a factor considered by Santow J in finding a breach of s 180.

41 White Paper, above n 2, 7-10. The National Greenhouse and Energy Reporting System ('NGERS') is to be used as the basis for this reporting system. A single report will be sufficient to satisfy obligations under both the NGERS and the CPR Scheme. It should also be noted that in 2006 the International Organization for Standardization ('ISO') published the ISO 14064 range of standards for greenhouse gas accounting and to provide government and industry with an integrated set of tools aimed at reducing greenhouse gas emissions, as well set the basis for emissions trading. These international standards have been adopted by Standards Australia through the EV-015 — Greenhouse Gas Measurement and Accounting Committee. The ISO 14064 series comprises the following three standards:
• ISO 14064-1:2006, Greenhouse gases — Part 1: Specification with guidance at the organization level for the quantification and reporting of greenhouse gas emissions and removals.
• ISO 14064-2:2006, Greenhouse gases — Part 2: Specification with guidance at the project level for the quantification, monitoring and reporting of greenhouse gas emission reductions and removal enhancements.
• ISO 14064-3:2006, Greenhouse gases — Part 3: Specification with guidance for the validation and verification of greenhouse gas assertions. Retrieved August 5, 2008, from http://www.standards. org.au/downloads/forums/emissions/standards_info.pdf

42 Ibid 7–35.

43 Carbon Pollution Reduction Scheme Bill 2009 (Cth) cl 324.

44 Carbon Pollution Reduction Scheme Bill 2009 (Cth) cl 325.

45 Pew Center on Global Climate Change, above n 9, 29.

46 DG Cogan, *Corporate Governance and Climate Change: The banking sector* (Ceres: 2008) 16.

47 JA Smith and M Morreale, 'Boardroom Climate Change', *New York Law Journal*, (New York) 16 July 2007, 1.

48 *Salsitz v Nasser* 208 FRD 589 (ED Mich 2002).

50 Above n 47.

51 KPMG International above n 1, 28. Kevin Healy (a partner at US law firm Bryan Cave LLP and co-chair of the Global Climate Change subcommittee of the New York State Bar Association) states that: 'Shareholders might argue that the duty of care of a company's directors includes the responsibility to take into account the impending business risks presented by climate change. While there are many uncertainties concerning the impact of climate change on shareholder value, the possibility of shareholder claims means that companies should prepare for the possible risks in order to avoid allegations of gross negligence.'

52 *Howard Smith Ltd v Ampol Petroleum Ltd* [1974] AC 821, 832.

53 *Corporations Act 2001* (Cth) s 180(3).

54 *Corporations Act 2001* (Cth) s 180(2).

55 [2007] NSWSC 124.

56 [2007] NSWSC 124 [1437].

56 *ASIC v Adler* [2002] NSWSC 171 [372].

57 *Greenhalgh v Arderne Cinemas Ltd* [1951] Ch 286, 291; see also *Ngurli Ltd v McCann* (1953) 90 CLR 425.

58 *Coleman v Myers* [1977] 2 NZLR 225.

59 S Bielefeld, S Higginson, J Jackson and A Ricketts, 'Directors' Duties to the Company and Minority Shareholder Environmental Activism' (2005) 23 *Company and Securities Law Journal* 28, 38.

60 A Lumsden, Saul Fridman, 'Corporate Social Responsibility: The case for a self-regulatory model' (2007) 25 *Company and Securities Law Journal* 147, 170.

61 In relation to corporate donations generally, see Elizabeth Klein and Jean J Du Plessis, 'Corporate Donations, the Best Interest of the Company and the Proper Purpose Doctrine' (2005) 28 *UNSW Law Journal* 69.

62 Department for Environment Food and Rural Affairs, *The Social Cost of Carbon and the Shadow Price of Carbon: What they are, and how to use them in economic appraisal in the UK,* (2007).

63 See for example, the announcement by National Grid plc concerning the adoption of a shadow carbon price: National Grid plc, *Policy Brief — Climate change* <http://www.nationalgrid.com/corporate/About+Us/climate/> at 12 November 2008. See also J Murray, 'Execs Urged to Adopt a "Shadow Carbon Pricing" Strategy' (News Release, 22 April 2008) <http://www.climatebiz.com/news/2008/04/22/execs-urged-adopt-shadow-carbon-pricing-strategy> at 12 November 2008.

64 For example, Rio Tinto's 2007 bid for Alcan included a premium reflecting the low carbon intensity of Alcan's aluminium production, secured through long term hydro power supply contracts. According to Rio Tinto's CEO, Tom Albenese: 'the access to long-term, low-cost, sustainable energy with essentially a zero-carbon footprint, very much puts the Alcan aluminium production in a competitive advantage'. PriceWaterhouse Coopers (NZ), *Carbon Value: Robust carbon management — A framework to protect and enhance shareholder value in response to climate change* (2008) 20 <http://www.pwc.com/en_NZ/nz/sustainability/carbon-value-report.pdf> at 4 August 2008.

65 Carbon Uncertainty, *Sydney Morning Herald* (26 May 2009) 24.

66 *Corporations Act 2001* (Cth) s 237(2).

67 *Corporations Act 2001* (Cth) s 249D.

68 K Bubna-Litic, 'Climate Corporate social responsibility: using climate change to illustrate the intersection between corporate law and environmental law' (2007) 24 *Environmental and Planning Law Journal* 253.

69 This occurred recently in relation to Exxon where these proposals were supported by 31% and 27% of the shareholders, respectively: see CL Slusarchuk, 'Low Carbon Resolutions on the Agenda — Shareholder activism And Exxon' (McCarthy Tétrault, Article, 8 August 2008) <http://www.mccarthy.ca/article_detail.aspx?id=4108> at 24 August 2009.

70 According to the Global Reporting Initiative, climate change resolutions accounted for 17% of all shareholders resolutions filed followed by resolutions dealing with political donations (15%). See: Global Reporting Initiative, 'Climate Change and Political Giving Top 2008 US Shareholder

Concerns' (News Release, 24 April 2008) <http://www.globalreporting.org/NewsEventsPress/LatestNews/2008/NewsApril08Riskmetrics.htm> at 17 December 2008.

71 A number of definitions of this phrase have been proposed, including:

- That a company acts in a manner which is socially responsible 'if it operates in an open and accountable manner, uses its resources for productive ends, complies with relevant regulatory requirements and acknowledges and takes responsibility for the consequences of its actions'. (Corporations and Markets Advisory Committee)

- '[A] concept whereby companies integrate social and environmental concerns in their business operations and in their interaction with their stakeholders on a voluntary basis'. (Commission of the European Union)

- '[T]he commitment of business to contribute to sustainable economic development, working with employees, their families, the local community and society at large to improve their quality of life'. (World Business Council for Sustainable Development)

See Corporations and Markets Advisory Committee, *The Social Responsibility of Corporations* (2006) iv, [2.1].

72 Ibid 40.

73 Ibid.

74 *Corporations Act 2001* (Cth) s 140(1).

75 Lumsden and Fridman, above n 60, 176.

76 *Companies Act 2006* (UK) c 46, s 172(1).

77 *Companies Act 2006* (UK) c 46, s 172(1).

78 Explanatory Notes, *Companies Act 2006* (UK) c 46 [327].

79 *Companies Act 2006* (UK) c 46, s 174.

80 Above n 76, [328].

81 Financial Services and the Treasury Bureau, *CO Rewrite: Rewrite of the Companies Ordinance* (2008) 20.

82 CAMAC, above n 71, [1.2.1].

83 Ibid [3.12].

84 Ibid.

85 MA Jones, SD Marshall, R Mitchell and I Ramsay 'Company Directors' Views Regarding Stakeholders' (University of Melbourne Legal Studies Research Paper No 270, 2007).

86 Ibid 29.

87 Ibid 33 and 35.

88 Pew Center on Global Climate Change, above n 9, 16–18.

89 See, eg, the 2008 Annual Report of Toll Holdings Ltd that details the approach of that company to developing and implementing a response to climate change. Toll Holdings Ltd, *Annual Report 2008* (2008).
See also the case studies of Entergy Corporation, The Travelers Companies Inc and Rio Tinto in Pew Center on Global Climate Change, above n 9, 19–27.

90 D Cogan, M Good, G Kantor and E McAteer, *Corporate Governance and Climate Change: Consumer and technology companies* (Ceres: 2008), 7. This report assessed how the senior management of 63 of the largest consumer and information technology companies (within the US and internationally) are addressing corporate governance systems to deal with climate risks and maximise climate business opportunities. It was found that:

- 15 of the 63 companies have established board level committees with responsibility for environmental oversight;

- 11 companies receive climate specific reports from management;

- 7 of the chief executive officers have adopted leadership roles in the environmental and climate change areas.

91 Ibid 4.

Labour Law and Climate Change Law

Victoria Lambropoulos

Many would wonder what labour law has to do with climate change and/or environmental law. It is not immediately apparent what connection there is between the two areas. However there has been recognition internationally, in particular by the United Nations, that there is an overlap between labour and the environment. The United Nations Environment Programme (UNEP) has explored how environmental sustainability can be promoted while protecting workers. They highlight three main areas within the existing legal framework common to most developed countries that can be used to promote environmental sustainability while protecting workers. They are first, to incorporate 'green friendly clauses' in collective workplace agreements. Second, the link between existing occupational health and safety laws (OHS) and environmental concerns can be strengthened and thirdly, corporate social responsibility initiatives can be used as a means of promoting standards in this area. This chapter will discuss these areas in the context of Australian law.

Introduction

The effect that climate change will have on the workplace has, until recently,[1] been omitted from the debates relating to climate change in Australia.[2] UNEP has explored how countries may implement reform within their existing legal frameworks in order to promote climate change initiatives and environmental sustainability.[3] It is clear that the reforms have a secondary purpose of protecting workers[4] and the security of their jobs. Protecting the environment and promoting sustainable development as a whole can contribute to protecting employment.[5] Much of the damaging impact of climate change, in particular the 'social flow-on'[6] effects that will impact work and workers, can be mitigated and possibly avoided if reform is implemented with this goal in mind.

The United Nations Environmental Programme

UNEP[7] outlines three main areas that can be used by countries like Australia, seeking to adapt their workplaces to the impacts of climate change and promote long-term sustainable development. They are:

- The use of the collective bargaining process and agreement making to incorporate clauses in workplace agreements, which are 'green friendly' and protect employees' rights.
- The strengthening of the link between OHS laws with practices that benefit the environment and promote sustainable development; and
- The strengthening of corporate social responsibility (CSR), laws and practices to ensure employers promote a sustainable workplace while protecting employees and adhering to international labour standards.

Collective Bargaining and Agreement Making

The collective bargaining process can be used to negotiate clauses in workplace agreements that promote environmentally sustainable production and other initiatives that reduce the production of carbon emissions within the workplace. The Australian labour law framework also includes individual agreement making, however, it is beyond the scope of this chapter to discuss individual agreements in detail. Generally, individuals will find it difficult, due to the unequal power imbalance between employers and most individual employees, to bargain for rights relating to the environment or climate change. Employers may still desire to promote such initiatives, however their preference is more likely to use CSR within their workplaces as these measures are largely voluntary, rather than binding themselves to enforceable obligations. This chapter will therefore confine its analysis to collective agreements. The advantages of registered collective workplace agreements under the *Workplace Relations Act 1996* (Cth)[8] is that they enjoy the substantial protections and remedies employed under the Act's statutory regime. The protections for employees are generally considered to be stronger and more effective than the general law.[9] Further, the chapter will only focus on common law employees as the *Workplace Relations Act* only permits employees to be covered by collective workplace arrangements.[10]

Potential clauses that are relevant and impact upon the employment relationship due to environmental sustainability initiatives include employee consultation clauses, investment in employee training and redundancy. These will be discussed below.

Restricted Content in Workplace Agreements in Australian Law

There is a significant legal limitation in Australia at the Federal level in using the collective bargaining process to incorporate clauses in workplace agreements that relate to the environment. The content of a workplace agreement is limited by the concept of 'matters pertaining to the employment relationship'. This concept was most recently considered by the High Court in *Electrolux Home Products Pty Ltd v Australian Workers' Union*.[11] Generally, it is considered that matters permitted in workplace agreements must have a 'direct, and not merely consequential', impact upon the employment relationship.[12] In *Electrolux*, a clause that related to the payment of trade union bargaining fees was held invalid, as it did not 'pertain to the employment relationship'. In the *Electrolux* decision, the union bargaining fees were prohibited because they did not directly relate to the relationship between employer and employee. The union was considered not to be a party to the employment relationship as defined in the relevant statutory provisions of the *Workplace Relations Act*. The test in *Electrolux* has been criticised as unclear and 'vague but rigid'.[13] *Electrolux* is particular to its facts and the trade union nexus to the clause, which was the focus of the dispute, was central to the decision made by the High Court.

The subsequent case law since *Electrolux* suggests that the door is open if clauses were phrased so as to impose a direct obligation on the employer and employees free of trade union interference.[14] Clauses that show a clear nexus to the employment relationship between the employer and the employee should be valid. For example, clauses which oblige the employer to train employees in new technologies implemented in the workplace by the employer to combat climate change should be upheld as they 'pertain to the employment relationship'. This is however conditional, providing that such technologies have a direct impact on the employment relationship. For example, general training courses that do not exhibit a clear link to the activities or reasonably foreseeable activities of the workplace could be considered too tenuous. However, if the training directly related to a specific workplace change that the employer was to introduce then the argument that the clause has a direct impact on the employment relationship is stronger. It is also common for many collective agreements, particularly union collective agreements, to include employee consultation clauses that relate to major structural change in the workplace. If changes to the workplace due to climate change were considered to be major structural changes then existing consultation clauses could enable consultation with employees on these matters.[15]

Employees are also concerned about job creation and a 'just transition'[16] for employees that will be made redundant due to changes in the workplace that will inevitably occur in some industries.[17] The collective bargaining process can also be used to enable employees in vulnerable sectors to be better prepared for transition into 'green jobs'.[18] Many collective agreements and awards throughout Australia already include redundancy payments for employees who will lose their jobs due to industrial restructuring.

Training clauses are permitted, given that they have a direct impact on how work is performed. This may extend to training for transition to carbon lean employment, especially if this involves a major structural change to the workplace. In the absence of this, there is no legal obligation upon the employer to provide training generally.[19] It is left to the bargaining process to decide this issue. There is evidence that the present reality in Australia is that employers do not invest in training or that the investment in training is low.[20] The evidence further suggests that the obligation for investment in training 'has shifted from employers to individual workers'.[21] Investment in training on this scale would need to be government sponsored in Australia or the law reformed to impose a positive obligation upon employers to provide such training.

Examples

The collective bargaining process has been used in other common law jurisdictions to address environmental concerns in the workplace with mixed success. An example of this is in Canada where the Canadian Auto Workers' Union (CAW) in 2004 experimented with the collective bargaining process in order to influence management to implement manufacturing processes that were environmentally sound. Environmentally friendly sources of employment are the premise upon which unions can ensure that their members have secure employment. In this instance, the CAW were campaigning for the introduction of Extended Producer Responsibility (EPR). EPR returns vehicles at the end of their life-cycle to the manufacturers for safe environmental disposal or for the recovery of usable parts. EPR has been established in Europe however in this instance the CAW did not have success in swaying car companies in Canada to implement the process.[22] It is likely that clauses requiring management to implement environmentally sound processes would not be permitted content in Australian workplace agreements. This is because it infringes the 'matters pertaining' formulation. This is discussed below later in this chapter.

In Australia, the National Tertiary Education Union (NTEU) is seeking agreement on environmental sustainability clauses in the latest round of negotiations with universities. The University of Ballarat University Union has been successful in including such a clause in its collective agreement which the Workplace Authority has approved.[23]

Even if these clauses passed the 'matters pertaining to the employment relationship' test, there is no compulsion on the part of employers to agree to be bound by such clauses. The inclusion of these clauses will largely depend on the bargaining process. It seems more likely that workplaces with a trade union presence, such as universities, will be able to negotiate environmental sustainability clauses in collective agreements.

Labour Law in Transition — Fair Work Bill

The Fair Work Bill 2008 (Cth),[24] once it is passed, will repeal the *Workplace Relations Act 1996* (Cth). The Fair Work Bill retains the 'matters pertaining to the employment relationship' jurisprudence.[25] The Explanatory Memorandum to the Fair Work Bill states that terms relating to 'corporate social responsibility and terms requiring an employer to commit to … climate change initiatives are not matters which pertain to the employment relationship'.[26] Presumably, this statement does not exclude clauses, which have a direct impact on the employment relationship, as it complies with the 'matters pertaining' formulation and complies with the 'permitted matters' requirements at s 172(1)(a) of the Bill, as discussed above. If this were the intent of the proposed sections then it would contradict existing understandings of the parameters set by the existing jurisprudence. As already noted with the University of Ballarat, the Workplace Authority has recently authorised an agreement, which contains content relating to climate change. It is likely that the statements in the Explanatory Memorandum refer to measures that oblige employers to change modes of production such as in the Canadian example already noted above. Generally, production decisions are considered to be the domain of managerial prerogative.[27] The relevant sections of the Fair Work Bill do not provide further detail and the reference to climate change in the Explanatory Memorandum is brief and vague.

The Fair Work Bill does provide for a new mandatory requirement to include a consultation clause in all collective (now called 'enterprise')[28] agreements requiring employers to consult with employees about major workplace change.[29] This will give employees throughout Australia some protection through the consultation process if major changes are made to a workplace because of climate change and/or general environmental law and policy. Significantly, consultation rights are now independent of employer agreement in the bargaining process and apply to all employees covered by a collective/enterprise agreement. Further, the right to redundancy payments will become a National Standard for many employees in the Federal system.[30] This will ensure that more employees throughout Australia will have some protection in the event of redundancy due to changes in workplace production.

The scope of the 'matters pertaining to the employment relationship' principle is unclear. As already noted, the delineation between managerial prerogative and content that pertains to the employment relationship is uncertain and appears to be ever evolving. However, even if clauses relating to climate change and environmental sustainability were permitted, the practical significance of these clauses in collective/enterprise agreements and in the wider enterprise bargaining framework is uncertain. Employees in some industries will likely be more concerned with wage rises than environmental sustainability, particularly in times of economic uncertainty. These observations suggest that the enterprise bargaining process has limitations, which cannot be ignored. However, a more detailed examination of collective agreement making and environmental sustainability clauses is required before any firm conclusions can be made.

Occupational Health and Safety Legislation and Protecting the Environment

'There is a strong connection between defending health and safety in the workplace and defending the local and wider environment.'[31] There are many instances where workplace production and activity have polluted the environment and caused a significant health hazard to workers and the community at large. [32] Unfortunately in the early years of the industrial revolution in the United Kingdom this was common.[33] The connection between environment and health and safety was exhibited by the fact that 'early environmental regulation was driven by public health concerns rather than environmental conservation.'[34] Since the 19th century, OHS legislation has promoted improved standards around the world but has developed separately to environmental law and policy.

OHS is predominantly a state responsibility in Australia.[35] The existing model of OHS in all states throughout Australia is generally similar although there are differences. The OHS laws impose broad general duties upon employers and controllers of workplaces and employees to ensure that the workplace is free of danger to health and safety. This is subject to the general principle of what is reasonably practical. The system is underpinned by the participation of health and safety representatives and committees at work. It is also enforced by state-appointed inspectors who can enter the work site at short notice to ensure that proper health and safety standards are maintained.[36] The current OHS model can promote environment standards in instances where health and safety in the workplace intersect with environment matters such as in 'major hazards facilities'.[37] However, outside of this, there are significant limitations in the present model if environmentally-sound work practices are to be maintained. The modern OHS model was developed in the late 1970s and 1980s in Australia,[38] at a time when the environmental agenda was not as prominent as it is now. Therefore it is not surprising that the OHS legislation makes no reference to the links between the workplace and the environment.

Workplace practices are legal if they meet the requirements under OHS laws. These requirements include a minimisation of risk in the workplace to health and safety as far as is reasonably practicable. However the practice may not be environmentally sound and further may contribute to global warming. This is evident in most of the high emission producing industries such as mining, gas and stationary energy. The production may continue as long as the employer puts in place precautions, which minimises the risk to health and safety of its workers. The precautions or measures, which the employer must put in place, are assessed against the concept of what is reasonably practicable. What is reasonably practicable is then assessed by reference to a number of factors relating to the particular context of the employer's business.[39] Therefore as long as the employer has minimised any risk to health and safety in the workplace, as far as is 'reasonably practicable', then the production can continue.[40]

The Place of the Environmental Delegate in the OHS Regime

The UNEP proposes the extension of rights and powers given to existing health and safety representatives or delegates so that they may also work on environmental matters in the workplace, they are also called 'environmental delegates'.[41] For example, the delegate may promote information and education on sustainable work practices amongst workers the way they do now in OHS. The challenge in introducing this in Australian law, and possibly in most other jurisdictions, is that the general OHS duty in legislation should be redefined so that consideration is given to the breadth of the duty and upon whom the duty will be imposed. The current legal duty is confined to the terms of health and safety and is 'based largely on the common law negligence standard of care'.[42] An environmental duty, which is imposed on controllers of the workplace, employers and employees, should be articulated and then legally defined. It may be appropriately described as an occupational or workplace environmental duty. Legal concepts and duties found in environmental law and jurisprudence can be examined for comparison and assistance in articulating an appropriately defined duty in the workplace context. UNEP acknowledges:

> The norms and standards of labour laws, which have a long tradition, were established in times when there was no environmental agenda. When, in the past few decades, the norms and standards that make up environmental law were developed, this took place without any consideration of labour rights.[43]

This is illustrated in the development of OHS laws, not just in Australia but also throughout the world. This is ironic given the genesis of early environmental legislation noted previously. The UNEP further acknowledges (at the time of writing) that no country has included in their domestic OHS laws the right of OHS representatives to effectively act as 'environmental delegates'.[44]

In spite of the lack of development in the field of domestic OHS laws, larger enterprises have conceded recognition of rights given to environmental delegates through the OHS model in their collective/enterprise agreements. This is as a result of bargaining and negotiation in the workplace between employers and employees represented by trade unions.

In Spain, there are examples of major enterprises negotiating wider powers for their occupational safety delegates in their collective agreements. The example given by UNEP is the recognition of such a right by the Michelin Company in Spain.[45] The inclusion of similar clauses in collective agreements in Australia is subject to the 'matters pertaining' requirement discussed above. Environmental delegates empowered under the auspices of OHS however must be limited to OHS concerns. The delegate is unlikely to be able to address wider concerns relating to sustainable production. Even UNEP does not envision a separation from general OHS concerns.

However, the current OHS laws do not preclude strengthening the link between OHS and environmental concerns. The machinery is in place for this to occur. Authorities that administer OHS laws such as Worksafe in Victoria can be used to educate the workplace on environmental matters that are linked to OHS. Like many OHS concerns of the past, education and training in this area is dependent upon the scientific knowledge of the time. As the scientific knowledge becomes available, this can be fed into workplaces through the OHS processes. Worksafe also has draft guidelines and codes for various industries to assist them to maintain OHS standards.[46] Sustainability guidelines linked to OHS can also be drafted to connect OHS with sound environmental workplace practices. This will strengthen the link between OHS, the environment and the activity of workplaces.

Environmental Health Officers in Australia: An Integrated Approach

Many companies and organisations in Australia currently employ environmental health officers.[47] The responsibilities of these positions vary depending on the organisation. However their general duties often include monitoring OHS in the workplace and environmental management. It is not unusual for organisations to combine OHS and environment in their risk management systems. In fact, an integrated approach to risk management is encouraged at the corporate level.[48] These initiatives have not been instigated as a consequence of mandatory legal obligation. Rather, it is seen as an effective management practice in companies and organisations. This indicates that organisations see advantages to an integrated approach between OHS and the environment. It has also been suggested that OHS officers (otherwise known as 'safety officers') in existing organisations will be 'coopted to manage the environmental portfolio in addition to their safety responsibilities,'[49] in response to the introduction of an emissions trading scheme. However, direct regulation may be necessary to enforce more widespread and effective change in workplaces.

Corporate Social Responsibility

What is Corporate Social Responsibility?

The concept of corporate social responsibility (CSR) has become a significant focus and concern in business and the wider community. It does not have a precise or fixed meaning.[50] The report prepared by the Australian Corporations and Markets Advisory Committee (CAMAC) in 2006, entitled 'The Social Responsibility of Corporations', noted that CSR essentially means the following:

> In essence, the focus of the issue of corporate social responsibility is on the way in which the affairs of companies are conducted and the ends to which their activities are directed, with particular reference to the environmental and social impact of their conduct. A responsible company, like a responsible individual, is one that acknowledges and takes responsibility for its actions.[51]

In Australia, CSR is not a mandatory legal obligation imposed upon companies. Rather, it is a voluntary code of conduct, which companies can choose to adopt. The primary legal basis, which governs the operation of CSR in Australia, is the duty imposed upon directors of companies to act in the best of the interests of the company. The duty is a common law fiduciary duty and a statutory duty in the *Corporations Act 2001* (Cth) s 181(1).[52] The concept of the best interests of the company is wide enough to encompass sustainable production, which promotes and protects employees. However, decisions based on this principle cannot be subordinate to the interests of shareholders. In other words, directors can have regard to such principles as long as their decisions are made in the best interests of the company, which generally means its shareholders.[53] Decisions that have this central focus have a long-term view of the company in mind.

> Directors are not confined in law to short-term considerations in their decision-making, such as maximising immediate profit or share price return. The interests of a company can include its continued long-term well-being.[54]

Equally however, in Australia, there is no case law, which precludes directors from taking a short-term view of their decision-making, especially if this will maximise profits and shareholder value.[55] This leaves directors with a great deal of discretion in assessing what they believe is in the best interests of the company in their particular context.

The Place of Sustainability in the Corporate Context

The CAMAC report states that the concept of sustainability has had increasing importance in corporate decision-making in Australia.[56] The Australian Institute of Company Directors and the Business Council of Australia have promoted the concept of sustainability to the business community.[57] It has been applied by companies in assessing their long-term viability, mainly in the context of voluntary non-financial reporting by companies, otherwise known as 'triple bottom line reporting'.[58] The purpose of triple bottom line reporting is to assess the company's performance against environmental, social and economic criteria and how these relate to the success of the business. Again this is voluntary in Australia but does relate to the company's mandatory reporting obligations under the *Corporations Act 2001* (Cth).[59]

Labour Standards and CSR

It has been observed that employees have received very little attention in the CSR and the wider corporate governance debate.[60] Much of the debate has focused upon the environmental impact of corporate behaviour. There are however various instruments which Australian companies may follow to accord with the best practice methods for the management of labour. An example of this is the Australian Stock Exchange's ten best practice corporate governance principles which recognise the value to companies of human capital. The principles have been adopted by many companies in Australia.[61] These voluntary guidelines bring some attention to employment or labour matters. Some companies such as BHP and Rio Tinto have adopted CSR principles relating to employment practices in their management strategies. However, some studies have shown that although there is a general commitment to these strategies, companies will depart from these principles if there is a strong business case to do so. There is little evidence of a deeper level of commitment in the actual workplace in these companies.[62]

CSR is sufficiently broad to allow managers of companies to adopt sophisticated strategies to simultaneously protect labour and the environment. It has been observed by some scholars that bad publicity associated with mass sackings as a consequence of restructures does not lead to public or shareholder approval.[63] In Australia, the Australian Workers' Union has been vocal in the media discussing the possibility of companies in Australia moving offshore to other countries that do not have an emissions trading scheme in place.[64] It is unclear whether this will affect shareholder value for companies that do decide to do this. It would certainly be an undesirable outcome for workers and Australia's economy. CSR can and should have the potential to influence corporate decision-making so that companies will keep their operations in Australia. There is evidence that the Australian community is willing to accept higher costs for goods and services in order to introduce an emissions trading scheme.[65] If the Australian community is willing to accept this, does this mean that shareholders are willing to sacrifice short-term profit in order to reduce carbon production? Given the current public concern surrounding climate change, will shareholders seek to invest in companies that promote long-term sustainable development and in turn, environmentally friendly sources of employment for their workers over other companies that do not do this? There is an incentive in the current climate for directors of companies to adopt management strategies that protect employment and the environment. These strategies are principally part of the long-term investment strategies and policies of a company. What will influence this decision-making will be the views of shareholders and what they are willing to view as being in the best interests of the company. What is difficult for directors is to gauge the changing mood of shareholders on such matters. Sentiments toward climate change policy will and have changed during economic crises.

Conclusion

This chapter has provided a preliminary analysis of what are complex issues in the debate between labour and the environment, in particular in the current context of climate change law and policy. UNEP has provided a useful starting point for policy and law-makers. Employers and employees in Australia will find it difficult to confidently bargain for environmental sustainability clauses given the uncertainty of the law in relation to permitted content of workplace agreements. This has been further complicated by the comments relating to climate change in the Explanatory Memorandum to the Fair Work Bill. The OHS laws need reform if they are to be used as a vehicle for the maintenance of environmental standards in the workplace as they do not directly require an employer to take into account the impact the activity in the workplace has on the environment. Further, future reform should take into account whether it is desirable to continue to regulate environmental standards within the context of environmental law and keep workplace health and safety separate. Perhaps it is CSR initiatives that will be undertaken by some companies to protect labour, particularly if this will improve shareholder value and the long-term viability of particular companies. Ultimately reform will be required in all three areas discussed in this chapter if the UNEP proposals are to be adopted in Australia.

Endnotes

1   S Hatfield-Dodds, G Turner, H Schandl and T Doss, Growing the Green Collar Economy: Skills and Labour Challenges in Reducing our Greenhouse Emissions and National Environmental Footprint. Report to the Dusseldorp Skills Forum (Canberra, CSIRO Sustainable Ecosystems, 2008).

2   C Lipsig-Mumme, 'Strategies for Work in a Warmer World', The Age, 25 February 2008.

3   See United Nations Environment Programme (UNEP), Labour and the Environment: A Natural Synergy (2007).

4   UNEP desire for all workers whether they are employees at law or not to be protected. This paper will concentrate on protection for common law employees.

5   See generally Hatfield-Dodds et al, above n 1; UNEP, above n 3; Australian Council of Trade Unions(ACTU), Principles and Policy on Global Warming, (Position Paper 2008); Trade Union Congress, A Green and Fair Future for a Just Transition to a Low Carbon Economy, (London, Trade Union Congress, London, 2008); European Trade Union Confederation, Report by European Trade Union Confederation and Social Development Agency, Climate Change and Employment:, Impact on Employment of Climate Change and CO_2 Emission Reduction Measure in the EU-25 to 2030 (European Trade Union Confederation and Social Development Agency, 2007). The documents discuss how protecting the environment can improve and protect employment with statistics in various for various industries.

6   See Lipsig-Mumme, above n 2.

7   See UNEP, above n 3.

8   At the time of writing, the act in force is the Workplace Relations Act 1996 (Cth). The Fair Work Bill 2008 (Cth) which repeals the Workplace Relations Act 1996 (Cth) has not passed the Senate.

9   See A Stewart, Stewart's Guide to Employment Law (2008) 8.1.

10 See B Creighton and A Stewart, Labour Law (4th ed, 2005) ch 11; see also R Owens and J Riley, The Law of Work (2005) ch 4 for a discussion of the different categories of worker. This will not change under the proposed legislation.

11 Electrolux Home Products Pty Ltd v Australian Workers' Union (2004) 221 CLR 309 ('Electrolux'). The doctrine of 'matters pertaining to the employment relationship' has been retained in the Fair Work Bill 2008 (Cth).

12 See Creighton and Stewart, above n 10, [4.30].

13 See A Frazer, 'Industrial Tribunals and Regulation of Bargaining', in C Arup, P Gahan, J Howe, R Johnstone, R Mitchell and A O'Donnell et al (eds), Labour Law and Labour Market Regulation: Essays on the construction, constitution and regulation of labour markets and work relationships (2006) 237.

14 See Owens and Riley, above n 10, 506–9; Creighton and Stewart, above n 10, [4.28]–[4.32].

15 See the discussion later in this paper discussion titled 'Labour Law in Transition'.

16 It is beyond the scope of this paper to discuss 'just transitions' in detail. See Trade Union Congress, above n 5, 6, 14. The concept of just transition came out of the United States.

17 See generally ACTU, above n 5; Trade Union Congress, above n 5.

18 See generally Hatfield-Dodds et al, above n 1.

19 See Creighton and Stewart, above n 10, [13.57].

20 M Jones and R Mitchell, 'Legal Origin, Legal Families and the Regulation of Labour in Australia'. In S Marshall, R Mitchell and I Ramsay (eds), *Varieties of Capitalism, Corporate Governance and Employees* (2008) 74–86.

21 Ibid 74.

22 UNEP, above n 3, 4, 12.

23 Clause included in the Appendix of this paper.

24 At the time of writing, the Fair Work Bill 2008 (Cth) is before the Senate.

25 Fair Work Bill 2008 (Cth) s 172.

26 See Explanatory Memorandum, Fair Work Bill 2008 (Cth) [673].

27 For a discussion on managerial prerogative see Creighton and Stewart, above n 10, [4.28]–[4.32].

28 See Fair Work Bill 2008 (Cth) s 12 (definition of 'Enterprise Agreement').

29 Fair Work Bill 2008 (Cth) s 205.

30 Fair Work Bill 2008 (Cth) s 119; Fair Work Bill 2008 (Cth) s 121 excludes employees of small business from redundancy rights.

31 UNEP, above n 3, 10.

32 See M Tooma, *Safety, Security, Health and Environment Law* (2008) ch 1; UNEP, above n 3, 65–82.

33 Ibid ch 1.

34 Ibid 5.

35 *Occupational Health and Safety Act 2000* (NSW); *Occupational Health and Safety Act 2004* (Vic); *Workplace Health and Safety Act 1995* (Qld); *Occupational Health Safety and Welfare Act 1986* (SA); *Workplace Health and Safety Act 1995* (Tas); *Occupational Health and Safety Act 1984* (WA); *Occupational Health and Safety Act 1989* (ACT); *Work Health Act 1986* (NT).

36 See Creighton and Stewart, above n 10, ch 19 for a summary of the OHS laws.

37 *Occupational Health and Safety Act (Major Hazard Facilities) Regulations 2000* (Vic).

38 R Johnstone, 'Regulating Occupational Health and Safety in a Changing Labour Market'. In C Arup, P Gahan, J Howe, R Johnstone, R Mitchell and A O'Donnell (eds) *Labour Law and Labour Market Regulation: Essays on the construction, constitution and regulation of labour markets and work relationships* (2006) 618.

39 *Occupational Health and Safety Act 2004* (Vic) ss 20–21.

40 *Occupational Health and Safety Act 2004* (Vic) s 21.

41 UNEP, above n 3, 35.

42 Johnstone, above n 38, 618.

43 UNEP, above n 3, 31.

44 Ibid 35.

45 Ibid.

46 *Occupational Health and Safety Act 2004* (Vic) ss 98–101, 149–51.

47 See Australian Government Department of Education, Employment and Workplace Relations, websites for job descriptions at <www.jobguide.thegoodguides.com.au/occupation/view/254313B> and <www.myfuture.edu.au/services/default.asp?FunctionID=5050&ASCO=254313B>.

48 See Tooma, above n 32, 1.

49 Ibid.

50 Corporations and Markets Advisory Committee ('CAMAC'), *Social Responsibility of Corporations Report* (December 2006) 13.

51 Ibid 15.

52 *Corporations Act 2001* (Cth) s 181(1).

53 See Richard Mitchell, Anthony O' Donnell and Ian Ramsay, Shareholder Value and Employee Interests: Intersections between Corporate Governance, Corporate Law and Labour Law (Research Paper No 128, CCLSR and CELR, University of Melbourne, 2005).

54 Ibid 84. See also *Provident International Corporation v International Leasing Corp Ltd* [1969] 1 NSWR 424, 440, noted in CAMAC, above n 50.

55 CAMAC, above n 50, 85.

56 Ibid 68.

57 Mitchell, O'Donnell and Ramsay, above n 53, 68; Business Council of Australia, *Towards Sustainable Development: How leading Australian and global corporations are contributing to sustainable development* (May 2001).

58 CAMAC, above n 50, 69.

59 *Corporations Act 2001* (Cth) s 299.

60 S Bottomley and A Forsyth, The New Corporate Law (Working Paper No 1, Corporate Law and Accountability Research Group, Monash University, 2006) 21.

61 For a more detailed list of instruments and guidelines see M Jones, S Marshall and R Mitchell, 'Corporate Social Responsibility and the Management of Labour in Two Australian Mining Industry Companies' (2007) 15 *Corporate Governance* 57, 58.

62 Ibid 64.

63 See S Deakin and F Wilkinson, *The Law of the Labour Market* (2005) 2.

64 B Norrington, 'Union Voices Fear on Carbon Trade', *The Australian* (Sydney), 24 July 2008; P Howes, Give Workers A Voice in Climate Change Solutions, *The Age* (Melbourne), 24 July 2008; see generally Australian Workers' Union, The National Emissions Trading Scheme, (Position paper, Australian Workers' Union, 23 July 2008).

65 See P Coorey, 'Voters Want Eco Action', *Sydney Morning Herald* (Sydney), 21 July 2008. The Herald/Nielsen poll, the first since the Green Paper on a proposed emissions trading scheme was introduced, finds that Australians are willing to pay the price for cutting carbon emissions, even though most do not have an understanding of how the scheme will work.

66 University of Ballarat, University of Ballarat Union Collective Agreement 2008–09 (2008) 60 <https://www.ballarat.edu.au/aasp/staff/personnel/uca_08_09.pdf>.

Appendix
University of Ballarat Union Collective Agreement 2008–09

Environmental Sustainability[66]

75.1 The parties are committed to improving the sustainability performance of the University through promoting a culture of sustainability, ensuring that the operations are managed in a manner that minimises the University's environmental and social impacts and through enabling the integration of sustainability principles and practices into teaching and learning, research and community engagement of the University.

75.2 The University will meet its commitment to sustainability through adopting the following principles:

75.2.1 Ensuring sustainability is reflected in the University's strategic documents;

75.2.2 Developing an annual sustainability reporting framework;

75.2.3 Developing education for sustainability through utilising the University's research, curriculum and administrative practice to enhance organisational sustainability;

75.2.4 Incorporating the principles of Ecologically Sustainable Design (ESD) into all capital and infrastructure projects;

75.2.5 Providing training and support for staff to adopt sustainability principles and practices to achieve sustainable behaviour change in the workplace;

75.2.6 Undertaking research which will develop and strengthen regional partnerships that address sustainable development;

75.2.7 Building capacity in the community by producing graduates who are sustainability literate through their experience at the University; and

75.2.8 Partnering with the community to respond to the regional needs for a sustainable future and provide leadership in our region.

75.3 Staff and unions will be kept regularly informed about the University's carbon emissions, water and energy consumption levels and this information will be reported publicly.

Climate Change Litigation

Hon Justice Brian J. Preston

In the absence of an international treaty to address global warming, climate change litigation is presenting itself as an attractive alternative path to address the effects of climate change. Environmental groups and affected individuals and groups have taken up the challenge and brought climate change related actions before the courts. Lawsuits have targeted unresponsive governments or companies that are major greenhouse gas emitters. This paper focuses on avenues that have been used, and other potential causes of action which could be used, to litigate issues relating to climate change. At the national level, plaintiffs have used tort law (public nuisance, negligence, civil conspiracy, misrepresentation), administrative law (merits review and judicial review proceedings) or constitutional law grounds (enforcement of a constitutional right). Environmental disputes have also been litigated before a range of international fora, including the International Court of Justice, the International Tribunal for the Law of the Sea or regional human rights courts.

Introduction

The Intergovernmental Panel on Climate Change (IPCC) found in its latest assessment report that climate patterns have changed significantly in the 20th century and the second half of the century brought the warmest years on record.[1] The causes of climate change are largely anthropogenic, leading many environmental groups to call on governments to tackle these causes. As some effects of climate change are already noticeable, such as increased coastal hazards, adaptive measures are also needed.

A comprehensive and action-forcing international treaty, ratified by all the major contributors to global warming, is regarded as the preferable choice to address the global warming phenomenon, as collective action taken by all states is what is required in order to meaningfully combat climate change.[2] However, international negotiations in this field are not advancing, hence presenting litigation as an attractive path, despite some drawbacks. While litigation might eventually force governments to take some action,[3] it might also mean that the results would be piecemeal. Ultimately, litigation is unlikely to have a great overall effect on climate change.[4] Despite this assessment of litigation, environmental groups and affected individuals and groups have nonetheless taken up the challenge and brought climate change-related actions before the courts. These lawsuits have mainly, but not solely, targeted unresponsive governments, through their agencies or departments,[5] or companies that are major greenhouse gas ('GHG') emitters, such as car manufacturers or power plants.[6]

Climate change litigation is a fairly new phenomenon. The first significant American court decision relating to climate change dates from 1990,[7] and the first Australian one from 1994.[8] Since then there has been an increase in the number of cases where issues relating to climate change are being litigated, more or less successfully. It is only in recent years that climate change as a phenomenon has been more widely accepted by the courts,[9] though there are still cases where the science of climate change is challenged.[10] Taking climate change into account when deciding upon the merits of a development proposal, is similarly a new development.[11]

Plaintiffs have used the legal avenues available to them to bring climate change-related actions before the courts, from actions based on the law of torts to domestic statutes and international conventions. These actions are not always based on environmental legislation, but also other provisions applicable nationally or internationally. This explains the variety of both causes of action which have been employed and fora which have heard climate change or air pollution actions.

This chapter provides a perspective on the avenues that have been used or could be used to litigate issues relating to climate change.

At the national level, plaintiffs have used tort law. Causes of action include the traditional actions of nuisance, both public and private, and negligence, as well as less common actions such as civil conspiracy.

Misrepresentation as to the environmental credentials of goods and services can give rise to claims by customers who have suffered loss by relying on the misrepresentation. Claims may be in the torts of deceit and negligence (negligent misstatement), in contract (including sale of goods) or in trade practices for false or misleading conduct or representations.

Litigants have also used administrative law to bring climate change issues before the courts. In particular, litigants have instituted judicial review and merits review proceedings to challenge administrative decisions or conduct relating to environmental issues, such as planning proposals and their impacts.

Constitutional law grounds have also been employed, through the enforcement of a constitutional right, including a right to life generally or a right to a clean and healthy environment in particular.

At the international level, environmental disputes have been litigated before a range of international fora, including regional human rights courts, the International Court of Justice, the World Trade Organization Appellate Body, the International Tribunal for the Law of the Sea, and the World Heritage Committee.

Tort

Tort actions, either at common law or in continental legal systems, are one of the ways in which remedies for environmental damage resulting from climate change could be sought. The causes of action employed or likely to be employed are in nuisance, negligence and conspiracy. Misrepresentation in tort, contract and trade practices is also likely to be employed.

Nuisance

The word nuisance derives from 'nocumentum' meaning hurt, inconvenience or damage. Nuisance generally covers acts unwarranted by law which cause inconvenience or damage to the public in the exercise of rights common to all subjects (public nuisance), acts connected with the occupation of land which injure another person in the use of land or interfere with the enjoyment of land or some right connected therewith (private nuisance), and acts or omissions declared by statute to be a nuisance (statutory nuisance).[12]

Actions in public nuisance have been brought in the United States by state and local governments. The targets of these suits have been large, industrial contributors to global warming, in the transportation, energy and power sectors.[13]

In *Connecticut v American Electric Power*,[14] 12 states, a municipality and three environmental, non-governmental organisations sued five electric power companies which, through their fossil-fired electric power plants, emitted around 10% of all carbon dioxide in the United States. The plaintiffs sought a permanent injunction requiring the defendants to cap their carbon dioxide emissions and to commit to yearly reductions over at least 10 years. The government plaintiffs sued on their own behalf to protect public lands (eg, the hardwood forests of the Adirondack park in New York) and as parens patriae on behalf of their citizens and residents to protect public health and wellbeing.

In *People of the State of California v General Motors*,[15] California sued six of the world's largest manufacturers of automobiles based on the alleged contributions (past and present) of their vehicles to climate change impacts in the state. The suit alleged these impacts constitute a public nuisance and sought monetary damages.

Both the General Motors and American Electric Power suits were dismissed by the respective District Courts on grounds of non-justiciability, the courts stating that it was impossible to decide the matters without making an initial policy determination of a kind clearly for non-judicial discretion.[16] The plaintiffs have appealed in both matters.

In *Kivalina v ExxonMobil*,[17] the native Inupiat village of Kivalina in Alaska has commenced a public nuisance suit against nine oil companies, 14 power companies and a coal company. The village suffers from the melting of Arctic ice which used to protect its coasts from severe weather and, hence, erosion. The current erosion of coastal areas means the village has to relocate or be abandoned. The plaintiffs seek monetary damages from the defendants for their contribution to climate change.

An action in private nuisance may also be available to an affected private person. Private nuisance involves an act or omission which is an interference with, disturbance of, or annoyance to a person in the exercise or enjoyment of his or her ownership or occupation of land or some easement, profit or right in connection with the land.[18]

Circumstances where a private nuisance might be caused include the undertaking of works to mitigate the effects of climate change. For example, a public authority might construct rock walls or levee banks to control or mitigate increased sea or water levels caused by climate change. If such works are poorly located, designed or constructed, they may exacerbate the problem they were intended to remedy or shift the problem to other locations.[19] An action in private nuisance by affected land owners may lie against the public authority, subject to any statutory immunities or defences.

In *Open Space Inst v American Electric Power Co*,[20] three environmental, non-governmental organisations who owned and preserved land in the State of New York sued five electric power companies which, through their fossil-fired electric power plants, were allegedly the five largest carbon dioxide emitters in the United States. The suit was joined with the *Connecticut v American Electric Power* suit discussed earlier and was also dismissed on grounds of non-justiciability.

Negligence

To succeed in negligence, a plaintiff must prove that: the defendant owed the plaintiff a duty, recognised by law, requiring the defendant to adhere to a certain standard of conduct; the defendant breached that duty; the plaintiff suffered loss; the loss was caused by the defendant's breach of duty; and the loss suffered by the plaintiff was not too remote.[21]

An action in negligence by a person who has suffered damage or loss by a climate change-induced event could potentially be brought in relation to a failure to mitigate or to adapt to climate change, although the former is more problematic than the latter.

In a negligence action for failure to mitigate, the defendants would likely fall into three categories: first, producers of fossil fuels whose combustion increases GHG emissions (including oil, gas and coal companies); second, the users of fossil fuels that cause GHG emissions (including electricity power generators, steel, aluminium and other metal mills and the transport industry); and third, manufacturers or marketers of products whose use contributes to climate change (such as automobile manufacturers). Any negligence action by a plaintiff against defendants of these kinds is, however, likely to face considerable hurdles.

The first hurdle is establishing the defendant owed the plaintiff a duty of care. Duties of care are not owed in the abstract.[22] Rather, a duty of care is relational; a duty is owed to another person or class of persons, not to the world at large:

A duty of care involves a particular and defined legal obligation arising out of a relationship between an ascertained defendant (or class of defendants) and an ascertained plaintiff (or class of plaintiffs).[23]

Because the obligation arises out of a particular relationship, the scope of the obligation may vary — be more or less expansive — depending on the relationship in question.[24] Hence, ascertaining the scope or content of the duty of care depends on ascertaining the particular relationship between the defendant and the plaintiff. Further, all duties of care, whatever their scope, impose an obligation to exercise reasonable care; they do not impose a duty to prevent potentially harmful conduct.[25]

Where there is a relationship between the defendant and the plaintiff, which falls into a category which the law has, in prior cases, held to give rise to a duty of care, a duty of care will exist. In cases that do not fall within established categories, the approach of the court is incremental and operates by analogy with established categories.[26] For other cases, not within an established category or analogous to cases in an established category, no single accepted test for determining when a duty of care will exist has emerged.[27] The court must engage in 'a multifactorial or "salient features" analysis'[28] to determine the existence of a duty of care in a particular case. At the minimum, there needs to be reasonable foreseeability: a defendant must know or ought reasonably to know that its conduct is likely to cause harm to the person or the tangible property of the plaintiff unless the defendant takes reasonable care to avoid the harm.[29] However, reasonable foreseeability may not be sufficient in itself to always give rise to a duty of care.[30]

The difficulty that has arisen is ascertaining what, if any, further requirements need to be satisfied before the law will impose a duty of care. Proximity or neighbourhood had been suggested as a determinant of a duty of care, but it has now been rejected as a determinant,[31] although it remains as a salient feature to be considered. Other factors have been suggested,[32] but no consensus has emerged.

In the climate change context, the relationship between potential defendants of the kinds earlier noted and persons who might suffer damage or loss as a result of climate change-induced events neither falls into any established category in respect of which the law has held a duty of care to arise, nor is analogous to cases in which a duty of care has been held to exist. An analysis of the salient features of the relationship is therefore required.

At the outset, there is a difficulty in actually identifying a relationship between an ascertained defendant (or class of defendants) and an ascertained plaintiff (or class of plaintiffs). As Hunter and Salzman note, 'climate change is essentially a global environmental tort'.[33] Defendants and plaintiffs are indeterminate. In relation to the defendant, no individual defendant of the kinds earlier identified is likely to have a relationship with any person who might suffer damage or loss by a climate change-induced event. It is, therefore, necessary to identify a class of defendants. However, this too would be difficult. At the broadest, the class would be all persons who cause GHG emissions (directly or indirectly). Such a class has a vast membership, including producers of fossil fuels, users of fossil fuels and manufacturers or marketers of products whose use contributes to climate change. Membership of the class is also distributed globally. Although members of the class are unified by being contributors to climate change, the differences in the nature and capabilities of members of this vast class would be immense.

Turning to the plaintiff, identifying in advance either any individual plaintiff or the class of which the plaintiff is a member would also be problematic. Climate change is liable to affect all of humanity to varying degrees. Hence, membership of the class would include all of humanity.

Next, problems are likely to arise in relation to reasonable foreseeability and proximity. What risks are reasonably foreseeable and which plaintiffs are within the zone of foreseeable risk? Even though the test of reasonable foreseeability is often expressed in the undemanding terms of not being far fetched or fanciful,[34] establishing the test may prove extremely hard. Is it reasonably foreseeable that a particular defendant's conduct, such as emitting GHGs by using fossil fuels in the course of its business, could result in a specific climate change-induced event, such as greater tidal inundation that, in turn, harms a resident or their property in a coastal community? And when did such foreseeability arise?[35] The criterion of proximity, whether expressed as spatial or temporal proximity, is also unlikely to be satisfied. The defendant and the plaintiff may be in spatially remote locations, and the cause (GHG emissions) and effects (damage from climate change-induced events) may be temporally distant. Other salient features of the relationship might also tend against establishing a relevant relationship between the defendant and the plaintiff.[36]

These difficulties in identifying a relationship between an ascertained plaintiff (or class of plaintiffs) and an ascertained defendant (or class of defendants), in turn, impede framing 'a particular and defined legal obligation' that could arise out of the relationship. The scope of the duty cannot be to prevent potential harmful conduct; it must be to take reasonable care in a particular defined way, arising out of the relationship in question. This cannot readily be formulated.

The second hurdle is establishing that the defendant breached any duty of care. Determining breach of duty customarily entails undertaking the task described by Mason J in *Wyong Shire Council v Shirt* of asking what a reasonable person in the defendant's position would have done by way of response to the foreseeable risk of harm to the plaintiff (or class of plaintiffs).[37] In an action for failure to mitigate, it would involve consideration of the magnitude of the risk of damage or loss by a climate change-induced event being caused by the defendant's GHG emissions, the degree of probability of its occurrence, along with the expense, difficulty and inconvenience of taking alleviating action (such as stopping all production causing GHG emissions or completely offsetting its GHG emissions) and any other conflicting responsibilities the defendant may have. Balancing such considerations may tend against any conclusion of breach of duty by the defendant.

The third hurdle is establishing that the defendant's particular breach of duty was, in a common sense way,[38] causative of the damage or loss suffered by the plaintiff. Causation for a climate change-induced event resulting in loss or damage to a plaintiff is confused by there being myriad and diffuse contributors of GHG emissions, distributed globally and over long timeframes, with delays between the emission of GHGs and the consequence of climate change and any particular adverse effect of climate change.

The fourth hurdle concerns remoteness of damage. There is a policy question raised by the spectre of defendants being exposed to 'a liability in an indeterminate amount for an indeterminate time to an indeterminate class', to use Cardozo J's words in *Ultramares Corp v Touche*.[39]

These hurdles in a negligence action for failure to mitigate were encountered in *Comer v Murphy Oil*.[40] Comer and 13 other individuals harmed by Hurricane Katrina in the United States sued nine oil companies, 31 coal companies and four chemical companies, including in negligence. They claimed the defendants had 'a duty to conduct their business in such a way as to avoid unreasonably endangering the environment, public health and public and private property, as well as the citizens of Mississippi' and that they breached their duty 'by emitting substantial quantities of greenhouse gases, knowing such emissions would unreasonably endanger the environment, public health, and public and private property interests'.[41] The court dismissed the claim summarily on the defendant's motion stating that the plaintiffs do not have standing and that the claims are non-justiciable pursuant to the political question doctrine.[42]

Actions in negligence in relation to a failure to adapt to climate change, however, might fare better. These actions would likely be based on more conventional ways of establishing liability and negligence, particularly against public authorities. An action in negligence might challenge:

(a) the appropriateness of development approvals in flood prone, coastal zone or bushfire prone areas;

(b) the adequacy of building standards to withstand extreme weather events, as their area and frequency increases;

(c) the responsibility for erosion, land slides, flooding, etc, resulting from extreme weather events;

(d) the adequacy of emergency procedures when more frequently put to the test and over a greater area;

(e) the failure to undertake disease prevention programs, as the area and frequency of diseases spread; and

(f) the failure to preserve public natural assets in the face of climate change.[43]

To date, there have not yet been any successful actions in negligence for damage or loss caused by climate change, but the potential exists.[44] Liability of public authorities for negligence can be limited by statute.[45]

Conspiracy

Conspiracy consists in the agreement of two or more persons to do an unlawful act, or to do a lawful act by unlawful means.[46] The tort of conspiracy takes two forms: conspiracy to use unlawful means; and conspiracy to injure. The second does, but the first does not, require a predominant purpose to injure.[47] Hence, conspiracy is actionable if the predominant purpose is disinterested harm, or unlawful means are used to secure it, but not if the predominant purpose is legitimate, such as to secure a larger market share.[48]

Prior to being used in climate change litigation, claims of civil conspiracy were used in lawsuits against tobacco companies, where it was argued that tobacco companies had conspired to deceive the public about the dangers of cigarettes.[49] The first climate change case to rely on conspiracy is *Kivalina v ExxonMobil*.[50] The plaintiffs have included 'claims for civil conspiracy and concert of action for certain defendants' participation in conspiratorial and other actions intended to further the defendants' abilities to contribute to global warming'. The plaintiffs argue that a number of defendants participated in an agreement to mislead the public with regard to the science of global warming. The purpose of this conduct was to delay public awareness of global warming and its effects, which delay would allow the defendants to pursue their activities as contributors to GHG emissions and global warming without being pressured into a costly change of their current behaviour. The plaintiffs also allege that the defendants are engaging in concert with each other over the creation of global warming, giving substantial assistance or encouragement to each other to so conduct themselves.

Misrepresentation

Businesses represent their products or services to be environmentally friendly — they put them in a 'green light'. Products or services are sometimes promoted as having 'low carbon emission', others as having had their carbon emission offset by the business promoting them, or in other cases, the consumer is given the opportunity to offset the emissions in a certain way organised by the seller. Such representations may be false or misleading and give rise to a claim by the customer against the representor in the torts of deceit or negligence or contract, as appropriate.

The tort of deceit involves making a false representation, knowing it to be false, or without honest belief in its truth, or recklessly, not caring whether it be true or false, with the intention that another should rely on the representation and act to their detriment and with the result that the other does so act.[51]

Negligent misstatement is a form of negligence. A person may owe a duty to take reasonable care not to cause purely economic loss by giving misleading information or advice. If another person, in a relationship with the giver of information or advice that the law recognises as sufficient, relies on that misleading information or advice and suffers loss as a result, the giver may be liable for that loss.[52]

Misrepresentation can also arise in contract. Misrepresentation is a false statement of fact, or of mixed fact and law, made by one party to the other party with the object, and having the result, of inducing the other to enter into a contract or similar relationship with the representor. It may be made by statement or other action, or by concealment but not by mere omission, silence or inaction except where such omission, silence or inaction would distort the natural inference from other facts or where, exceptionally, there is a positive duty to disclose all relevant facts.[53]

Trade Practices

Misrepresentations as to the environmental credentials of products and services may also be actionable under trade practices law. The *Trade Practices Act* 1974 (Cth), for example, regulates:

1. Misleading or deceptive conduct under s 52(1):

   > A corporation shall not, in trade or commerce, engage in conduct that is misleading or deceptive or is likely to mislead or deceive.

2. False or misleading representations under s 53:

   > A corporation shall not, in trade or commerce, in connexion with the supply or possible supply of goods or services or in connexion with the promotion by any means of the supply or use of goods or services:
   >
   > (a)   falsely represent that goods are of a particular standard, quality, value, grade, composition, style or model or have had a particular history or particular previous use;

    (aa) falsely represent that services are of a particular standard, quality, value or grade;

    (c) represent the goods or services have sponsorship, approval, performance characteristics, accessories, uses or benefits they do not have.

3. False or misleading representations about land under s 53A(1):

> A corporation shall not, in trade or commerce, in connexion with the sale or grant, or the possible sale or grant, of an interest in land or in connexion with the promotion by any means of the sale or grant of an interest in land:
>
> (b) make a false or misleading representation concerning the nature of the interest in the land, the price payable for the land, the location of the land, the characteristics of the land, the use to which the land is capable of being put or may lawfully be put or the existence or availability of facilities associated with the land.

The Australian Competition and Consumer Commission (ACCC), the regulatory body with responsibility for administering the *Trade Practices Act*, has investigated and reported on environmental claims of business. It has released two reports this year Green Marketing and the *Trade Practices Act*[54] and *Carbon Claims and the Trade Practices Act*.[55]

The Green Marketing Report contains guidelines aimed at businesses using environmental claims as part of their marketing campaigns. The purpose of the report is to educate businesses about their obligations under the *Trade Practices Act* to avoid misleading and deceptive conduct and false or misleading representations as to the environmental credentials of their products. Consumers are to be provided with accurate information in order to make informed decisions.

The Carbon Claims Report addresses issues surrounding carbon offset and neutrality claims. The purpose of the report is to inform businesses and consumers as to their obligations and rights under the *Trade Practices Act* in relation to such claims. The report emphasises the need to provide accurate, clear, substantiated information about their products, so as to develop a credible and transparent carbon offset market by having good marketing practices.

The ACCC has been active in scrutinising environmental claims made by businesses to ensure that they comply with the *Trade Practices Act*. Three examples are the ACCC actions concerning the environmental claims by De Longhi, Goodyear and Saab that their products are climate change friendly.

The ACCC challenged the unqualified claim by De Longhi that the refrigerant component gas R290 used in its portable air-conditioners was 'environmentally friendly'. De Longhi provided court-enforceable undertakings that it would refrain from using unqualified claims and amended its advertising to specify that 'R290 is the most environmentally friendly refrigerant gas currently available for use in domestic portable air conditioners'.[56]

The ACCC challenged a number of unsubstantiated representations made by Goodyear during 2007 and 2008 in relation to the Goodyear Eagle LS2000 tyre, including that the Eagle LS2000 is 'environmentally-friendly', has 'minimal environmental impact', its production process results in reduced carbon dioxide emissions and that the BioTRED technology 'increases the life' of the tyre and 'improves fuel economy'. Goodyear gave a court-enforceable undertaking that it would offer partial refunds as compensation to customers who relied on the unsubstantiated environmental claims during 2007 and 2008 with regard to this range of tyres. Goodyear also issued a corrective notice.[57]

The ACCC has challenged the 'green' claims by GM Holden Ltd in the advertising of Saab cars. Saab advertised that 'Grrrrrreeen, Every Saab is green', 'Carbon emissions neutral across the entire Saab range' and 'Switch to carbon neutral motoring' to promote the green credentials of its motor vehicles. The advertisements also stated that Saab would plant 17 native trees in the first year after a Saab vehicle purchase as a carbon offset. ACCC considers such claims to be misleading because:

(a) there would, in fact, be a net release of carbon dioxide into the atmosphere by the operation of any motor vehicle in the Saab range;

(b) planting 17 native trees would not provide a carbon dioxide offset for any period other than a single year's operation of any motor vehicle in the Saab range; and

(c) Saab vehicles do not have any attribute or attributes which contribute to reduced carbon dioxide emissions by those vehicles compared with Saab vehicles supplied prior to the publication of the advertisement.[58]

The ACCC commenced proceedings in the Federal Court and obtained declarations that GM Holden had breached ss 52 and 53(c) of the *Trade Practices Act*.[59] In addition, GM Holden accepted a court-enforceable undertaking to refrain from republishing the original advertisements and to train its marketing staff in relation to misleading and deceptive green marketing claims.[60] GM Holden advised the ACCC that it will plant 12500 native trees which it believes to be sufficient to offset the carbon emissions for the life of all Saab motor vehicles sold during the advertising campaign.[61]

The ACCC also pursued V8 Supercars Australia Pty Ltd after the company introduced its Racing Green Program claiming that it would plant 10 000 native trees to fully offset the carbon emissions from the V8 Championship Series.[62] The ACCC considered that the claims were unclear, suggesting that the trees would quickly absorb the carbon emission when it was likely that it would take a number of decades for one year's racing emissions to be absorbed by the specially planted trees.[63] The ACCC accepted a court enforceable undertaking from V8 Supercars Australia Pty Ltd that:

- any future claims that it publishes about 'green marketing' will be first considered by a solicitor with experience in trade practices law to ensure that it complies with the Act,
- any future claims that it makes about trees being planted to offset carbon emissions will include an explanation about the time before those emissions will be offset, and
- an acknowledgement of the ACCC's concerns and the undertaking will be placed on its Racing Green pages of the V8 Supercars website.[64]

Administrative Law

Issues relating to climate change can arise in judicial review and merits review proceedings.

Judicial Review

Standing

The legality or validity of administrative decisions and action may be reviewed by the courts on numerous grounds relating to climate change issues. However, at the outset, the person seeking review must have standing to sue. Plaintiffs have had mixed success in establishing standing to sue in climate change litigation, particularly in the United States.[65]

A breakthrough came with the US Supreme Court's decision last year in *Massachusetts v EPA*.[66] The State of Massachusetts, together with 11 other states, three cities, two United States territories and several environmental groups sought review of the denial by the Environment Protection Agency (EPA) of a petition to regulate the emissions of four GHGs, including carbon dioxide, under s 202(a)(1) of the *Clean Air Act*. Section 202(a)(1) of the Clean Air Act requires that the EPA shall by regulation prescribe standards applicable to the emission of any air pollution from any class of new motor vehicles which, in the EPA's judgment, causes or contributes to air pollution reasonably anticipated to endanger public health or welfare.

The Supreme Court held that Massachusetts had standing to challenge the EPA's denial of their rulemaking petition. The Supreme Court applied the three part test for standing in *Lujan v Defenders of Wildlife*,[67] namely:

(a) The plaintiff has suffered 'an injury in fact' which is both concrete and particularised, and actual and imminent, as opposed to conjectural or hypothetical.

(b) The injury is fairly traceable to the challenged action of the defendant.

(c) There is a likelihood that the injury can be redressed by a favourable decision, as opposed to this being merely speculation.

The Supreme Court held that Massachusetts had suffered an injury in fact as owner of the state's coastal land which is and will be affected by climate change-induced sea level rise and coastal storms.[68] The fact that other states suffered similar injuries did not disqualify Massachusetts.[69]

In relation to causation, the EPA did not contest the link between GHG emissions and climate change. However, the EPA argued that its decision not to regulate GHG emissions from new motor vehicles contributes so insignificantly to the petitioner's injuries that it cannot be challenged in court.[70] The Supreme Court held against the EPA stating that:

> Its argument rests on the erroneous assumption that a small, incremental step, because it is incremental, can never be attacked in a federal judicial forum. Yet accepting that premise would doom most challenges to regulatory action. Agencies, like legislatures, do not generally resolve massive problems in one fell regulatory swoop.[71]

The Supreme Court found that reducing domestic automobile emissions, a major contributor to GHG concentrations, is 'hardly a tentative step'.[72]

In relation to redressability, the Supreme Court held that while the remedy sought by the plaintiffs, regulating motor vehicle emissions, would not reverse global warming, it might slow down or reduce its effects.[73]

The issue of standing does not present the same procedural barrier in states where statutes have open standing provisions. For example, in New South Wales, many planning and environment statutes have open standing provisions — any person may bring proceedings to remedy or restrain a breach of the statute.[74] Judicial review of administrative action relating to climate change is thereby facilitated.

Grounds of review

Administrative decisions and action raising issues relating to climate change could conceivably be challenged on many of the grounds of judicial review. However, the more likely grounds would be:

(a) under the rubric of illegality, misdirection as to the applicable law or failure of the repository of power to have a required state of mind before exercising the administrative power;

(b) under the rubric of irrationality, failure of the repository of power to consider relevant matters or making a manifestly unreasonable decision; and

(c) under the rubric of procedural impropriety, failure of the repository of power to comply with some procedure in the statute, such as a requirement for environmental impact assessment, or for consultation.

Error of Law

Judicial review for error of law will lie where the administrative decision-maker misinterprets or misdirects itself as to the applicable law or question to be determined.

In *Massachusetts v EPA*, the US Supreme Court found that the EPA's reading of the applicable statutory provision, s 202(a)(1) of the *Clean Air Act*, was erroneous. GHGs are 'air pollutants' and the statutory provision authorised the EPA to regulate GHG emissions from new motor vehicles in the event that it formed the judgment that such emissions contribute to climate change.[75] The court held that the administrator must determine whether or not emissions of greenhouse gases from new motor vehicles cause or contribute to air pollution which may reasonably be anticipated to endanger public health or welfare, or whether the science is too uncertain to make a reasoned decision.[76]

In response to the Supreme Court's ruling, in April 2009, the EPA proposed two findings and submitted them for public debate. These are, first, an endangerment finding, being a proposal to find that the current and projected concentrations of the mix of six key greenhouse gases — carbon dioxide (CO_2), methane (CH_4), nitrous oxide (N_2O), hydrofluorocarbons (HFCs), perfluorocarbons (PFCs), and sulfur hexafluoride (SF6) — in the atmosphere threaten the public health and welfare of current and future generations.[77] Second, a cause or contribute finding, being a further proposal to find that the combined emissions of CO_2, CH_4, N_2O, and HFCs from new motor vehicles and motor vehicle engines contribute to the atmospheric concentrations of these key greenhouse gases and hence to the threat of climate change.[78]

Failure to Have Requisite State of Mind

The statute reposing power may require, as a condition precedent, that the decision-maker consider certain facts and form some opinion, satisfaction or belief that such facts exist. Failure to do so will entitle the court to review the decision-maker's decision as being *ultra vires*.

An example is the provision of an environmental planning instrument that a consent authority not grant consent unless satisfied that carrying out the development is consistent with the objectives of the zone. The zone objectives might be to prevent development which would adversely affect, or be adversely affected by, coastal processes. Development of land that is susceptible to coastal erosion, exacerbated by climate change, may not be consistent with such zone objectives.[79]

Failure to Consider Relevant Matters

A decision-maker will be bound to take into account matters that the statute expressly or by implication from the subject matter, scope or purpose of the statute require the decision-maker to consider.[80]

Examples of express relevant matters are provisions in local environmental plans requiring consideration of the effect of coastal processes and coastal hazards and potential impacts, including sea level rise, on a proposed development or arising from a proposed development.[81]

More commonly, the statute does not expressly state the matters relating to climate change and it is necessary to ascertain, from the subject matter, scope and purpose of the statute, whether the statute impliedly requires consideration of matters relating to climate change. Most of the climate change litigation has involved this task.

In *Australian Conservation Foundation v Latrobe City Council*,[82] the Victorian Civil and Administrative Tribunal held that the environmental effects of GHG emissions that were likely to be produced by use of the Hazelwood Power Station were relevant to the proposed amendment to the planning scheme to facilitate mining coal fields to supply coal for the power station.[83]

In *Gray v Minister for Planning*,[84] the Land and Environment Court held that GHG emissions from downstream use (burning) of coal mined from the proposed coal mine were relevant matters to be considered in the environmental assessment of the mine[85] and in the Director-General's decision to accept the environmental assessment as adequately addressing the environmental assessment requirements of the Director-General.[86]

In *Walker v Minister for Planning*,[87] the Land and Environment Court held that climate change flood risk for a project for the subdivision and residential development of land on a flood constrained coastal plain was a relevant matter to be considered by the Minister in determining to approve a concept plan for the project.[88]

On appeal, the New South Wales Court of Appeal, although reversing the Land and Environment Court's decision to void the Minister's decision in that case, nevertheless held that the Minister must consider the public interest in fulfilling functions under the *Environmental Planning and Assessment Act 1979* (NSW) (*EPA Act*).[89]

The Court of Appeal held that 'in respect of a consent authority making a decision in accordance with s 79C of the *EPA Act*, and a court hearing a merits appeal from such a decision, consideration of the public interest embraces ESD'.[90]

Further, the Court of Appeal held 'that the principles of ESD are likely to come to be seen as so plainly an element of the public interest, in relation to most if not all decisions, that failure to consider them will become strong evidence of failure to consider the public interest and/or to act bona fide in the exercise of powers granted to the Minister, and thus become capable of avoiding decisions'.[91]

In *Natural Resources Defense Council v Kempthorne*,[92] the court held that data about climate change that may adversely affect a threatened species of fish and its habitat was a relevant matter to be considered in the biological opinion of the United States Fish and Wildlife Service.[93]

Weight to be Attributed to Objects or Relevant Matters

Although generally the weight to be attributed to objects or relevant matters is within the discretion of the decision-maker, this general rule is subject to any statutory indication of the weight to be given. In an environmental context, statutes are increasingly providing an indication of the weight that a decision-maker is required to give to certain relevant considerations or the priority that should be accorded to certain objects of the statute or certain purposes for which the power may be exercised.[94]

Where the statute does indicate the weight to be given to a relevant consideration or the priority that should be accorded to a certain object or purpose, a reviewing court can intervene to set aside a decision if the decision-maker fails to accord the required weight or priority.

Non-Compliance With Procedural Requirements

Many planning and environment statutes require, as a precondition to the exercise of power to approve a development, compliance with certain procedures. These include undertaking an environmental impact assessment (EIA) of the proposed development. The EIA may be inadequate for failure to consider the impact of a proposed development on climate change or the impact climate change might have on a proposed development. A failure to comply with such procedural requirements may be judicially reviewed on the ground of procedural impropriety. Considerable climate change litigation has seized on this aspect of procedural impropriety.

Two examples are in Australia. In *Minister for the Environment & Heritage v Queensland Conservation Council,*[95] a relevant impact of a controlled action under the *Environment Protection and Biodiversity Conservation Act 1999* (Cth) was broadly interpreted to mean the influence or effect of an action. The impact could readily include the indirect consequences of an action and might include the results of acts done by persons other than the principal actor.[96] Applied to the facts of that case, the relevant impacts of the proposed action of constructing the Nathan Dam on the Dawson River in Queensland could include the impacts of the use of the water impounded by the dam for growing and ginning cotton downstream of the dam.[97] In *Gray v Minister for Planning,*[98] both direct and indirect effects of mining and subsequent use of the coal from the proposed coal mine were required to be considered in the environmental assessment.[99]

In four North American decisions, courts have held environmental impact assessments to be inadequate for failure to consider climate change impacts. In *Border Power Plant Working Group v Department of Energy*, the EIA for proposed electricity transmission lines was held inadequate for failure to discuss the carbon dioxide emissions from new power plants in Mexico, which would be connected by the proposed electricity transmission lines with the power grid in southern California.[100]

Mid States Coalition for Progress v Surface Transportation Board[101] concerned an EIA for a proposed rail line. The line would provide a less expensive route by which low-sulphur coal could reach electricity power plants and hence it would likely be utilised more than other routes. This would increase the supply of coal to the power plants and hence their consumption of coal. Greater consumption of coal by the power plants would increase the adverse effects of burning coal, including greenhouse gas emissions and climate change. The court held the EIA to be inadequate for failure to consider the possible effects of an increase in coal consumption.[102]

In *Center for Biological Diversity v NHTSA*,[103] the EIA of making a rule setting the corporate average fuel economy standard for light-duty trucks was held inadequate for failure to consider the effect of greenhouse gas emissions from light duty trucks on climate change.[104]

In *Pembina Institute for Appropriate Development v Attorney General of Canada*,[105] the Federal Court of Canada upheld a judicial review challenge to a Joint Review Panel's report on the EIA of the Kearl oil sands mine in northern Alberta. The court held that the Panel failed to explain in its report why the potential impacts of greenhouse gas emissions of the project will be insignificant and also failed to provide any rationale as to why the intensity-based mitigation proposed to be adopted would be effective to reduce greenhouse gas emissions, equivalent to 800 000 passenger vehicles, to a level of insignificance.[106]

Merits Review

Merits review involves a court (or tribunal) re-exercising the power of the original decision-maker. The court is not confined to the evidentiary material that was before the original decision-maker but may receive and consider fresh evidence in addition to or substitution of the original material.

Courts in merits review appeals have considered the effects a proposed development might have on climate change and the effects climate change might have on a proposed development.

In *Charles & Howard Pty Ltd v Redland Shire Council*,[107] the Planning and Environment Court of Queensland held that the impact of climate change on sea levels on an area of flood prone land proposed to be filled for residential development justified a condition requiring the proposed dwelling to be relocated to an area less prone to tidal inundation. In *Northcape Properties Pty Ltd v District Council of Yorke Peninsula*,[108] the Environment Resources and Development Court of South Australia held that changes in flood patterns and sea levels by global warming would erode a buffer zone and prevent public access to the coast, making coastal land subdivision unacceptable. In *Gippsland Coastal Board v South Gippsland Shire Council*,[109] the Victorian Civil and Administrative Tribunal held that the likely increase in severity of storm events and sea level rise due to the effects of climate change created a reasonably foreseeable risk of inundation of the land and proposed dwellings, which was unacceptable.

In a different context, courts in planning appeals have weighed in the balance the public interest in addressing climate change against narrower private interests, both in carrying out development or objecting to development.

In *Taralga Landscape Guardians Inc v Minister for Planning and RES Southern Cross Pty Ltd*,[110] the Land and Environment Court of New South Wales approved a large wind farm. Local residents of a nearby village, Taralga, and its surrounds had objected to the proposed wind farm on a variety of grounds, including visual impact and noise.

The wind farm was, however, beneficial in providing renewable energy with no greenhouse gas emissions, which could be substituted in part for non-renewable, fossil fuel energy with greenhouse gas emissions. The conflict was between the geographically narrower concerns of the residents and the broader public good of increasing the supply of renewable energy.[111] The court noted that increasing the supply of renewable energy involved promoting sustainable development, including intergenerational equity.[112] On balance, the court concluded that 'the overall public benefits outweigh any private disbenefits either to the Taralga community or specific landowners'.[113]

In a similar case, *Perry v Hepburn Shire Council*, the Victorian Civil and Administrative Tribunal also approved a wind farm. The tribunal took into account 'the benefits to the broader community of renewable energy generation as well as the contribution of the proposal to reducing greenhouse gas emissions'.[114]

Constitutional Law

Constitutions or statutes may provide for certain rights such as a right to life or right to a healthy environment. Such rights may provide a source for climate change litigation.

In India, the constitutional right to life (art 21) has been held to include the right to enjoy pollution-free water and air for full enjoyment of life and as providing a basis for sustainable development and intergenerational equity.[115] In Pakistan, the constitutional right to life (art 9) has been held to include a right to have a clean atmosphere and unpolluted environment.[116] In Kenya, the constitutional right not to be deprived of life save by court sentence (s 71(1)) has been held to include a denial of a wholesome environment in which to live.[117] In the Philippines, the right to a balanced and healthful ecology in accord with the rhythm and harmony of nature (art II, s 16) has been held to be a deduction from, if not a reiteration of, the constitutional right to life provision (art III, s 1).[118]

Such constitutional rights have the potential to found a challenge by an affected citizen against the government (or its instrumentalities) responsible for contributing to climate change and its effects. This is a vertical challenge. Rarely do such constitutional rights enable a horizontal challenge by the affected citizen against another citizen (including private industry) responsible for contributing to climate change and its effects.[119]

While there have been a number of actions, based on constitutional rights to life, addressing the effects of air pollution,[120] there has not yet been constitutional litigation focused on GHG emissions or climate change, although there is the potential.[121]

International Human Rights

Human rights under international conventions and instruments may provide a source for climate change litigation. Environmental litigation has occurred under two such instruments, the European Convention for the Protection of Human Rights and Fundamental Freedoms, before the European Court of Human Rights (ECtHR),[122] and the American Convention on Human Rights, before the Inter-American Commission for Human Rights (IACHR).

ECtHR Decisions

There have been three cases before the European Court of Human Rights concerning the infringement of human rights by air pollution. None of them addressed climate change, but these cases illustrate the potential for climate change litigation.

In *Lopez Ostra v Spain*,[123] the applicant lived metres away from and suffered for three years from smells, noise and polluting fumes caused by a sewerage plant treating liquid and solid waste. The responsible municipal and other authorities adopted a passive attitude to her entreaties. The ECtHR held Spain had breached art 8 (right to respect for private and family life) in that the authorities did not strike a fair balance between the town's need for a sewerage plant and the applicant's right under art 8.[124] The ECtHR held that the actions of the authorities in resisting judicial decisions and otherwise prolonging the situation amounted to a breach of the applicant's right to respect for private and family rights.[125]

In *Fadeyeva v Russia*,[126] the applicant alleged that the operation of a steel plant (the largest iron smelter in Russia) in close proximity to her home endangered her health and wellbeing due to the state's failure to protect her private life and home from severe environmental nuisance from the plant, in violation of art 8 of the Convention. The ECtHR held that while the Convention does not contain a right to nature preservation as such, art 8 could apply if the adverse effects of the environmental pollution had reached a certain minimum level. This threshold had been reached as the average pollution levels were way over the safe concentrations of toxic elements and local courts had recognised the applicant's right to resettle.[127] The ECtHR held Russia to be in breach of art 8 and awarded damages and costs.[128]

In *Okyay v Turkey*,[129] the applicants sought to stop the operation of three thermal power plants situated in the Aegean region of Turkey. The plants used low quality lignite coal. Sulphur and nitrogen emissions from the sites affected the air quality of a large area, while activities incidental to the plant's operation adversely affected the region's biodiversity. The applicants brought proceedings in local courts seeking to stop the operation of the plants, arguing that the plants did not have the required licences to function lawfully. They relied on the right to a healthy, balanced environment in art 56 of the Turkish Constitution, as well as provisions of the *Environment Act* requiring authorities to prevent pollution or ensure its effects are mitigated. The local courts upheld their appeal, finding that the plants did not have the required licences and ordered the plants to stop operating. The Turkish authorities refused to enforce the local court decisions. The applicants complained to the ECtHR that their right to a fair hearing under art 6 of the Convention had been breached by the authorities' failure to enforce the local courts' decisions to halt the operation of the power plants. The ECtHR found Turkey had violated art 6 and awarded the applicants compensation.[130]

IACHR decision

In 2005, the Inuit, Indigenous people in the Arctic region, filed a petition against the United States alleging human rights violations resulting from the US's failure to limit its emissions of GHGs and therefore reduce the impact of climate change. The petitioners invoked the right to culture, the right to property, the right to the preservation of health, life and physical integrity. The Inter-American Commission for Human Rights rejected the petition in 2006 without giving reasons. However, on the request of the petitioners, the Commission agreed to a hearing of the matter in 2007.[131]

International Law

Transboundary Environmental Damage

Many types of environmental damage do not stop at national boundaries or respect other states' sovereignty. In recent years, climate change has proved to be one of the best examples. Examples of transboundary environmental incidents causing air pollution or other environmental damage in more than one state include:

(a) the air pollution leading to the Trail Smelter arbitration[132] between Canada and the United States in 1906;

(b) the Chernobyl incident which took place in 1986, when a radioactive leak from a nuclear power plant in the former USSR (currently the Ukraine) was carried off over a number of European States harming human health and damaging ecosystems;[133] and

(c) haze from Indonesian forest fires noticeable in a number of neighbouring countries and going as far as Australia in 1999.[134]

Climate change is a form of transboundary environmental harm. Various fora have been put forward as having a role to play in resolving disputes arising out of climate change, including the International Court of Justice, the World Trade Organization Appellate Body, the International Tribunal for the Law of the Sea, and the World Heritage Committee.

International Court of Justice

The International Court of Justice ('ICJ'), the judicial body of the United Nations, had an important role in shaping the law of this area. It first held, in the 1949 Corfu Channel case,[135] that a state had a duty 'not to allow knowingly its territory to be used for acts contrary to the rights of other states'.[136] This was interpreted to be the recognition of a duty on a state to warn others when the danger is located on the state's territory.[137] The case, however, did not deal with air pollution, but with the loss of an English ship due to mines found in Albanian territorial waters.

The ICJ has not had many opportunities to decide on environmental matters. The main cases so far[138] have stopped short of giving a detailed exposition on environmental law.[139]

In 2002, after the United States of America and Australia refused to ratify the Kyoto Protocol to the United Nations Framework Convention on Climate Change, the Pacific Island-State of Tuvalu threatened to take action in the ICJ against countries who have not ratified the treaty.[140] Tuvalu never commenced proceedings. However, this potential litigation highlighted some of the difficulties which arise in bringing proceedings before the ICJ. These difficulties stem mainly from the way in which the rules of the court are framed, in accordance with principles of public international law. Only state parties to the United Nations Charter can bring disputes before the

ICJ.[141] This means that individuals or organisations need to persuade a government to bring a claim in their name, which is not always an achievable task. Also, the parties to a dispute have to accept the jurisdiction of the court.[142] This requirement makes states such as the United States, which has rescinded its acceptance of the compulsory jurisdiction in the 1980s, unlikely to be easily brought before this court.[143] Alternatively, the parties have to agree to bring the dispute before the court,[144] which is not always a likely outcome on issues as sensitive as GHG emissions constituting a violation of international environmental law.[145] Finally, parties can agree to bring disputes before the ICJ under a treaty which is in effect between them.[146] In Tuvalu's case, its only treaty with the United States, the *Treaty of Friendship*,[147] does not contain such a clause.

World Trade Organization (WTO)

The World Trade Organization Appellate Body has had an opportunity to deal with some environmental matters. In particular, art XX of the General Agreement on Tariffs and Trade (GATT) provides parties with the opportunity to raise environmental issues as justification for not complying with an obligation under GATT. Article XX provides:

> Subject to the requirement that such measures are not applied in a manner which would constitute a means of arbitrary or unjustifiable discrimination between countries where the same conditions prevail, or a disguised restriction on international trade, nothing in this Agreement shall be construed to prevent the adoption or enforcement by any contracting party of measures:
>
> ... (b) necessary to protect human, animal or plant life or health;
>
> ... (g) relating to the conservation of exhaustible natural resources if such measures are made effective in conjunction with restrictions on domestic production or consumption.

While this provision has not been employed in respect of climate change, it has proven a source of environmental litigation before the WTO.[148] In the shrimp/turtle dispute,[149] the chapeau as well as paragraphs (b) and (g) to art XX were accepted as a basis for the imposition of a unilateral ban by the United States on shrimp imports from certain south Asian countries ostensibly to protect an endangered species of sea turtle, listed under CITES.[150] However, the ban failed the chapeau on grounds of discrimination. In the case of climate change litigation, it would seem more likely that paragraph (b) of the chapeau to art XX would be invoked.

One of the arguments that has been suggested that parties to a dispute could use, would be that failure to ratify the Kyoto Protocol to the UNFCCC constitutes a state subsidy to the firms registered in that state, in breach of the relevant provisions of the GATT.[151]

International Tribunal for the Law of the Sea

Another possible international forum for climate change litigation is the International Tribunal for the Law of the Sea (ITLOS), established under the United Nations Convention on the Law of the Sea (UNCLOS).[152] This body has also seen some environmental-related litigation.[153] Under the UNCLOS, another agreement was negotiated in 1995, the Agreement for the Implementation of the Provisions of the United Nations Convention on the Law of the Sea 10 Dec 1982 Relating to the Conservation and Management of Straddling Fish Stocks and High Migratory Fish ('UNFSA'). This is a more likely legal framework in which to bring disputes relating to climate change.

Fish are very susceptible to changes in the temperature of oceans,[154] which means that they, and consequently the commercial fisheries sector, may be profoundly affected by climate change.[155] The agreement has a broad application given the number of parties that have adhered to it — 71 — including large GHG emitters such as the United States and India, and the fact that it covers one fifth of the total marine catch.[156] The advantage that the UNFSA presents is that it provides for a binding dispute resolution mechanism. While the Agreement's main objective is the long-term conservation and sustainable use of straddling fish stocks and highly migratory species, it also envisages other activities that would imperil conservation.[157] One argument that could be put forward is that activities resulting in the emission of GHGs could be considered to hinder conservation efforts because of the link between GHG concentrations in the atmosphere and climate change, and climate change's impact on fish stock conservation. However, such an argument has not yet been presented in a dispute under the UNFSA.

Heritage Committee

Numerous cases have also been commenced before the International Committee for the Protection of the Cultural and Natural Heritage of Outstanding Universal Value (World Heritage Committee), part of the United Nations Educational, Scientific and Cultural Organisation (UNESCO). These petitions requested that various designated world heritage sites be placed on the list of world heritage sites in danger owing to the impact of climate change on these sites. This was a bid to ensure that the parties to the World Heritage Convention, including large GHG emitters like the United States of America or Canada, abide by their obligation under the Convention and 'do all [they] can ... to the utmost of [their] own resources'[158] to protect and conserve natural heritage within its boundaries, namely that they reduce their GHG emissions in order to limit climate change and its impacts on natural heritage.[159] The sites covered in the petitions are the Waterton-Glacier International Park (USA/Canada), Sagarmatha National Park (Nepal), Belize Barrier Reef Reserve System (Belize), Huascaran National Park (Peru), and the Great Barrier Reef (Australia).[160]

In 2005, the World Heritage Committee recommended that a group of experts analyse the situation and report to the following annual meeting of the Committee on their findings in regards to the effects of climate change on world heritage sites.[161] At the following session, in 2006, the committee limited itself to endorsing the 'Strategy to assist State Parties to implement appropriate management responses'[162] and requested that State Parties implement the strategy.[163] The strategy refers generally to mitigating and adaptive measures that can be taken to limit the effects of climate change on world heritage sites, but does not present as a document that imposes any particular actions to be taken,[164] to the disappointment of the petitioners.[165]

Conclusion

Even if many of the attempts to litigate climate change are unsuccessful, there is a consensus among commentators that there is value in the attempts themselves. While courts are bound by domestic or international norms in their activity, and cannot bring about dramatic change, climate change litigation has proved to be a vehicle through which matters that are important to communities are being brought to the attention of the governments and, hence, act as a catalyst for executive action.[166] Another important effect of litigation is that such actions raise the defendants' and the public's awareness of the implications of climate change[167] and, sometimes, solutions are reached at a much faster pace by commencing proceedings.[168]

Not only will there be a more frequent use of the avenues of litigation covered in this chapter, but it is likely that the avenues used to litigate climate change-related matters will continue to expand. As governments are likely to implement new legislation to tackle climate change, such as carbon emissions trading schemes, this could also provide litigants with new ways in which to challenge climate change-inducing actions.

Endnotes

1   Intergovernmental Panel on Climate Change (IPCC), *Climate Change 2007 Synthesis Report: Summary for policymakers* (2007) 2.

2   E Posner, Climate Change and International Human Rights Litigation: A critical appraisal (2007) 155 *University of Pennsylvania Law Review* 1925, 1925.

3   JL Sax, *Defending the Environment: A handbook for citizen action* (1971) xviii, 152.

4   Posner, n 2, 1925-6. See also R Meltz, Report for Congress: Climate change litigation: A growing Phenomenon (Congressional Research Service, 2007) 33; A Huggins, Is Climate Change Litigation an Effective Strategy for Promoting Greater Action to Address Climate Change? What other legal mechanisms might be appropriate? (2008) 13 *Local Government Law Journal* 184.

5   *Massachusetts v EPA* 549 US 1 (2007); *Walker v Minister for Planning* (2007) 157 LGERA 124.

6   *Connecticut v American Electric Power* 406 F Supp 2d 265 (SDNY, 2005); *Genesis Power Ltd v Greenpeace New Zealand Inc* [2008] 1 NZLR 803.

7   *City of Los Angeles v National Highway Traffic Safety Administration* 912 F 3d 478 (DC Cir, 1990) in Meltz, above n 4, 15.

8   *Greenpeace Australia v Redbank Power Company* (1994) 86 LGERA 143 in T Bonyhady, 'The New Australian Climate Law' in T Bonyhady and P Christoff (eds), Climate Law in Australia (2007) 11.

9   *Environmental Defence Society (Inc) v Auckland Regional Council and Contact Energy Ltd* [2002] 11 NZRMA 492; *Gray v Minister for Planning* (2006) 152 LGERA 258; *Taralga Landscape Guardians Inc v Minister for Planning and RES Southern Cross Pty Ltd* (2007) 161 LGERA 1; *Walker v Minister for Planning* (2007) 157 LGERA 124.

10   *Re Xstrata Coal Queensland Pty Ltd* [2007] QLRT 33.

11   *Walker v Minister for Planning* (2007) 157 LGERA 124; *Gray v Minister for Planning* (2006) 152 LGERA 258.

12   DM Walker, *The Oxford Companion to Law* (1980) 894.

13   See M Pawa and K Krass, Global Warming as a Public Nuisance (2004-05) 16 Fordham Environmental Law Review 207; K Alek, A Period of Consequences: Global warming as public nuisance (2007) 26A/43A *Stanford Journal of International Law* 77; and D Hunter and J Salzman, Negligence in the Air: The duty of care in climate change litigation (2007) 155 *University of Pennsylvania Law Review* 1741, 1788–94.

14   406 F Supp 2d 265 (SDNY, 2005).

15   2007 WL 2726871 (NDCal, 2007).

16 *Connecticut v American Electric* 406 F Supp 2d 265 (SDNY, 2005) at 274; People of the *State of California v General Motors* 2007 WL 2726871 (NDCal, 2007) 5-13; Alek, above n 13, 91; Meltz, above n 4, 24.

17 08-CV-1138 (NDCal, filed 26 February 2008).

18 A Dugdale and MA Jones (eds), Clerk and Lindsell on Torts (19th ed, 2006) [20-01] 1162; and *Sedleigh-Denfield v O'Callaghan* [1940] AC 880, 903.

19 J McDonald, 'A Risky Climate for Decision-Making: The liability of development authorities for climate change impacts' (2007) 24 *Environment and Planning Law Journal* 405, 414.

20 04-CV-05670 (SDNY, filed 21 July 2004).

21 *Kenny & Good Pty Ltd v MGICA* (1992) Ltd (1997) 77 FCR 307, 322.

22 *Roads and Traffic Authority of NSW v Dederer* (2007) 238 ALR 761 [43] (Gummow J) with whose reasons on the nature and extent of the duty of care Hayden J ([283]) and Callinan J ([270]) agreed. See also *Palsgraf v Long Island Railroad Company* 162 NE 99 (NY, 1928) 99 (Cardozo J).

23 *Roads and Traffic Authority of NSW v Dederer* (2007) 238 ALR 761 [44].

24 *Roads and Traffic Authority of NSW v Dederer* (2007) 238 ALR 761 [43]–[44].

25 *Roads and Traffic Authority of NSW v Dederer* (2007) 238 ALR 761 [18], [43], [51].

26 *Sutherland Shire Council v Heyman* (1984) 157 CLR 424, 481; Perre v Apand Pty Ltd (1999) 198 CLR 180 [93]–[94].

27 *Perre v Apand Pty Ltd* (1999) 198 CLR 180 [76], [93].

28 See the summary in the joint judgment in *Sullivan v Moody* (2001) 207 CLR 562 [50]-[51]. See also *Perre v Apand Pty Ltd* (1999) 198 CLR 180 [27], [201], [333], [406]; Graham Barclay Oysters Pty Ltd v Ryan (2002) 211 CLR 540 [149], [236]–[237]; *Hunter Area Health Service v Presland* (2005) 63 NSWLR 22 [9].

29 *Perre v Apand Pty Ltd* (1999) 198 CLR 180 [70]; see also s 5B(1)(a) of the Civil Liability Act 2002 (NSW).

30 See, for example, *Sullivan v Moody* (2001) 207 CLR 562 [42].

31 Hill v Van Erp (1997) 188 CLR 159, 176–7, 210, 237–9; *Perre v Apand Pty Ltd* (1999) 198 CLR 180 [27], [74], [280], [330]–[331]; *Sullivan v Moody* (2001) 207 CLR 562 [48]; *Vairy v Wyong Shire Council* (2005) 223 CLR 422 [28], [66].

32 See *Perre v Apand Pty Ltd* (1999) 198 CLR 180 [10]–[15], [133], [201], [259], [335], [406].

33 Hunter and Salzman, above n 13, 1748.

34 *Wyong Shire Council v Shirt* (1980) 146 CLR 40, 47.

35 See *Hunter and Salzman*, above n 13, 1745, 1769.

36 See further *Hunter and Salzman*, above n 13, 1745-9.

37 *Wyong Shire Council v Shirt* (1980) 146 CLR 40, 47–8, affirmed as still being the proper approach in *NSW v Fahy* (2007) 81 ALJR 1021.

38 March v E & MH Stramare Pty Ltd (1991) 171 CLR 506, 515, 522, 524.

39 *Ultramares Corp v Touche* 174 NE 441 (NY, 1931) 444, cited in Bryan v Maloney (1995) 182 CLR 609, 618.

40 *Comer v Murphy Oil* No 1:05-CV-436 (SD Miss, 18 April 2006); Hunter and Salzman, above n 13, 1754.

41 The claims are quoted in *Hunter and Salzman*, above n 13, 1754-5.

42 *Comer v Murphy Oil* No 1:05-CV-436 (SD Miss, 18 April 2006) dismissed on 30 August 2007.

43 P England, 'Heating Up: Climate change law and the evolving responsibilities of local government' (2008) 13 *Local Government Law Journal* 209, 217.

44 Hunter and Salzman, above n 13; Z Lipman and R Stokes, 'Shifting Sands: The implications of climate change and a changing coastline for private and public authorities in relation to waterfront land' (2003) 20 *Environmental and Planning Law Journal* 405; J McDonald, 'The Adaptation Imperative: Managing the Legal Risks of Climate Change Impacts' in Bonyhady and Christoff, above n 8, 124; McDonald, above n 19; England, above n 43.

45 In New South Wales, the liability of local councils is limited through the Local Government Act 1993 (NSW) with respect to coastal hazards (s 744) and pt 5 of the Civil Liability Act 2002 (NSW).

46 *Mulcahy v The Queen* (1868) LR 3 HL 306, 317.

47 Dugdale and Jones, above n 18, 1611.

48 Walker, above n 12, 276.

49 S Faris, Climate Conspiracy?, *Australian Financial Review* (Sydney) 27 June 2008, 9.

50 08-CV-1138 (NDCal, filed 26 February 2008).

51 Walker, above n 12, 341; and see also Dugdale and Jones, above n 18, Ch 18.

52 LexisNexis, Halsbury's Laws of Australia, 300 Negligence, '2 Duty of Care' [300-10].

53 Walker, above n 12, 844.

54 Australian Competition and Consumer Commission (ACCC), Green Marketing and the *Trade Practices Act* (2008).

55 ACCC, Carbon Claims and the *Trade Practices Act* (2008).

56 ACCC, *De Longhi Alters 'Environmental Friendly' Claims* (Media release 112/08, 30 April 2008).

57 ACCC, *Goodyear Tyres Apologise, Offer Compensation for Unsubstantiated Environmental Claims* (Media release 181/08, 26 June 2008).

58 ACCC, *ACCC Takes Action Against GM Holden Ltd over Saab 'Green' Claims* (Media release 8/08, 18 January 2008).

59 *Australian Competition and Consumer Commission v GM Holden Ltd* (ACN 006 893 232) [2008] FCA 1428 [16], [10].

60 *Australian Competition and Consumer Commission v GM Holden Ltd* (ACN 006 893 232) [2008] FCA 1428 [6], [10] setting out the undertaking given to ACCC by GM Holden for the purposes of s 87B of the *Trade Practices Act* 1974 (Cth). See [22]-[23] of the undertaking.

61 *Australian Competition and Consumer Commission v GM Holden Ltd* (ACN 006 893 232) [2008] FCA 1428 [6], [10] setting out the undertaking given to ACCC by GM Holden for the purposes of s 87B of the *Trade Practices Act* 1974 (Cth). See [17] of the undertaking. See also Australian Competition and Consumer Commission (ACCC), 'Saab "Grrrrrreen" Claims Declared Misleading by the Federal Court' (Media release 267/08, 18 September 2008).

62 ACCC, 'V8 Supercars corrects carbon emissions claims' (Media release 265/08, 18 September 2008).

63 Ibid.

64 Ibid.

65 Examples where standing has been denied include *Center for Biological Diversity v Abraham* 218 F Supp 2d 1143 (ND Cal, 2002) and *Korsinsky v US EPA* 2005 US Dist LEXIS 21778 affirmed *Korsinsky v US EPA* 2006 US App LEXIS 21024 (2d Cir NY, 2006) while standing was upheld in *City of Los Angeles v National Highway Traffic Safety Administration* 912 F2d 478 (CADC, 1990); *Friends of the Earth Inc v Watson*, 2005 WL 2035596 (ND Cal, 2005); 35 Envtl L Rep 20,179; and *Natural Resources Defense Council v EPA*, 464 F 3d 1 (CADC, 2006).

66 549 US 1 (2007).

67 504 US 555 (1992).

68 *Massachusetts v EPA* 549 US 1 (2007) 19-20.

69 *Massachusetts v EPA* 549 US 1 (2007) 19.

70 *Massachusetts v EPA* 549 US 1 (2007) 20.

71 *Massachusetts v EPA* 549 US 1 (2007) 21.

72 *Massachusetts v EPA* 549 US 1 (2007) 21-2.

73 *Massachusetts v EPA* 549 US 1 (2007) 22.

74 For example, see Environmental Planning and Assessment Act 1979 (NSW) s 123.

75 *Massachusetts v EPA* 549 US 1 (2007) 25-6, 29-30.

76 US Environment Protection Agency, 'Proposed Endangerment and Cause or Contribute Findings for Greenhouse Gases under the Clean Air Act' (17 April 2009) <http://epa.gov/climatechange/endangerment.html> at 5 June 2009.

77 Ibid.

78 Ibid.

79 Eg, see Byron Local Environmental Plan 1988 cl 9(3).

80 See *Minister for Aboriginal Affairs v Peko Wallsend Ltd* (1986) 162 CLR 24, 39–40, 55.

81 See Standard Instrument (Local Environmental Plans) Order 2006 cl 5.5 (2).

82 (2004) 140 LGERA 100.

83 *Australian Conservation Foundation v Latrobe City Council* (2004) 140 LGERA 100 [43]–[47].

84 (2006) 152 LGERA 258.

85 *Gray v Minister for Planning* (2006) 152 LGERA 258 [100], [125].

86 *Gray v Minister for Planning* (2006) 152 LGERA 258 [115], [126], [135].

87 (2007) 157 LGERA 124.

88 *Walker v Minister for Planning* (2007) 157 LGERA 124 [166].

89 *Minister for Planning v Walker* (2008) 161 LGERA 423 [39].

90 *Minister for Planning v Walker* (2008) 161 LGERA 423 [42].

91 *Minister for Planning v Walker* (2008) 161 LGERA 423 [56].

92 506 F Supp 2d 322 (EDCal, 2007)

93 *Natural Resources Defense Council v Kempthorne* 506 F Supp 2d 322 (EDCal, 2007) 368–70.

94 Examples where there is a statutory indication of the weight or priority to be given to aspects of ecologically sustainable development are Coastal Protection Act 1979 (NSW) s 37A; National Parks and Wildlife Act 1974 (NSW) s 2A(1); Water Management Act 2000 (NSW) s 9(1); Fisheries Management Act 1994 (NSW) s 3(2); and Sydney Regional Environmental Plan (Sydney Harbour Catchment) 2005 cl 2(1).

95 (2004) 134 LGERA 272. Full Federal Court decision upholding Kiefel J's decision in *Queensland Conservation Council Inc v Minister for the Environment and Heritage* [2003] FCA 1463.

96 *Minister for the Environment and Heritage v Queensland Conservation Council* (2004) 134 LGERA 272 [53]–[55].

97 *Minister for the Environment and Heritage v Queensland Conservation Council* (2004) 134 LGERA 272 [60]. See DE Fisher, The Meaning of Impacts — The Nathan Dam Case on appeal (2004) 21 *Environmental and Planning Law Journal* 325; C McGrath, Key concepts of the *Environment Protection and Biodiversity Conservation Act 1995* (Cth)' (2005) 22 *Environmental and Planning Law Journal* 20, 36; N Sommer, *Queensland Conservation Council Inc v Minister for the Environment and Heritage* [2003] FCA 1463 ('*Nathan Dam Case*') (2004) 9 AJNRLP 145.

98 (2006) 152 LGERA 258.

99 See also *Australian Conservation Foundation v Latrobe City Council* (2004) 140 LGERA 100, 110.

100 *Border Power Plant Working Group v Department of Energy* 260 F Supp 2d 997 (SD Cal, 2003) [42].

101 345 F 3d 520 (8th Cir, 2003).

102 *Mid States Coalition for Progress v Surface Transportation Board* 345 F 3d 520 (8th Cir, 2003) [29].

103 508 F 3d 508 (9th Cir, 2007).

104 Center for Biological Diversity v NHTSA 508 F 3d 508 (9th Cir, 2007) [20]–[22].

105 2008 FC 302 (5 March 2008).

106 *Pembina Institute for Appropriate Development v Attorney General of Canada* 2008 FC 302 (5 March 2008) [73]–[75], [78], [79].

107 (2007) 159 LGERA 349.

108 [2008] SASC 57.

109 [2008] VCAT 1545.

110 (2007) 161 LGERA 1. A subsequent application to modify the development was dealt with by the *New South Wales Land and Environment Court in RES Southern Cross v Minister for Planning and Taralga Landscape Guardians Inc* [2008] NSWLEC 1333.

111 *Taralga Landscape Guardians Inc v Minister for Planning and RES Southern Cross Pty Ltd* (2007) 161 LGERA 1 [3].

112 *Taralga Landscape Guardians Inc v Minister for Planning and RES Southern Cross Pty Ltd* (2007) 161 LGERA 1 [73], [74].

113 *Taralga Landscape Guardians Inc v Minister for Planning and RES Southern Cross Pty Ltd* (2007) 161 LGERA 1 [352].

114 *Perry v Hepburn Shire Council* (2007) 154 LGERA 182 [27].

115 MC *Mehta v Union of India* AIR 1988 SC 1037; *Vellore Citizens Welfare Forum v Union of India* AIR 1996 SC 2715; (1996) 5 SCC 647; AP *Pollution Control Board v Prof MV Nayudu* (ret'd) [1999] 1 LRI 185; and MC *Mehta v Kamal Nath* AIR 2000 SC 1997.

116 *Shehla Zia v WAPDA PLD* 1994 SC 693; General Secretary, West Pakistan Salt Miners Labour Union, *Khewral, Jhelum v Director of Industries and Mineral Development*, Punjab 1994 SCMR 2061.

117 *Waweru v Republic* (2006) 1 KLR (E&L) 677.

118 *Minors Oposa v Factoran*, Secretary of the Department of Environment and Natural Resources 33 ILM 173 (1994); 224 SCRA 792 (1994).

119 See P Alston (ed), Human Rights Law (1996) xii–xiii, and for an international human rights perspective see H Steiner, P Alston and R Goodman (eds), *International Human Rights in Context: Law, politics, morals* (2008) 58–9.

120 See, for example, MC *Mehta v Union of India and Shriram Food and Fertiliser Industries* AIR 1987 SC 965 (Oleum Gas Leak case I); AIR 1987 SC 982 (Oleum Gas Leak case II); AIR 1987 SC 1026 (Oleum Gas Leak case III); AIR 1987 SC 1086 (Oleum Gas Leak case IV); *Indian Council for Enviro-Legal Action v Union of India* (1996) 3 SCC 212; AIR 1996 SC 1446; MC *Mehta v Union of India* WP13381/1984 (30 December 1996); AIR 1997 SC 734 (Taj Trapezium case).

121 L Horn, 'Climate Change Litigation Actions for Future Generations' (2008) 25 *Environmental and Planning Law Journal* 115, 131.

122 Similar cases, based on the European Convention for Human Rights (EHCR), can arise before the European Court of Justice (ECJ), the judiciary arm of the European Community. The ECJ has long recognised human rights and the ECHR specifically as being part of European Community Law (case 11/70, *Internationale Handelgesellschaft mbH v Einfuhrund Vorratselle fr Getreide und Futtermittel* [1970] ECR 1125; case 4/73, Nold, Kohlen und *BarstoffgroBhandlung v Commission of the European Communities* [1974] ECR 491 at 507; case 36/75, *Roland Rutili v Minister for the Interior* [1975] ECR 1219, 1232). Also, once the Charter of Fundamental Rights of the European Union enters into force, the ECJ will also have recourse to a right to environmental protection (art 37 of the Charter), a provision which has no counterpart in the ECHR.

123 Application No 16798/90 (European Court of Human Rights, Chamber, 9 December 1994).

124 *Lopez Ostra v Spain* (European Court of Human Rights, Chamber, 9 December 1994) 58.

125 *Lopez Ostra v Spain* (European Court of Human Rights, Chamber, 9 December 1994) 56.

126 Application No 55723/00 (European Court of Human Rights, Chamber, 30 November 2005).

127 *Fadeyeva v Russia* Application No 55723/00 (European Court of Human Rights, Chamber, 30 November 2005) 80, 84, 86.

128 *Fadeyeva v Russia* Application No 55723/00 (European Court of Human Rights, Chamber, 30 November 2005) 134, 138, 149–150.

129 Application No 36220/97 (European Court of Human Rights, Chamber, 12 October 2005).

130 Okyay v Turkey Application No 36220/97 (European Court of Human Rights, Chamber, 12 October 2005) 74–75, 79.

131 Earthjustice, Nobel Prize Nominee Testifies about Global Warming (Press release, 1 March 2007) Retrieved 30 March 2009 from http://www.earthjustice.org/news/press/007/nobel-prize-nominee-testifies-about-global-warming.html

132 Trail Smelter case (*United States v Canada*) [1941] 3 RIAA 1907; 3 UN Rep Int; 1 Arb Awards (1941). Sulphur-dioxide fumes from a British Columbia (Canada) smelter damaged apple crops in the State of Washington (United States). In 1930, for example, 300-50 tonnes of sulphur were emitted from the smokestacks of the smelter.

133 D McClatchey, Chernobyl and Sandoz One Decade Later: The evolution of state responsibility for international disasters, 1986–1996 (1996) 25 *Georgia Journal of International and Comparative Law* 659.

134 AK-J Tan, Forest Fires of Indonesia: State responsibility and international liability (1999) 48 *International and Comparative Law Quarterly* 826; Laode M Syarif, *Orang-utans Can't Wear Smoke Masks: Indonesia's legal response to forest fires* (Paper presented at the Second Asian Law Institute Conference, The Challenge of Law in Asia: From Globalisation to Regionalisation, Bangkok, Thailand, 26-27 May 2007). For a discussion on possible legal avenues to engage the responsibility of the Indonesian State in international law, see Laode M Syarif, *Regional Arrangements for Transboundary Atmospheric Pollution in ASEAN Countries* (Doctor of Philosophy thesis, University of Sydney, 2006) 218-26, 233-9.

135 Corfu Channel case (*UK v Albania*) 1949 ICJ 4 (1949).

136 McClatchey, above n 133, 664.

137 Ibid.

138 Judgment in the Case Concerning the Gabcíkovo-Nagymaros Project (*Hungary v Slovakia*) 1997 ICJ 3 (September 25), reprinted in 37 ILM 162 (1998); Legality of the Threat or Use of Nuclear Weapons (Advisory Opinion) 1996 ICJ 226 (July 8), reprinted in 35 ILM 809 (1996); Nuclear Tests (*New Zealand v France*) 1995 ICJ 228 (September 22).

139 P Sands, 'International Environmental Litigation and Its Future' (1999) 32 *University of Richmond Law Review* 1619, 1633.

140 For more articles on Tuvalu and global warming, see Tuvalu Islands, 'Tuvalu and Global Warming' <http://www.tuvaluislands.com/warming.htm> at 30 March 2009.

141 *Statute of the International Court of Justice* arts 34(1) and 35(1).

142 *Statute of the International Court of Justice* art 36.

143 A Strauss, 'The Legal Option: Suing the United States in International Forums for Global Warming Emissions' (2003) 33 *ELR* 10185, 10185-6.

144 *Statute of the International Court of Justice* art 36(1).

145 Strauss, above n 143, 10186.

146 Ibid 10186.

147 *Treaty of Friendship between the United States of America and Tuvalu*, opened for signature 9 February 1979, 2011 UNTS 79 (entered into force 23 September 1983).

148 Environmental or human health issues were the basis for non-compliance with GATT and for the subsequent challenge by the trade affected party in the EC-Hormones case (European Communities — Measures concerning meat and meat products (hormones), WT/DS26&48/AB/R (Appellate Body Report adopted by the Dispute Settlement Body (DSB) on 13 February 1998)); France-Asbestos case (European Communities — Measures affecting asbestos and asbestos-containing products, WT/DS135/AB/R (Appellate Body Report adopted by the DSB on 5 April 2001)); United States-Reformulated Gasoline case (United States — Standards for reformulated gasoline (Appellate Body Report adopted by the DSB on 20 May 1996)); Australia-Salmon case (Australia - Measures affecting the importation of salmon, WT/DS18 (Appellate Body Report and Panel Report as modified adopted by the DSB on 6 November 1998; Compliance Panel Report adopted by DSB on 20 March 2000)); and Brazil-Retreaded Tyres case (Brazil — Measures affecting imports of retreaded tyres, WT/DS332/AB/R (Appellate Body Report adopted by the DSB on 19 December 2007)); Japan — Apples case (Japan — Measures affecting the importation of apples, WT/DS245/AB/R (Appellate Body Report adopted by the DSB on 10 December 2003; Appellate Body and Panel Compliance Reports adopted by the DSB on 20 July 2005)).

149 United States — Import prohibition of certain shrimp and shrimp products, WT/DS58/AB/R. Appellate body report adopted on 6 November 1998.

150 Sands, above n 139, 1635-6.

151 W Burns, The Exigencies that Drive Potential Causes of Action for Climate Change at the International Level (2004) 98 *American Society for International Law Proceedings* 223, 227.

152 Strauss, above n 143, 10188 and, drawing on that article, see also Burns, above n 151.

153 Southern Bluefin Tuna Cases (*Australia v Japan; NZ v Japan*) (1999).

154 W Burns, 'A Voice for the Fish? Litigation and potential causes of action for impacts under the United Nations Fish Stocks Agreement' (2008) 48 *Santa Clara Law Review* 605, 617.

155 Burns, above n 151, 607.

156 Ibid 608.

157 Ibid 635–6.

158 *Convention Concerning the Protection of the World Cultural and Natural Heritage*, opened for signature 16 November 1972, 15 UNTS 511 (entered into force 17 December 1975) art 4

159 See E Thorson, A Stasch, C Scott, K Gibel and K McCoy, *Petition to the World Heritage Committee Requesting Inclusion of Waterton-Glacier International Peace Park on the List of World Heritage in Danger as a Result of Climate Change and for Protective Measures and Actions* (2006) vii–viii.

160 Meltz, above n 4, 30.

161 World Heritage Committee, Decision 29 COM 7B.a in Decisions adopted at the *29th session of the World Heritage Committee*, WHC-05/29.COM/22 (Durban, 2005) 36, 7.

162 World Heritage Committee, Decision 30 COM 7.1 in Decisions adopted at the *30th session of the World Heritage Committee*, WHC-05/29.COM/22 (Vilnius, 2006) 7, 6.

163 World Heritage Committee, Decision 30 COM 7.1 in Decisions adopted at the *30th session of the World Heritage Committee*, WHC-05/29.COM/22 (Vilnius, 2006) 7, 8.

164 World Heritage Committee, 'Strategy to Assist State Parties to Implement Appropriate Management Responses', WHC-06/30.COM/7.1 (adopted on 17 July 2006) 1-2, B.9, C.10.

165 See Climate Justice, *World Heritage Committee Fails to Act* (Press release, 20 July 2006). Retrieved 30 March 2009 from http://www.climatelaw.org/cases/country/intl/unescobelize/2006Jul20/

166 Sax, above n 3, xviii, 152. See also BJ Preston, The Role of Public Interest Environmental Litigation (2006) 23 *Environmental and Planning Law Journal* 337, 339.

167 B Harper, Climate Change Litigation: The federal common law of interstate nuisance and federalism concerns (2006) 40 *Georgia Law Review* 661, 697.

168 See, eg, ibid 697.

Liability in Tort for Damage Arising From Human-Induced Climate Change

Peter Cashman and Ross Abbs

In his provocative book *Liability: The Legal Revolution and its Consequences*,[1] United States tort law 'reform' advocate Peter Huber analyses the historical expansion of the boundaries of tort liability.

As he notes, historically, the law of nuisance, like negligence, 'kept most of the world out of court most of the time'.[2] Isolated individual litigation, relying on the law of nuisance or the principles arising out of *Rylands v Fletcher*,[3] arose where aggrieved individuals were affected adversely by water, smoke, vibration, barking dogs and the fouling of water by cattle.

Gradually, increasing concern developed with respect to the possible effects of exposure to various toxic chemicals in food, water, cosmetics and pharmaceuticals. In many instances, toxic time bombs began to explode, or at least tick louder. As Huber observes, '[t]he moment had come for personal injury law, in its relentless and unending expansion, to invade the formerly sleepy kingdom of nuisance law'.[4]

Until this point, little legal attention had been paid to widely dispersed but apparently low-level injuries and 'nuisances'. On Huber's assessment, the prevailing view had been that it was 'better for 10,000 people to suffer the minor harms of stinging eyes and sooty window sills than for 200 workers to suffer the major harm of unemployment'.[5] This view was shared by employers and employees and acquiesced by the public.

Subsequent environmental litigation, and legislation, arose out of the discovery that activities such as chemical manufacturing, waste disposal, the production of

This chapter is based on a paper presented at the Climate Change Litigation Roundtable, University of Sydney, 28 February 2009. The primary focus of the paper is on the law of negligence, but nuisance-based litigation is also considered. We are grateful to Professor Barbara McDonald and Dr Tim Stephens for their helpful comments on an earlier draft.

nuclear power, and the transport of chemicals were extremely hazardous, and potentially injurious to the health of a significant number of people. By this time, as Huber notes, the *second stage* of the 'new tort revolution' was underway:

> The demise of the old threshold of harm, itself rather like a bursting reservoir, flooded the legal landscape with the waters once placidly confined within the narrow Rylands enclosure. The courts have taken the limited legal theories of the past, meant to apply to front yard sorts of environmental mischief, and stretched them to cover the inner space of intimate contractual relations; now they extended them to the outerspace of the public square, with its myriad low level mass contacts. Public risks and environmental torts, once all but excluded from the torts system, quickly became the vibrant arm of a whole new field of litigation.[6]

As Huber goes on to point out:

> No matter how strict our religious observance or civic hygiene, we are all bombarded on a molecular level with bits of pig, dog and rodent, as well as biological matter even more vile. Each camp fire or radium painted watch dial, each filling station pump or can of turpentine, sends out chemical ripples that spread and reverberate indefinitely, touching an ever-increasing number of people ever more softly. The environment knows no bounds.[7]

Correspondingly, the environmental law suit appeared to know no bounds and 'the modern history of toxic tort litigation has been the record of an ever widening circle'.[8] Huber identifies three generations of cases over a relatively short historical period:

- the *first generation* of cases encompassed product liability litigation arising out of the use of therapeutic drugs and devices. Early cases involved thalidomide, diethystlbestol (DES) and high dose x-ray therapy;
- the *second generation* of cases, on Huber's analysis, arose out of the workplace, where occupational exposure to hazardous substances (such as asbestos, dust, radiation, toxic chemicals, pesticides and environmental tobacco smoke) occurred, often at a high level over long periods of time. As Huber notes,'[i]n moving from drugs to the workplace the toxic tort clientele grew dramatically';[9]
- the *third generation* of cases arose out of familiar substances, such as asbestos, dioxin and radiation, which had spread over huge geographical areas including school buildings (asbestos insulation), towns (dioxin road spraying) and entire states or regions (fall-out from nuclear testing). In Huber's words, 'the toxic tort once bounded by the human body or the factory walls now encompassed rivers, municipal water supplies, lakes, outdoor air, and entire communities, cities and water sheds'.[10]

In a relatively short period of time, small-scale legal proceedings between individual parties had spawned complex, large-scale group or class action litigation. This was facilitated, in part, as Huber notes, by the expansion of the boundaries of tort liability.

But is this trend likely to extend to a *fourth generation* of tort cases arising out of personal injury, property loss and environmental damage caused by human-induced climate change?

Recently, the Chief Justice of New South Wales concluded that the 'imperial march' of the law of negligence had come to an end. He noted that:

> From the 1960s to the 1990s, a long-term trend of judicial decision making can be discerned by which liability and damages expanded … However, that trend has, in recent years, been decisively stopped and reversed.[11]

The degree to which the general contraction of the law of negligence in recent times militates against the recognition of liability in respect of the impacts of climate change remains to be seen. Such impacts may potentially be devastating, and have been summarised by Durrant in the following terms:

> Changes to our climate system are predicted to result in rising sea levels, rising temperature and higher incidences of severe storms. Climate related harm could include loss of homes, livestock and other property, damage to public infrastructure and to coastal settlements, impaired agricultural yield, loss of livelihoods and population displacement. The human health impacts could involve thermal stress and heat-related deaths and illnesses, proliferation and geographical shifts of infectious diseases, impaired nutrition and other adverse mental and physical health risks.[12]

Climate change presents new challenges for the legal system in terms of regulatory policy and both preventive and remedial litigious strategies. The threats presented by climate change have spurred a renewed interest in use of litigation on the part of 'ideological' plaintiffs and organisations seeking to secure objectives perceived to be in the 'public interest'. Such litigation has already been brought in Australia in the planning and administrative law context,13 and it is highly likely that an 'ideological' plaintiff will, at some point, seek to invoke the law of negligence in order to achieve particular climate change-related objectives. Of course, most tort litigation arises after the event, is narrowly focused and is concerned primarily to achieve a pecuniary benefit for the individual litigant. However, this ought not to be taken to suggest that such litigation is incapable of achieving broader objectives considered to be in the 'public interest'. The prospect of tort litigation and judicially-imposed remedies may, even if only indirectly, play an important role in deterring hazardous conduct and improving standards of safety.[14] In some respects, tort law is also effectively 'punitive'.[15]

Having considered administrative law and other litigation directed to compelling particular governmental action, Hsu (writing in the North American context) contends that 'seeking direct civil liability against those responsible for greenhouse gas emissions' is the only litigation strategy 'that holds out any promise of being a magic bullet'[16] in connection with climate change. Hsu observes:

> By targeting deep-pocketed private entities that actually emit greenhouse gases (or, in the case of automakers, produce the means of omitting greenhouse gases), a civil litigation strategy, if successful, skips over the potentially cumbersome, time-consuming and politically perilous route of pursuing legislation and regulation. The civil litigation strategy is potentially a means of regulation itself, as a finding of liability could have an enormous ripple effect and send greenhouse gas emitters scrambling to avoid the unwelcome spotlight.[17]

Any plaintiff seeking to use tort law for either preventive or remedial purposes in connection with damage arising out of human-induced climate change would face formidable legal, logistical, evidentiary and financial obstacles. Some of these are addressed below. The picture is undoubtedly complicated by underlying uncertainties in the law of negligence itself, some of which have been created or exacerbated by recent legislative 'reforms'. Negligence-based climate change cases presently lie in the realm of the hypothetical. Given that a factually diverse array of such cases may be imagined and because in some respects we are dealing with the fringes of the law of negligence, our exploration of the law will proceed at an appropriate level of generality, and our conclusions should be taken as provisional rather than categorical. For the purposes of illustration, we will refer to the law applicable in New South Wales, but we will endeavour to draw attention to different rules applicable in other Australian jurisdictions where the circumstances warrant it.

The Factual and Scientific Background

Establishing Parameters

It is impossible to consider the potential application of negligence law in the context of human-induced climate change without first having some appreciation of its nature. In particular, it is important to have regard to the actual and projected impacts of climate change, for the obvious reason that compensable loss or damage lies at the heart of any successful tort action. Because we are concerned with legal proceedings under Australian law, we will focus upon impacts having a specific Australian connection.

While we are well aware that there is much that is uncertain about the likely impacts of climate change, we are in no position to evaluate the vast body of scientific literature on the subject, and it is not the purpose of this chapter to do so. Rather, as our concern is to consider one particular aspect of the legal dimension of climate change at a relatively abstract level, we will be content to rely on a highly schematic factual model, notwithstanding any imprecision that this may entail. In developing such a model, we have depended heavily upon the Fourth Assessment Report of the Intergovernmental Panel on Climate Change,[18] which (at the risk of incurring the wrath of climate change skeptics) we take to be broadly representative of the mainstream of scientific opinion. We have also drawn upon the *The Garnaut Climate Change Review — Final Report (Garnaut Report)*,[19] which deals specifically with projected impacts of climate change in the Australian context. Our purpose is not to reproduce the detail contained in these reports, which are readily available to interested persons, but rather to draw attention to certain points that are essential to the legal exploration that will follow.

The Nature of Climate Change

The dynamics of climate are complex, and 'climate change', where it occurs, is the product of a confluence of factors. As the Intergovernmental Panel on Climate Change (IPCC) has written, '[t]he climate system evolves in time under the influence of its own internal dynamics and due to changes in external factors that affect climate'.[20] Of these 'external factors', the most significant, for present purposes, are those arising from human activities that have affected, and continue to affect, the atmospheric concentration of certain gases, known as greenhouse gases (GHGs). The IPCC identifies such activities as the 'dominant force' behind a net warming of the earth's atmosphere over the past five decades.[21]

The most significant GHGs in terms of atmospheric warming are carbon dioxide (CO_2), methane (CH_4) and nitrous oxide (N_2O), with halocarbons also having some historical importance.[22] The atmospheric concentration of these gases is a function of both natural processes and human activities, including the production and combustion of fossil fuels and other agricultural and industrial practices.[23] There has been a dramatic increase in the atmospheric concentration of each gas since 1750, a fact which the IPCC links to 'human activities in the industrial era'.[24] In each case, a number of anthropogenic factors are said to have contributed to the increase. For present purposes, it will suffice to note that the IPCC identifies the principal causes as fossil fuel combustion (eg, in transportation and power generation), agricultural activities (eg, fertiliser use), industrial activities (for example, cement and nylon manufacture) and landfill.[25] It should be clear from the foregoing that individual sources of GHG emissions are huge in number and global in distribution.

Net atmospheric warming is expected to continue for many decades and even centuries to come, with the rate of change dependent upon the success of efforts to reduce the atmospheric concentration of GHGs.[26]

The Impacts of Climate Change

Climate change is a global phenomenon, and while its central manifestation may consist in net atmospheric warming, the direct and indirect consequences of such warming are far from uniform, and, in particular, are variable from location to location. This is attributable to the fact that the 'impacts' of climate change are produced by the interaction of broad climatic shifts with specific human and natural systems.[27]

Climate change impacts need not, therefore, represent the end results of simple, linear chains of cause and effect. The IPCC acknowledges that in some circumstances, climate change will combine with other stressors to produce adverse consequences, noting that '[f]or example current stresses on some coral reefs include marine pollution and chemical runoff from agriculture as well as increases in water temperature and ocean acidification'.[28] Moreover, the severity of symptoms of climate change will be to a considerable extent dependent upon the adaptive capacity of human and natural systems,[29] and upon efforts made to mitigate resultant damage.

Because of the complicated and contingent character of the sequence of events that will precede the realisation of most possible climate change impacts, it is difficult for scientists to make specific predictions about their likelihood and probable magnitude. In specific connection with its projections concerning climate change impacts on Australia, the IPCC acknowledges that:

> Assessment of impacts is hampered because of uncertainty in climate change projections at the local level (eg in rainfall, rate of sea-level rise and extreme weather events) ... Other uncertainties stem from an incomplete knowledge of natural and human system dynamics, and limited knowledge of adaptive capacity, constraints and options.[30]

For the sake of convenience, it is useful to consider damage incidental to the projected impacts of climate change in terms of two general categories, although we acknowledge that the division is to some extent artificial:

- gradual damage attributable to the sustained effect of predicted climate change impacts such as altered rainfall patterns, shifts in the range of agricultural pests, the degradation of ecosystems and sea level rise ('gradual damage'); and
- more acute and immediate damage attributable to what the IPCC terms 'extreme weather events', such as storms, cyclones, heat waves and droughts ('acute damage').

Gradual Damage — Long-Term Change

Projected long-term shifts in climatic patterns will conceivably have significant consequences for many sectors of the Australian economy. For example, the IPCC predicts that climate change is likely to result in water availability problems, which will in turn affect the sustainability of farming in some areas:

> Climate change is likely to change land use in southern Australia, with cropping becoming non-viable at the dry margins if rainfall is reduced substantially, even though yield increases from elevated CO_2 partly offset this effect ... Land degradation is [also] likely to be affected by climate change.[31]

The IPCC also forecasts a climate-related decline in the quality of some crops[32] and changes to the distribution of some agricultural pests.[33] The *Garnaut Report* suggests that without 'effective global mitigation' of climate change, agriculture will no longer be possible in the Murray-Darling Basin region by 2100, with depopulation the result.[34]

The *Garnaut Report* further notes that climate change could negatively affect the 'natural resource-based tourism' industry in various ways.[35] For instance, incremental changes in the Australian climate are expected to result in a loss of biodiversity in some ecologically sensitive areas.[36] While the gravity of such an outcome from an environmental point of view is obvious enough, it is important to recognise that it could also decimate businesses that rely, in one way or another, on the ecological quality of the areas in question. By way of example, the IPCC points out that tourism in Queensland will suffer severe detriment if predictions about the degradation of the Great Barrier Reef in consequence of thermal stress and ocean acidification (attributable to increased atmospheric CO_2 concentration) come to fruition.[37] The *Garnaut Report* identifies further tourist areas considered to be at particular risk from climate change impacts, including the Australian alpine regions, which may suffer economically if snow coverage patterns continue to be affected by rising temperatures.[38] The collapse or diminution of a local tourism industry will inevitably have flow-on effects for other local businesses, particularly where the economy of a region is principally dependent upon tourism-related income.

Another gradual consequence of global climate change is a general rise in sea levels attributable to 'thermal expansion of the oceans ... and the loss of land-based ice due to increased melting'.[39] Although Australia is better-placed to mitigate the threats that rising sea levels pose than less-developed countries,[40] the economic costs of an increasing risk of inundation and erosion[41] could nevertheless be severe. This is particularly so because of the concentration of population centres along the eastern seaboard of Australia. As the IPCC notes:

> Over 80% of the Australian population lives in the coastal zone, with significant recent non-metropolitan growth. About 711,000 addresses ... are within 3 km of the coast and less than 6 m above sea level.[42]

There is also the potential for the distribution of vectors of disease to be altered as a result of changing weather patterns. For example, the IPCC points out that there is a likelihood that the geographical range of dengue fever in Australia will expand.[43]

Where damage occurs gradually, it is unlikely to lead to sudden, catastrophic economic loss. However, significant costs may be involved in undertaking adaptation measures to limit the extent of the damage as it is occurring.[44] It should also be noted that some of the gradual impacts of climate change discussed above may in fact result in large-scale environmental and social disruption if certain tipping points are reached.[45] A tipping point refers to the potential for a relatively small change in a climatic variable such as oceanic or atmospheric temperature to generate a rapid and large change in a climatic feature, such as the Asian monsoon.

Acute Damage — Extreme Weather Events

The IPCC defines an 'extreme weather event' as '[a]n event that is rare within its statistical reference distribution at a particular place.[46] The *Garnaut Report* lists as examples hot days and nights (including heatwaves), cold days and nights (including frosts), heavy rainfall events, droughts, floods, hail and thunderstorms, tropical cyclones, bushfires and extreme winds.[47] Scientific data supports the proposition that climate change affects the *probability* that certain types of extreme weather event will occur in particular places.[48] As the *Garnaut Report* explains:

> Single events, such as an intense tropical cyclone or a long-lived heatwave, cannot be directly attributed to climate change. Climate change may, however, affect the factors that lead to such events. It may make certain events, like the heatwave that occurred in Adelaide in the summer of 2007-08, *more likely.*[49]

In specific connection with Australia, the IPCC notes that '[h]eatwaves and fires are virtually certain to increase in intensity and frequency ... [and f]loods, landslides ... and storm surges are likely to become more frequent and intense'.[50] The IPCC also anticipates more widespread drought.[51]

Self-evidently, extreme weather events of all varieties are likely to have direct and indirect effects leading to extensive and multifaceted detriment, in terms of damage to property and infrastructure,[52] personal injury[53] and purely economic loss. Unlike the creeping, gradual damage discussed above, harm attributable to extreme weather events is liable to be of an immediate, catastrophic character, although such events may of course also cause damage that is revealed only in the longer term. As the IPCC points out, the social and economic costs of extreme weather events 'spread from directly impacted areas and sectors to other areas and sectors through extensive and complex linkages'.[54] By way of example, the IPCC notes that the 'major projected impacts' of increased tropical cyclone activity are likely to include (but are not limited to) an increase in crop damage, the uprooting of trees, reef damage, power outages, disruption of the public water supply, traumatic deaths and injuries and post-traumatic stress disorder, increased risk of certain diseases, and other social and economic disruption and property damage.[55]

Some indication of the magnitude of the economic impact of an increased incidence of extreme weather events is provided by the IPCC's comment that:

> In Australia, around 87% of economic damage due to natural disasters (storms, floods, earthquakes, fires and landslides) is caused by weather-related events. From 1967 to 1999, these costs averaged US$719 million [per annum], mostly due to floods, severe storms and tropical cyclones. [56]

Conclusion

Some salient points to be drawn from the above are that: (a) in so far as human activities drive climate change, they do so principally by altering the atmospheric concentration of GHGs; (b) the range of individual activities (both historical and ongoing) combining to produce the aggregate result of an increased atmospheric concentration of GHGs is incredibly vast. Entire societies are presently essentially dependent on such activities (particularly those involving fossil fuel combustion) to sustain 'conventional' ways of life, and to that extent, at least, are complicit in their continuation;[57] and (c) the likely effects of climate change are difficult to predict with any degree of certainty, and the shape which they take depends upon many variables. However, there is little doubt that they could be extremely widespread, and in many respects catastrophic.

Some may regard climate change impacts as lying principally in the realm of the hypothetical, and to a significant extent the severity of many such impacts will depend upon events yet to occur. However, the *Garnaut Report* points out that some climatic changes attributable to anthropogenic warming have already occurred, citing as examples 'the [general] increase in average temperatures since the middle of the 20th century' and '[u]p to 50% of the reduction in rainfall in the south-west of Western Australia, and of the decline in snow cover in the south-east Australian alpine region'.[58] Moreover, as the *Report* notes, 'the CSIRO and the Bureau of Meteorology ... have concluded that the drought in many parts of [Australia] is linked to, or at least exacerbated by, global warming'.[59] In this light, it is reasonable to conclude that some of us may, unwittingly, already be paying the costs of adapting to climate change'.[60]

We therefore consider that it is valid to consider whether recovery in respect of negligent contributions to climate change may be possible in the near term under the law as it presently stands. Would, or should, the law of negligence provide a remedy for, say:

- A farmer whose land is no longer arable (or at least viable to farm) due to climate change-related degradation?
- A charter boat operator whose business suffers due to climate change-related damage to the Great Barrier Reef?
- A householder whose property is damaged by a tropical cyclone in a locality where there is thought to be an increased likelihood of such events due to climate change?

Subject to issues of jurisdiction, the defendant/s in a climate change-based tort action would likely comprise corporate entities whose activities could be shown to have made a significant direct or indirect contribution to the increased atmospheric concentration of GHGs, in circumstances where there was a reasonable basis for arguing that such activities were, as a matter of law, 'negligent'. The following sections will explore some of the theoretical and practical problems that a plaintiff seeking to recover damages in respect of climate change-related harm in the nature of personal injury, property damage or economic loss would be required to confront.

Climate Change and the Law of Negligence

A Duty of Care

It is clear, based on ordinary principles, that a duty of care does not arise in all contexts and that even where a duty exists, it may be of limited scope.

Donoghue v Stevenson[61] was notable for Lord Atkin's attempt to resolve cases decided under the hitherto prevailing 'categorial' approach to negligence, whereby a duty of care was taken to arise in various species of cases, into a consistent scheme anchored by the so-called 'neighbour principle',[62] and thus develop what might be termed the 'general part' of negligence law.[63] The purpose of adopting such an approach was presumably to afford greater certainty to the law, and thus assist in the resolution of future cases. Unfortunately, perhaps, and despite numerous attempts to advance it,[64] it would seem that the endeavour to rationalise the law in this respect has collapsed as a consequence of the range and diversity of circumstances with which the principles of negligence are expected to deal. Simply put, no formulation as to when a duty will exist has been deemed satisfactory to accommodate all possible cases that may arise. While there is a broad range of established relationships in which the law will recognise the existence of a duty of care as a matter of course,[65] difficulties arise when considering novel situations, particularly where the relationship between plaintiff and defendant is not readily susceptible of comparison to that which subsists in a known class of case, or is apparently tenuous. In general, judges appear to have reverted to a conservative, incremental approach founded on analogical reasoning by reference to principles espoused in decided cases.[66] In this context, it is arguable that Lord Atkin's foundational assumption — that the duty of care 'must logically be based upon some element common to the cases where it is found to exist'[67] — cannot be sustained. As has been suggested by Callaway JA, '[t]he search for principle is worse than looking for a needle in a haystack. The needle is not there.'[68]

At the heart of the conceptual problem lies the notion that it is not sufficient to support a duty of care that a plaintiff is reasonably foreseeable to a defendant as a potential victim of harm consequent upon the latter's negligence.[69] Self-evidently, the application of such a meagre standard alone and of itself could result in the ambit of a defendant's potential liability in respect of a single act or omission being extraordinarily wide.[70] While foreseeability of harm remains a prerequisite to the imposition of a duty of care in respect of particular damage,[71] the courts can frequently be seen to be reaching for some additional factor said to justify or exclude the imposition of a duty in a particular case or class of cases. However, the various pronouncements made in this respect have not crystallised into anything resembling a neat 'checklist' of factors to which it may be profitable to advert in circumstances of novelty.

The most recent authoritative pronouncement of the High Court in this connection may be found in *Sullivan v Moody*,[72] in which joint reasons were delivered by all five members of the bench. The court roundly rejected the idea that the demonstration of some element of 'proximity' as between the plaintiff and the defendant would necessarily found a duty of care.[73] In the court's view, the term 'proximity', although often used in this area of law, represents little more than a conclusory label.[74] The Court was also critical of approaches tending to focus attention on the question of what is 'fair, just and reasonable' in the particular case.[75] For the Court, such an incantation is 'capable of being misunderstood as an invitation to formulate policy [on a discretionary basis] rather than to search for principle'.[76] However, the Court was quick to point out that the apparent failure of the search for unifying doctrine 'does not mean that novel cases are to be decided by reference only to some intuitive sense of what is "fair" or "unfair"'.[77] Rather, it emphasised that 'the law of tort develops by reference to principles, which must be capable of general application, not discretionary decision-making in individual cases'.[78] It noted:

> Different classes of case give rise to different problems in determining the existence and nature or scope, of a duty of care. Sometimes the problems may be bound up with the harm suffered by the plaintiff, as, for example, where its direct cause is the criminal conduct of some third party. Sometimes they may arise because the defendant is the repository of a statutory power or discretion. Sometimes they may reflect the difficulty of confining the class of persons to whom a duty may be owed within reasonable limits. Sometimes they may concern the need to preserve the coherence of other legal principles, or of a statutory scheme which governs certain conduct or relationships. The relevant problem will then become the focus of attention in a judicial evaluation of the factors which tend for or against a conclusion, to be arrived at as a matter of principle.[79]

In *Sullivan* itself, the critical consideration was that those said to be subject to a duty of care were also subject to other statutory obligations, with which the duty of care contended for was said to be inconsistent.[80] The Court also referred to 'a question as to the extent, and potential indeterminacy, of liability',[81] but without really exploring its implications in the particular case. Precisely where *Sullivan* leaves the law as to duty of care is less than clear. In truth, the High Court's reasons provide little in the way of prospective guidance as to whether and when a duty of care will be imposed upon a defendant in a case having no direct precedent. While the apparent intractability of the duty of care concept doubtless renders highly individuated ad hoc analysis of novel cases an attractive and pragmatic option for judges, such an approach can hardly operate as anything other than a disincentive to potential plaintiffs contemplating bringing proceedings hinging on the recognition of a new category of duty. This is particularly so given that some judges have candidly acknowledged that decision-making in this area is a product of 'trade-offs and value judgments',[82] and requires them to address 'questions of fairness, policy, practicality, proportion, expense and justice'.[83] As Trindade et al have noted, even long-used terms like 'reasonable foreseeability' and 'proximity' encapsulate 'value judgments by courts as to when it is appropriate to impose liability for negligent conduct'.[84] Although the High Court has ostensibly eschewed the notion that discretionary considerations as to fairness and justice ought to inform the recognition of duties of care, it would seem that to some extent liability in negligence continues to be 'based upon a general public sentiment of moral wrongdoing for which the offender must pay'.[85]

In light of the above, it is difficult to anticipate how a court would approach a hypothetical case in which a plaintiff sought to establish negligence in respect of particular defendants' alleged contributions to climate change. Whether a court would be prepared to recognise a duty of care in such a case would doubtless turn upon the specific circumstances involved, and particularly upon the character of the harm suffered. However, it is probable that a plaintiff would face formidable hurdles in this respect.

In the first place, it may be very difficult to show that the harm said to have eventuated was a reasonably foreseeable consequence of a particular defendant's activities. Because of the global nature of climate change, a plaintiff would be unlikely to be able to rely on any direct or specific relationship with the defendants involved as supporting the existence of a duty.[86] Rather, a suit would be brought on the basis of such defendants' activities with respect to the world at large, and those activities may be both geographically and temporally divorced from the adverse consequences said to have been suffered by the plaintiff. Moreover, satisfying the court that it would be *appropriate* to recognise a duty as a matter of legal policy would probably be problematic. The incremental approach is inherently conservative, and one of the reasons for its adoption was to curb the expansionary tendency of methodologies reliant on stating the law in general terms. Recognising that negligence may potentially be found in the fact of emitting GHGs that feed into a global phenomenon having worldwide effects would have incalculable consequences, and would arguably represent a revolutionary rather than an evolutionary development of the law.

Much of what has been written in recent times with respect to the outer limits of the duty of care concept has arisen from the context of claims for purely economic loss, in which categories of liability are still being developed and refined. This is particularly relevant for present purposes, not only because much of the harm that may be suffered as a result of climate change is likely to be exclusively financial in character, but also because of the insight it provides into the type of considerations to which courts may advert when confronted with a novel claim as to the existence of a duty of care, some of which may also be applicable, by analogy, well beyond the field of purely economic loss.

Of particular significance is the case of *Perre v Apand*,[87] decided by the High Court in 1999. *Perre* is something of a difficult case to deal with. The Court's reasons ultimately consisted of seven different judgments manifesting various levels of semantic and substantive divergence, notwithstanding that there was significant overlap between their reasoning in many respects, and that all of the judges agreed in the conclusion that a duty existed. However, in a very lucid analysis, Mulheron notes that:[88]

- each of the judges stated a preference for an 'incremental' approach when determining the existence of a novel duty of care, and rejected any attempt to invoke or formulate some kind of overarching 'unifying principle';[89]

- there was no question that the loss suffered by the plaintiffs (the particulars of which are not here relevant) was reasonably foreseeable, but no judge considered this sufficient to support a duty of care of itself; and

- it therefore became necessary for each judge to address specifically the 'factors/principles/salient features/policy considerations' that he/she considered supported the imposition of a duty in the instant case.[90]

Mulheron then identifies some 15 factors referred to by various judges in *Perre* as being relevant to their respective conclusions that a duty of care existed in the circumstances. For example:[91]

- whether the plaintiff was particularly 'vulnerable' to suffering harm by reason of the defendant's conduct;
- whether the defendant had 'actual knowledge' of the risk of the harm that eventuated, and of its magnitude;
- whether the imposition of a duty of care would raise the spectre of indeterminate liability, and/or whether such imposition would burden a defendant 'out of all proportion to his wrong';
- whether the alleged negligence had the effect of interfering with the plaintiff's use or ownership of land; and
- whether the harm that materialised would have been 'relatively easy to avoid'.

In *Perre*, these factors were discussed with specific reference to purely economic loss, and it may be that some would have diminished (or no) significance if removed from this context. If nothing else, however, *Perre* provides some indication of the manner in which a court would probably go about addressing whether or not a duty of care exists as between an emitter of GHGs and a person said to have suffered loss of some sort in consequence of climate change. In all likelihood, a similar kind of factorial approach would be adopted.[92] While the fact-dependent character of such an approach makes it difficult to draw conclusions with respect to hypothetical claims, we would be pessimistic about the prospect of establishing that an emitter of GHGs owes any kind of duty to any particular person to prevent purely economic harm resulting from climate change. It was critical to the court's reasoning in *Perre* in several respects that the type of harm the subject of the plaintiffs' claim was inherently limited. However, the establishment of a duty in respect of climate change-related harm of some kind would raise the very real prospect of indeterminate liability, because realistically the potential effects of climate change are themselves indeterminate, and are not restricted either geographically or temporally. The cases suggest that a court would be extremely reluctant to impose such a duty.[93] Put simply, the sheer breadth that any such duty would potentially have, and the fact that it would not be anchored to any specific relationship nor subject to any other practical limitation, would render it unlikely to be recognised. Moreover, the diversity of agents to whose activities climate change may be attributed will render it arguable that the imposition of liability on any one of them would burden the specific defendant 'out of all proportion to his wrong'.

In general, it can safely be said that the inherent conservatism of the 'incremental' approach presently in favour with respect to the recognition of novel categories of duty of care would militate against the recognition of such a category in respect of climate-change related harm, particularly where the plaintiff and the defendant are linked only by the fact that such harm has come to pass as a matter of fact.

Standard of Care/Breach of Duty

Even if it could be established that a duty to avoid climate change-related harm existed in a particular case, it may be difficult, if not impossible, to establish that a defendant was actually in breach of that duty. The issue of breach of duty is now governed primarily by s 5B of the *Civil Liability Act 2002* (NSW) ('*CLA*'), which provides that:

(1) A person is not negligent in failing to take precautions against a risk of harm unless:

    (a) the risk was foreseeable (that is, it is a risk of which the person knew or ought to have known), and

    (b) the risk was not insignificant, and

    (c) in the circumstances, a reasonable person in the person's position would have taken those precautions.

(2) In determining whether a reasonable person would have taken precautions against a risk of harm, the court is to consider the following (amongst other relevant things):

    (a) the probability that the harm would occur if care were not taken,

    (b) the likely seriousness of the harm,

    (c) the burden of taking precautions to avoid the risk of harm,

    (d) the social utility of the activity that creates the risk of harm.

Section 5B appears in general to be affirmative of the antecedent common law,[94] albeit that some of its language differs slightly from that used in *Wyong Shire Council v Shirt*,[95] previously regarded as the leading authority on breach of duty in Australia. The *Ipp Report* expressed concern that the mere fact of foreseeability of risk was being taken as giving rise to a breach of duty without reference to the potentially countervailing considerations now enumerated in s 5B(2).[96]

Thus, in determining whether there has been a breach of an established duty of care, a court is required to consider whether a reasonable person, in the defendant's position, would have foreseen that his or her conduct involved a 'not insignificant' risk of injury to the plaintiff or to a class of persons including the plaintiff. If so, the court must then consider what a reasonable person would have done by way of response to that risk. As Mason J commented in *Shirt*:

> The perception of the reasonable man's response calls for consideration of the magnitude of the risk and the degree of probability of its occurrence, along with the expense, difficulty and inconvenience of taking alleviating action and any other conflicting responsibilities which the defendant may have.[97]

In the context of climate change litigation, a court would be required to address itself to the risk that the particular type of harm concerned would come to pass as a result of climate change, rather than the risk of climate change occurring in its generality. Specifically, the court would need to consider whether a reasonable person in the position of the defendant would, or should, have anticipated the creation of such a risk. Self-evidently, this would be likely to present a multiplicity of case-specific problems, which would differ according to the species of harm in question. However, as Durrant has noted, climate change is an area in which 'knowledge of the risk of harm has developed over time',[98] and as such there would be a temporal aspect to the question of breach of duty; the relevant standard of care may have heightened over time as scientific knowledge has increased and awareness of the potential consequences of climate change has become more widespread. Durrant points out that there have been a succession of international agreements and reports which may be taken to have significance in this respect.[99] To some extent, additionally, the issue may depend upon the position and character of the particular defendant/s against whom action is brought. The fact that the particular damage in issue might have been inflicted as a result of an unpredictable chain of specific events need not constitute a barrier to a finding of breach of duty; if the general species of risk that ultimately materialised should have been anticipated, the unlikelihood of the actual damage involved will not necessarily be problematic in connection with breach of duty. This may be of particular relevance to litigation dealing with extreme weather events, which are invariably brought about by a combination of factors.

If the specific risk concerned is deemed to have been foreseeable for the purposes of s 5B(1)(a), the court must then address the question of whether the risk was 'not insignificant'. The precise connotation of this phrase is difficult to discern,[100] and although it may not appear to involve a particularly rigorous standard, it is arguably more substantial than the 'not far fetched or fanciful' standard formulated in *Shirt*.[101] Self-evidently, 'significance' may only be evaluated with a particular risk in mind.

The issue of whether the defendant/s ought to have taken precautions against the risk of harm is required to be addressed prospectively, that is, by 'look[ing] forward to identify what a reasonable person would have done, not backward to identify what would have avoided the injury'.[102] It is in this context that the factors identified in s 5B(2) would be considered, with a focus on the reasonableness of the conduct of the defendant/s. The idea that 'the social utility of the activity that creates the risk of harm' must be considered may be particularly problematic in the context of climate change-related litigation. It is, of course, difficult to know how a court is generally expected to go about evaluating such 'utility'.[103] However, it would presumably be arguable that many, if not most, activities resulting in the emission of GHGs have some degree of objective social utility, and the real issue would be whether precautions could, and should, have been taken to reduce their adverse effects at the relevant time. The reasonableness of taking the particular precautions that the plaintiff alleges ought to have been taken will be critical in this context. As Durrant has written, '[a] balance needs to be identified between the risk and the reasonable steps that could be taken to minimise emissions'.[104] In some areas, it is conceivable that the only 'precaution' actually open to a defendant may have been to cease certain activities altogether. In such circumstances, the 'burden' of taking relevant precautions would have been considerable. In other cases, it may be arguable that by failing to moderate or modify those activities in some way (eg, by implementing available technological improvements), particular defendant/s failed to adhere to an acceptable standard of care.

One significant issue may relate to the degree to which society has become dependent on particular GHG-emitting activities. A court may be reluctant to hold that an emitter should have taken steps to reduce GHG emissions where the efficacy of the activities producing those emissions would thereby have been compromised, to the detriment or inconvenience of large sections of the population reliant on those activities. Durrant points out that it may also be necessary to consider 'any relevant statutory or customary standards'.[105] She observes that '[i]ndustries worldwide have historically emitted unabated [GHGs] since the time of the industrial revolution. Customarily, there have been no limits or restrictions on those emissions'.[106] Moreover, where a particular sector has been subject to government regulation in terms of emissions, it may be more difficult to argue that an entity which has complied with the relevant regulatory regime, but nevertheless continued activities that contribute to the process of atmospheric warming, has breached any relevant duty of care. In particular, the existence of, and fact of compliance with, such a regime may have a significant bearing on the evaluation of the foreseeability of the risk said to have materialised, and/or the reasonableness of the defendant's response to such risk. This point is likely to have particular salience if litigation is brought with respect to emissions generated by a defendant while the defendant was acting in accordance with government regulation of carbon emissions, such as the Commonwealth's Carbon Pollution Reduction Scheme (the legislation for which is currently before Parliament).

Causation

Australian authors Smith and Shearman observe that '[e]stablishing legal causation in climate change actions — that is, proving that a defendant's actions caused the harm suffered by the plaintiff — will pose the greatest obstacle for a majority of plaintiffs'.[107]

From a plaintiff's point of view, causation is likely to raise two distinct problems. The first, which will be addressed in some detail below, is doctrinal, and concerns the attribution of causal responsibility to a particular defendant or defendants in accordance with legal principles. However, it is important to recognise that this issue will be somewhat academic unless the second, more practical problem can be overcome. This requires the establishment by evidence of some kind of causal link between relevant negligence and the damage suffered. The difficulties that this second problem could pose should not be underestimated. Even in the relatively simple context of ordinary toxic tort and product liability litigation, the question of whether there is a causal connection between use of or exposure to a particular product or substance and the development of the particular disease or condition contracted by a plaintiff may be complex, and the subject of conflicting, wide-ranging and expensive expert evidence from persons with diverse medical and scientific backgrounds. The case of climate change is likely to present as significantly more complicated from an evidentiary point of view. As observed above, the science of climate change is possessed of inherent complexity, and in some respects involves significant uncertainties.[108] A plaintiff would probably be forced to rely upon expert evidence said to establish a statistical association between the negligence alleged and the harm said to have been suffered.[109] Mank has observed that:

> [P]roving specific causation between GHGs and climate change [impacts] is … difficult because the impacts involve intensification of existing climatic phenomena, such as more frequent storms, rather than the creation of unique 'signature diseases' such as

asbestosis, which is caused only by exposure to asbestos, or clear cell adenocarcinoma, which caused only by the drug DES. Because climate is affected by several factors interacting in complex ways, it is difficult for scientists to tease out what percentage of any climate change is affected by GHGs, and it is even more difficult to determine what percentage is affected by a specific polluter or group of polluters ... Climate is a chaotic system affected by natural fluctuations in frequency and severity, so it is difficult to determine to what extent human actitvities, such as producing GHGs, affect those frequencies or variations.[110]

Scientific uncertainty is not a subject that will be canvassed in detail in this chapter, but its omission should not be taken to downplay its formidable practical significance. Because litigation against an emitter of GHGs in respect of climate change-related damage would inevitably be viewed (if not conceived) as something of a test case, it is reasonable to suppose that the defendants would invest significant resources in challenging the scientific foundation of the claim made against them. If an 'ideological' plaintiff was contemplating running such a test case, it would be critical to identify an area in which the scientific basis of a claim would be particularly strong. It might be the case that in some areas, the science of climate change is simply insufficiently developed to support the commencement of legal proceedings, although as the IPCC emphasises throughout its *Fourth Assessment Report*, human understanding of climate change is continually improving.[111] Of course, if the scientific evidence presents insurmountable difficulties, then the legal principles and problems discussed below will be largely academic.

The 'Elements' of Causation

In New South Wales,[112] s 5D of the *CLA* provides, in part, that:

(1) A determination that negligence caused particular harm comprises the following elements:[113]

(a) that the negligence was a necessary condition of the occurrence of the harm (factual causation) and

(b) that it is appropriate for the scope of the negligent person's liability to extend to the harm so caused (scope of liability).

(2) In determining in an exceptional case,[114] in accordance with established principles, whether negligence that cannot be established as a necessary condition of the occurrence of harm should be accepted as establishing factual causation, the court is to consider[115] (amongst other relevant things) whether or not and why responsibility for the harm should be imposed on the negligent party.[116]

...

(4) For the purpose of determining the scope of liability, the court is to consider (amongst other relevant things) whether or not and why responsibility for the harm should be imposed on the negligent party.

Precisely how the enactment of these provisions affects the common law is an issue with which, we would suggest, Australian courts, and by implication Australian lawyers and litigants,[117] are yet to come to grips. Previously, causation in negligence was taken to revolve around an holistic 'commonsense' approach, articulated by Mason CJ in *March v Stramare (E & MH) Pty Ltd*[118] in terms specifically rejecting recourse to a structured, two-stage inquiry such as that now apparently embodied in s 5D:

> Commentators subdivide the issue of causation in a given case into two questions: the question of causation in fact — to be determined by the application of the 'but for' test — and the further question whether a defendant is in law responsible for damage which his or her negligence has played some part in producing ... It is said that, in determining this second question, considerations of policy have a prominent part to play, as do accepted value judgments ... However, this approach ... (a) places rather too much weight on the 'but for' test to the exclusion of the 'common sense' approach which the common law has always favoured; and (b) implies, or seems to imply, that value judgment has, or should have, no part to play in resolving causation as an issue of fact. As Dixon CJ, Fullagar and Kitto JJ remarked in *Fitzgerald v Penn*:[119] 'it is all ultimately a matter of common sense' and '[i]n truth the conception in question [ie, causation] is not susceptible of reduction to a satisfactory formula'.[120, 121]

Nevertheless, in New South Wales, it has been held in several cases that the principles 'embodied' in s 5D of the *CLA* are, as a matter of fact, in accord with the common law.[122] The common thread linking most of these cases has been the uncritical acceptance of certain dicta of Ipp JA to that effect in *Ruddock v Taylor*.[123] His Honour there stated that 'there are two fundamental questions involved in the determination of causation in tort', being whether causation exists in a factual/historical sense, and whether the defendant ought to be held liable in a normative sense.[124] He continued:

> The approach to causation that I have set out [the 'two stage' approach] forms the basis of s 5D of the *Civil Liability Amendment (Personal Responsibility) Act 2002*. This Act does not govern the present action but, in my view, *the principles it embodies in regard to causation are in accord with the common law*.[125]

It is difficult know whether Ipp JA was referring to s 5D in its substantive or its structural aspect, or both. However, in any event, it is extremely doubtful that s 5D is fundamentally consistent with the pre-existing law of negligence. As is well-known, Ipp JA was also the chairman of a panel whose Report[126] (*Ipp Report*) contained proposals which, broadly speaking, are reflected in s 5D. However, the perfunctory manner with which the Panel addressed itself to the state of the common law is curious,[127] and Bartie and McDonald have convincingly demonstrated that in various respects, its proposals (and therefore, by extension, s 5D) are essentially at odds with *March*.[128] So much is apparently confirmed by various remarks of members of the High Court in *Travel Compensation Fund v Tambree*[129] to the effect that the 'two stage' approach described in *Ruddock* was incompatible with *March* and related High Court authority.[130]

The conclusion seems inescapable that s 5D of the *CLA* is not, in fact, altogether harmonious with the common law. At the very least, it seems readily apparent that s 5D compels the judicial use of a different explanatory model to that which previously prevailed. Whether the divergence is confined to the level of the semantic, as decisions to this point seem to have assumed, is unclear. While we propose to structure our consideration of causation in the context of climate change around it, we emphasise that the introduction of s 5D may have consequences that are unpredictable, and yet to be fully worked out. Aside from the fundamental issue of how the identification of 'elements' of causation by s 5D affect the shape of the causation inquiry, we would suggest that the provision also gives rise to relevant uncertainty with respect to: (a) the compass of the 'exceptional case' concept in s 5D(2); and (b) whether or not 'remoteness of damage' is preserved as a separate 'element' of negligence liability, or whether it is subsumed within the 'scope of liability' criterion — as the authors of the *Ipp Report* seem to have intended.[131] We will return to both of these matters in due course.

Factual Causation

Somewhat inelegantly, s 5D(1)(a) of the *CLA* identifies as an 'element' of causation a criterion which is subsequently made conditionally dispensable by s 5D(2). In dealing with an issue of causation by reference to the Act, one must first apparently consider whether the negligence of a defendant was a 'necessary condition' of the occurrence of the harm concerned. However, in an 'exceptional case', this criterion may be ignored. The word 'exceptional' is of questionable utility in this context, given that any case in which s 5D(2) comes into play is of its nature 'exceptional' simply by virtue of the fact that s 5D(1)(a) has been bypassed.[132] Presumably because of this, the rider 'in accordance with established principles' is deployed to provide some direction to a decision-maker, although the 'established principles' being referred to are not identified.

Turning to the issue at hand, because the individual agents responsible for the increased atmospheric concentration of GHGs are so numerous and diffuse, it is difficult to imagine that a plaintiff could ever show that the specific negligence of a particular defendant was a 'necessary condition' of the occurrence of whatever climate-change related harm was said to have been suffered.[133] Even a large-scale, long-term emitter of GHGs would probably be able to construct a compelling argument to the effect that its activities had made but a nominal or marginal relative contribution to global climate change, and that anthropogenic warming (and its incidents) would have occurred to a similar degree even had its activities not taken place. This being the case, a plaintiff would probably be required to establish the fact of an 'exceptional case' in accordance with s 5D(2).

To the extent that s 5D(1) encapsulates the 'but for' test,[134] the idea that it may not necessarily be determinative of causation in its factual aspect appears broadly consistent with the common law.[135] Problematically, however, s 5D(2) is extremely ambiguous.[136] It establishes that the 'but for' test *can* be circumvented, but does little to properly explain the circumstances in which that is permissible. Section 5D(2) was 'explained' with considerable imprecision in the Second Reading Speech to the Civil Liability (Personal Responsibility) Bill 2002 ('the Second Reading Speech') as follows:

> The rules for factual causation are set out, including the very limited exception to the 'but for' test. This exception was developed by the court for those rare cases, often in the dust diseases context, where there are particular evidentiary gaps. By including this exception in the bill it is not intended that the bill extend the common law in any way. Rather, it is to focus the courts on the fact that they should tread very carefully when considering a departure from the but for test. [137]

In general, an 'evidentiary gap' has traditionally been said to exist where two (or more) putative tortfeasors are implicated in loss or damage suffered by the plaintiff in circumstances where the evidence is incapable of establishing the nature or extent of their respective causal contributions. The *Ipp Report* identified two distinct species of case in which there may be such a gap:

- a case in which harm 'is brought about by the cumulative operation of two or more factors, but which is indivisible in the sense that it is not possible to determine the relative contribution of the various factors to the total harm suffered', such as *Bonnington Casting v Wardlaw*,[138] which was said to establish the principle that 'any of the contributory factors can be treated as a cause of the total harm suffered, provided it made a "material contribution" to the harm';[139] and

- a case in which harm has resulted from one (or more) of two (or more) separate acts of negligence, but it is impossible to determine which negligent act/s actually caused the harm as a matter of fact. In *Fairchild v Glenhaven Funeral Services Ltd*,[140] the House of Lords held that in an appropriate case, any defendant whose conduct 'materially increased the risk' of the harm could be held liable.[141]

The report continued:

> The 'material contribution to harm' and 'material contribution to risk' principles both allow negligent conduct to be treated as a factual cause of harm even though it cannot be proved on the balance of probabilities that there was in fact a causal link between the conduct and the harm. ... The Panel's opinion is that, in certain types of cases, bridging the evidentiary gap in this way would be widely considered to be fair and reasonable. ... The major difficulty ... is to define those cases in which the normal requirements of proof of causation should be relaxed. ... The Panel believes that detailed criteria for determining this issue should be left for common law development [but] we consider that it would be useful to make explicit the normative character of the issue by [enacting] a provision that, in deciding whether proof that conduct that materially contributed to, or materially increased the risk of, harm should suffice as proof of causal connection, it is relevant to consider whether (and why) responsibility for the harm should be imposed on the negligent party, and whether (and why) the harm should be left to lie where it fell (that is, on the plaintiff) ...[142]

Section 5D(2) accords with the Panel's proposal at least in so far as it wholly abdicates responsibility for developing principles for dealing with 'evidentiary gap' cases to the courts. However, its reference to 'established principles' is somewhat ambiguous, particularly given that the *Ipp Report* specifically noted that the *Fairchild* principle was of uncertain standing in Australia.[143] For the purposes of this chapter, we will assume that s 5D(2) leaves it open to the courts to draw upon and further develop either of the lines of jurisprudence identified in the *Ipp Report*. Both may potentially be of use for a plaintiff bringing action in respect of climate change-related harm. But it is not inconceivable that s 5D(2) may have the effect of opening the door to the development of more inventive strands of jurisprudence, depending upon how exactly the reference to 'established principles' is interpreted. Indeed, certain comments in *Fairchild* suggest that the capacity of the law for innovation in this area is yet to be exhausted.

Whether or not the *Bonnington* principle should be regarded as anything 'exceptional' is perhaps unclear. Arguably, it has long been absorbed into the mainstream of the law of negligence. The *Ipp Report* specified that in a strict sense, the *Bonnington* principle meant that 'a defendant may be liable for the *total harm* suffered by a plaintiff even though it cannot be said that, but for the conduct of the defendant, the plaintiff would not have suffered the *total harm*'.[144] However, Mendelson notes that whatever the specificity of its origins, the 'material contribution' concept 'has ... been applied indiscriminately in any legal context'.[145] Indeed, explanations of causation at common law following *March* have customarily eschewed reference to the 'but for' test altogether and instead invoked the language of *Bonnington*. For example, in *Henville v Walker*, McHugh J said:

> If the defendant's breach has 'materially contributed' to the loss or damage suffered, it will be regarded as a cause of the loss or damage, despite other factors or conditions having played an even more significant role in producing the loss or damage. As long as the breach materially contributed to the damage, a causal connection will ordinarily exist even though the breach without more would not have brought about the damage.[146]

This formulation, not unorthodox under the common law, would appear difficult to reconcile with the *CLA*, given the latter's privileging of the 'but for'/'necessary condition' test. A breach of duty may clearly 'materially contribute' to the infliction of particular harm without being a 'necessary condition' of the occurrence of that harm. Under the 'common sense' test, a 'material contribution' of this nature could readily have been recognised as being legally consequential in an appropriate case. Now, such recognition would be contingent upon shoehorning the circumstances into the ambiguous 'exceptional case' category established by s 5D.

It seems clear, at least, that s 5D(2) was intended to refer to the principles developed in *Fairchild*, which have now been qualified by the decision of the House of Lords in *Barker v Corus UK Ltd*.[147] *Fairchild* arose out of circumstances in which the plaintiffs, who had suffered asbestos-induced mesothelioma, sued two former employers who had required them to work directly with asbestos. Lord Bingham of Cornhill described the essential question in the case as follows:

> If ... C [the plaintiff] cannot (because of the current limits of human science) prove, on the balance of probabilities, that his mesothelioma was the result of his inhaling asbestos dust during the employment by A or during his employment by B or during his employment by A and B taken together', is C entitled to recover damages against either A or B, or against both A and B?[148]

The Court of Appeal had held that the plaintiffs could not succeed against either defendant:

> [A]pplying the conventional 'but for' test of tortious liability, it could not he held that C had proved against A that his mesotheliomas would probably not have occurred but for the breach of duty by A, nor against B that his mesothelioma would probably not have occurred but for the breach of duty by B, nor against A and B that his mesothelioma would probably not have occurred but for the breach of duty by both A and B together. So C failed against both A and B.[149]

According to Lord Bingham of Cornhill, the critical issue on appeal was whether the 'special circumstances' of the case were such that 'principle, authority or policy' justified a modified approach to proof of causation.[150] The House of Lords was unanimous in holding that they were, with the result that the plaintiffs were entitled to succeed against both defendants. The decision was said to be limited to the 'special circumstances' of the particular cases before the House.

The decision in *Fairchild* has been extensively analysed elsewhere.[151] For present purposes, it will suffice to say that it may be of limited relevance in the context of climate change litigation. *Fairchild* was therefore a case in which the harm concerned could have been caused by the negligence of either or both of two different tortfeasors, but it was impossible to discern the actual scientific chain of causation that led to the plaintiff's injuries. Crucially, each agent's negligence was, of itself, theoretically sufficient to cause the harm. Moreover, all of the possible sources of risk to the plaintiffs were tortious.[152] By way of contrast, in a climate change case, assuming that any scientific uncertainty could be overcome to the satisfaction of the court, the situation would be that a GHG emitter, as defendant, could be shown to have contributed in some finite degree to the generalised global process of climate change, which in turn could be shown to have caused some form of harm to the plaintiff, or at least increased the risk that such harm would be inflicted. However, it could never be established or assumed that the activities of that particular emitter were of themselves sufficient to set in motion the process which resulted in the harm, even in theory.

The question of causation would, therefore, revolve around whether making some contribution to a *process* initiated as a result of the cumulative effect of a multitude of such contributions (as well as natural causes) could or should be taken as a legal cause — not whether a possible cause of harm should be deemed to be a cause-in-fact for reasons of justice and fairness.[153] If a sound scientific case could be advanced as to the aetiology of the damage, the causal contribution of a GHG-emitter to climate change-related harm *may* be relatively clear — it would not be a case, like *Fairchild*, in which the defendant may have caused the harm by itself, *may* have caused the harm in conjunction with another party, or *may not* have contributed to the harm at all. The real difficulty in a climate change case would, in our view, concern the magnitude of any particular defendant's contribution, and whether it was sufficiently substantial to support a finding of causation in a manner consistent with s 5D(2).

Much, we would suggest, would depend upon the significance that a court assigned to the 'material contribution' idea, and the degree to which a court was prepared to 'relax' the test for causation in light of the language of the *CLA*. Some extension of existing principles might arguably be required depending on the particular case, but as Lord Hoffmann noted in *Fairchild*, the law of causation, like the law governing the imposition of duties of care, is ultimately derived from conceptions of what is 'just and reasonable' in the circumstances,[154] and *Fairchild* itself stands as evidence that courts may be inclined to avoid undue rigidity in applying that law if the circumstances justify it.

Scope of Liability

Section 5D(1)(b) of the *CLA* recognises that demonstration of a causal link between negligent conduct and harm in a factual or technical sense will not of itself suffice to establish causation in the legal sense; there is an additional normative dimension. The distinction between scientific and legal conceptions of causation was adverted to by the High Court in *March v Stramare*, where to Mason CJ observed:

> In philosophy and science, the concept of causation has been developed in the context of explaining phenomena by reference to the relationship between conditions and occurrences. In law, on the other hand, problems of causation arise in the context of ascertaining or apportioning legal responsibility for a given occurrence. [155]

The requirement in s 5D(1)(b) that a decision-maker consider whether it is 'appropriate' to impose liability on a particular defendant would appear, on its face, to leave an extraordinarily wide variety of factors open for consideration. The *Ipp Report* observed that the 'scope of liability' idea 'tends to be seen as one that has to be answered case-by-case rather than by the application of detailed rules and principles',[156] and specified that it encompasses 'issues, other than factual causation, referred to in terms such as "legal cause", "real and effective cause", "commonsense causation", "foreseeability" and "remoteness of damage"'.[157] Although the *Ipp Report's* apparent agglomeration of what some would consider to be a number of quite discrete concepts is difficult to reconcile with the pre-existing common law in several respects,[158] it is apparent, at least, that its authors intended that all manner of what might loosely be referred to as 'policy' considerations should inform the normative aspect of the causation inquiry. One particular concern of the authors of the *Ipp Report* was that the role of such considerations was being insufficiently explained by judges in individual cases.[159] It remains to be seen whether the legislative provisions will lead to any improvement in this connection.[160]

For the purpose of this chapter, we will assume that s 5D has the effect of incorporating considerations as to 'remoteness' into the 'scope of liability' inquiry, as the authors of the *Ipp Report* seem to have intended. In general, at common law, damages were not recoverable in respect of harm considered to be too 'remote' from the negligence to which it is attributable in a causal sense. Trindade et al write that:

> Rules of remoteness of damage are chiefly concerned with cases in which the damage suffered is of an unexpected or unusual nature, or where it occurs in a surprising or extraordinary way. But damage may be too remote just because it is widely separated in time from the defendant's tort, or because the causal chain between the damage and the tort has many links ...[161]

The basic principle with regard to remoteness was identified in *Overseas Tankship (UK) Ltd v Morts Dock & Engineering Co Ltd (The Wagon Mound)*[162] as being bound up with the issue of reasonable foreseeability of damage. There it was held that the 'essential factor in determining liability is whether the damage [that occurred] is of such a kind as the reasonable man should have foreseen'.[163] What is necessary is that the general 'type' of harm suffered (eg, physical injury) was foreseeable in the circumstances; if it was so foreseeable, the courts will not concern themselves with arguments to the effect that its extent could not have been predicted, or that the precise causal chain leading to the harm was unusual or unpredictable.[164] Trindade et al comment that ideas such as 'foreseeability' are used, in this context, 'to give effect to the (typically unexpressed) value judgments about how the consequences of tortious conduct ought to be distributed as between tortfeasors and tort victims'.[165]

Some matters which may be relevant to the consideration of the appropriate 'scope of liability' in climate change-related litigation include:

- the significance of the causal contribution of the defendant to the phenomenon of climate change, relative to other causative factors (both anthropogenic and natural). If (as would often be the case) the defendant's negligent activities have made but a minor contribution to that phenomenon, it would be open to a court to conclude (to use language adverted to in the *Ipp Report*) that those activities were not a 'real' or 'effective' or 'commonsense' cause of the particular damage alleged. It might also be considered that natural phenomena or the activities of other GHG emitters ought to be taken as having broken whatever chain of causation could be said to exist as a matter of fact;

- the degree to which the end result in the chain of causation (that is, the damage) is removed from the defendant's initial negligence. If the damage is distant from that negligence in space or time, or has been produced in an unpredictable or improbable manner, a court may be inclined to view such damage as lying beyond the scope of the defendant's liability;

- other policy considerations, including some of those which might also arise in connection with determining whether there has been a breach of duty, particularly those concerned with the broader consequences of imposing liability in the specific type of case involved.

In sum, the fact that a defendant's GHG-emitting activities (even where negligent) are likely to have only a relatively tenuous factual connection to any *particular* example of climate-change related harm is likely to present a major problem for a plaintiff seeking to establish causation for the purposes of obtaining redress in respect of such harm. Even if a plaintiff could find some means of circumventing the 'necessary condition' test and establishing what s 5D of the *CLA* refers to as 'factual causation', the 'scope of liability' criterion has the potential to throw up all manner of problems. We agree with Durrant that '[o]verall, it is highly probable that the Court would conclude that it is not appropriate to impose liability for the emission of [GHGs] and the resulting ... harm'.[166]

Further Issues

The obstacles presented by the substantive principles of negligence in their current form do not exhaust the difficulties facing a prospective proponent of a climate change-related lawsuit, which are too numerous to be examined comprehensively by this chapter. Nevertheless, several related issues will be briefly considered below.

Remedies

While we do not intend to deal exhaustively with the issue of remedies, it is important to note that for all the effort likely to be expended in pursuing a claim in respect of climate change-related harm, it may be that any redress potentially available would be relatively nominal and/or ineffective to check the activities of GHG emitters. Of course, the significance of this concern would depend upon the particular objectives of the plaintiff in bringing the relevant proceedings.

In relation to damages, the *CLA* provides that in so far as claims for economic loss or damage to property are concerned, a proportionate liability regime applies, such that:

> the liability of a defendant who is a concurrent wrongdoer in relation to that claim is limited to an amount reflecting that proportion of the damage or loss claimed that the court considers just having regard to the extent of the defendant's responsibility for the damage or loss. [167]

Given any individual emitter's negligent contribution to human-induced climate change (and any consequent actionable damage) would probably be held to have been relatively inconsequential; it is difficult to imagine a court awarding any significant monetary damages to a plaintiff where such a provision is applicable. In any event, precisely how a court ought to go about determining the extent to which a specific defendant should be held responsible for particular damage by reason of such contribution is less than clear.

There has been some discussion at an academic level of how damages awarded in respect of climate change-related harm might be allocated as between emitters. In *Fairchild* and *Barker*, several members of the House of Lords referred to the development of 'market share' liability in the United States, in circumstances where a drug (DES) was known to have caused particular harm, but it was not possible to ascertain which individual manufacturer (of some 300) had been responsible for the drug involved in any individual case. [168] In *Barker*, Lord Hoffman noted that courts in California and New York had apportioned liability according to the national market share of each company. Thus, defendants had been held liable according to the *chance* that their drug had caused the injury in any individual case. Lord Hoffmann referred to the following passage from the decision of the Supreme Court of California in *Brown v Superior Court (Abbott Laboratories)*:

> In creating the market share doctrine, this court attempted to fashion a remedy for persons injured by a drug taken by their mothers a generation ago, making identification of the manufacturer impossible in many cases. We realised that in order to provide relief for an injured DES daughter faced with this dilemma, we would have to allow recovery of damages against some defendants which may not have manufactured the drug that caused the damage ... Each defendant would be held liable for the proportion of the judgment represented by its market share, and its overall liability for injuries caused by DES would approximate the injuries caused by the DES it manufactured. A DES manufacturer found liable under this approach would not be held responsible for injuries caused by another producer of the drug. The opinion acknowledged that only an approximation of a manufacturer's liability could be achieved by this procedure, but underlying our holding was a recognition that such a result was preferable to denying recover [sic] altogether to plaintiffs injured by DES. [169]

Some commentators have suggested that 'market share' principles of liability, or something analogous, may provide an appropriate framework for determining damages in climate change litigation. Thus, as Smith and Shearman note,[170] Penalver has contended that defendants should be liable for damage caused by human induced climate change on a proportional basis according to market share for fossil fuels.[171] The example provided by Smith and Shearman is that a defendant responsible for 3% of global carbon dioxide production via fossil fuel would be held liable for 3% of the relevant damage, before a discount to allow for the possibility of 'natural causes' having caused the damage.

Even if palatable in principle, as Smith and Shearman note, such a formulaic resolution would, 'in practice, be a costly and difficult task'.[172] However, more serious objections can also be raised. Presumably, if 'market share' principles were somehow to be applied, a court would need to be satisfied that each and every entity whose 'market share' was to be taken into account had, at least on the face of things, engaged in negligent conduct in emitting GHGs. Showing as much would be a vastly complicated exercise, particularly where, on the face of things, the conduct for which the 'defendants' were to be sanctioned consisted, on the face of things, in having engaged in lawful, regulated, activity (commercial, 'profit-making' or otherwise). Moreover, unless it could be accepted that each entity had been 'negligent' (and was therefore culpable) in the same degree, it would be strongly arguable that the blunt instrument of 'market share' calculation would serve only to distort their relative culpability and almost certainly offend basic principles of justice.

Depending upon the circumstances, it may be that a plaintiff would be more interested in seeking to restrain particular GHG-producing activities of a defendant which are said to be unlawful by means of a prohibitory injunction than in extracting damages. The jurisdiction of a court to grant an injunction is discretionary. It would probably be particularly difficult to persuade a court to grant any kind of interlocutory injunction in proceedings against an emitter of GHGs given that the plaintiff would need to show that his or her prospects of ultimate success in the action were sufficient to justify restraining relevant activities of the defendant, taking specific account of the practical consequences of doing so.[173] Such consequences are likely to be significant from the defendant's point of view. Moreover, a plaintiff seeking interim relief would ordinarily be required to provide an undertaking as to damages in connection with the effects of any prohibitory order sought, and if he or she will not be in a position to honour the undertaking then such relief will likely be refused.

Costs

The costs involved in maintaining large-scale legal proceedings are liable to present an intractable problem for all but the most well-resourced 'ideological' plaintiffs, particularly where those costs dwarf whatever pecuniary interest the particular plaintiff has in the outcome of the case. Considering: (a) the complexity of the issues that would be involved in any novel climate change-related proceedings; (b) the corresponding need to compile a substantial volume of high-quality scientific and other expert evidence; (c) the likelihood that such proceedings would be trenchantly contested by the defendant/s; and (d) the likelihood that such defendant/s would appeal if unsuccessful at first instance, it can be surmised that the cost of mounting and maintaining such proceedings would potentially be enormous, even if long-term pro bono legal representation could be secured. Moreover, if a 'public interest' or 'ideological' organisation was the proponent of such proceedings, it is unlikely that any costs could be saved by early recourse to alternative dispute resolution mechanisms, because the primary objective of such a plaintiff would not be the recovery of money, and settlement would probably be strategically unattractive. Thus, any such organisation would need to be prepared to devote significant resources to the proceedings over a potentially lengthy period of time.

Even more problematically, the plaintiff would run the risk of having to pay the costs of the defendant/s if ultimately unsuccessful. In Australia, the ordinary rule that costs follow the event may be displaced at the discretion of the court. However, courts have customarily been cautious in circumventing the rule in favour of litigants claiming to be acting in the 'public interest',[174] and in practice such litigants are rarely insulated from the risk of having to pay the other parties' costs if they lose. In *Oshlack v Richmond River Council*, Kirby J noted that:

> Courts, whilst sometimes taking the legitimate pursuit of public interest into account, have also emphasised, rightly in my view, that litigants espousing the public interest are not thereby granted an immunity from costs or a 'free kick' in litigation. [175]

In large part, the reluctance of the courts to grant costs concessions to 'public interest' litigants would appear to arise out of the amorphous quality of the 'public interest' concept, and the difficulty of consistently justifying exceptions to the ordinary rule.[176] It has often been noted that many cases involve a public interest of some kind,[177] which raises the spectre of arbitrariness where such an exception is made. Moreover, it has been recognised that even where a litigant genuinely and conscientiously seeks to achieve objectives perceived to be in the public interest, regard must also be had to the position of the defendant/s unwillingly dragged into the litigation, and that 'sympathy is not a legitimate basis to deprive a successful party of his or her costs'.[178]

It would seem, in general, that only a confluence of special circumstances is likely to convince a court to depart from the ordinary costs rule.[179] Even though climate change-related negligence litigation would undoubtedly raise issues of some novelty, it is questionable whether a court would be likely to regard such litigation as compelling the reversal of that rule. First, such litigation would be determined under the general law of negligence, and although that law is arguably unsettled in some respects, it may be difficult to persuade a court that the testing or clarification of its fringes is in the public interest. Second, the litigation would be brought under private law (and perhaps against private parties in the form of corporations responsible for large-scale GHG emissions), whereas comments of various judges tend to suggest that courts are more amenable to entertaining submissions as to the reversal of the usual costs rule on 'public interest' grounds where a genuine question as to the operation of public law has been raised. Third, where the circumstances are such as to make it inferable that the proceedings were a mere vehicle for ventilating concerns extrinsic to the specific factual situation on which they are founded, a particularly unsympathetic judge might be inclined to view those proceedings as actually being hostile to the public interest, in the sense of illegitimately occupying the time of the court. The danger that a judge would take this view would be compounded where the plaintiff's case, as framed, had poor prospects of success.

In any event, given the likely difficulty of assessing a plaintiff's chances of success in advance of any trial, it would simply have to be assumed that the risk of an adverse costs order would be of significant magnitude, and unfortunately, in Australia, there is no mechanism by which the attitude of the court to the question of costs may be antici-pated,[180] despite recommendations by the Australian Law Reform Commission[181] and (more recently) the Victorian Law Reform Commission[182] to the effect that such a mechanism ought to be established.

In sum, one effect of the current costs regime in Australia is to discourage the com-mencement of speculative proceedings directed to testing the boundaries of the existing law, and a climate change-related negligence action would almost certainly fall into that category. Even with a strong substantive case and exemplary motives, a plaintiff must be prepared to accept the risk of an adverse costs award in the event of failure. That a plaintiff bringing an analogous case in the United States would not face such a severe disincentive suggests that that jurisdiction might present more fertile soil for initial climate change-related tort litigation.

A Class or Representative Action?

In theory, proceedings seeking compensation for climate change-related harm, which is of its nature likely to be widely dispersed, would seem ideally suited to commence-ment under class or representative action rules.[183] However, this may be a less attrac-tive option where the extraction of money from the defendant/s does not represent the primary motivation of the proponents of such proceedings. In particular, the com-mencement of the proceedings in class or representative form would compound the complexity of the action, present the defendant/s with additional non-substantive grounds of attack upon the proponent's case, increase the likelihood that resolution of the proceedings would be inordinately delayed by reason of interlocutory disputation and inflate the costs that would be bound up in the outcome of the litigation.[184]

On the other hand, it may be that a 'public interest' or 'ideological' organisation would see some kind of strategic benefit in vesting the proceedings with a 'participa-tory' character. Moreover, the aggregation of a large number of claims under the aegis of unitary proceedings may have the effect of raising the stakes for the defendant/s, although it is difficult to say whether this would be likely to be of any net benefit to the proponent. In reality, the question of whether class or representative proceedings would be a suitable vehicle for a climate change-related negligence lawsuit could only be answered with a particular set of circumstances in mind.

Reasonable Prospects of Success

One feature of the civil litigation landscape in New South Wales that is assuming increasing importance is the requirement, introduced by the *CLA*, and now contained in the *Legal Profession Act 2004* (NSW), that a legal practitioner not commence a claim for damages unless he or she has a 'reasonable belief' that the claim has 'reasonable prospects of success'.[185] Breach of the relevant provision has the potential to result in professional sanctions and/or costs orders being made against the practitioner.[186] This places a significant burden on a practitioner considering whether or not to represent a particular plaintiff, and it would be quite understandable if such a practitioner was inclined to err on the side of caution when evaluating reasonable prospects of success.

For this reason, the effective 'certification' requirement imposed by the *Legal Profession Act* has considerable potential to deter the commencement of cases in which a plaintiff would be seeking to test the boundaries of existing doctrine, such as a climate change-related negligence case, or at least to dissuade legal practitioners from becoming involved. The deterrent effect would doubtless be even more pronounced where the plaintiff was at odds with, or sought to overturn, existing authority. In the High Court, Callinan J has drawn attention to the likelihood that the requirement will stunt the growth of the common law by impeding the commencement of exploratory, but arguably meritorious, claims — noting that 'the common law has often owed its development to, and has benefited from, the adventurousness and ingenuity of counsel'.[187] One would like to assume that a court would not exercise its discretion[188] to sanction a practitioner in respect of a novel case brought with a genuine public interest in mind, assuming that the legal argument submitted was not altogether frivolous. Relevantly, in *Lemoto v Able Technical Pty Ltd*,[189] McColl JA commented that '[a] statutory provision denying to the community legal services in a particular class of litigation cannot be intended to stifle genuine but problematic cases'. Nevertheless, the fact remains that the personal risks involved in 'certification' may operate in practice to do precisely that.

Nuisance

The tort of private nuisance is concerned with serious and unreasonable interference with the use of, or rights over, land. In particular, a plaintiff may allege that a defendant has either indirectly caused physical injury to land in which he or she has a particular interest, or has substantially impeded its enjoyment.

Nuisance cases have often been brought in circumstances where emissions from the land of a defendant (eg, smoke, dust, fumes, other pollutants) have inflicted 'damage' of some kind on the occupants of nearby properties.[190] However, as Trindade et al write, the law of nuisance has rarely ventured beyond cases in which there is an element of geographical propinquity as between the impugned activities and the affected land:

> In social terms, nuisance law is most relevant to issues of environmental pollution and, in particular, its effects on the use and enjoyment of land. However, the effects of pollution are so widespread, pervasive and serious that since the mid-nineteenth century, anyway, government regulation — in the form, for instance, of statutory controls and over the use of land and the emission of polluting substances into air and water — has played a much more significant role than tort law in attempts to manage the negative environmental impact ... of industrial activity and social life. ... *Nuisance law plays its most characteristic role in the resolution of small-scale local disputes between neighbouring landowners into which regulators cannot or will not intervene.*[191]

Similarly, Conaghan and Mansell refer to 'the relative impotency of nuisance law in the face of industrial development', although they note that such impotency is not easily explained by reference to the shape of common law doctrine.[192]

A dispute concerning climate change-related harm would be far removed from the paradigm case of private nuisance, and it would be difficult to plead given the kind of language used to express existing nuisance doctrine.[193] The cases often refer to 'neighbourhood' in a far more literal sense than that in which it was used in *Donoghue v Stevenson*.[194] While it is not the purpose of this chapter to undertake a detailed examination of the law of nuisance, a number of specific theoretical problems would seem to present themselves:

- in the usual case, nuisance cases concerning pollutants involve the direct transmission of such pollutants to the land concerned. In a climate change case, an additional step would be interposed, as the pollutants would have fed into a global phenomenon said to have effects on that land. Whether the law of nuisance can be stretched to cover such a case, in which the 'nuisance' and the 'interference' may not be geographically or otherwise proximate in any relevant sense, and the defendant/s, realistically, would have genuine 'control' over the nuisance, is highly questionable;

- the requirement that interference said to amount to nuisance be 'unreasonable' calls for something of a value judgment in the particular circumstances of the case, and the courts have referred to a number of different factors in making judgments as to reasonableness.[195] It is probable that in a climate change-related case a number of the 'policy' considerations referred to above as militating against, say, the recognition of a duty of care to avoid climate change-related harm in a negligence case would tend also to dissuade a court from holding that GHG-emitting activities were 'unreasonable' in the nuisance context;

- inherent in the law of nuisance is the idea that owners or occupiers of land must have a certain tolerance for interferences or inconveniences that are incidental to land use in a particular area. Moreover, the interference alleged must be 'substantial'. Certainly, the law does not recognise an absolute right to freedom from activities amounting to nuisance. Given that climate change is global in its effects, it is questionable whether it could be said that GHG-emitting activities amount to a nuisance in respect of their ultimate 'effects' in connection with particular land. Subjection to regional manifestations of a broad global phenomenon is arguably an ordinary incident of any land use.

Thus, while the law of nuisance might be viewed as having the potential to assist a plaintiff to circumvent some of the doctrinal problems that the law of negligence presents, it could also raise a number of difficulties of its own. Indeed, we would suggest that the doctrine of nuisance would need to be the subject of aggressive renewal before it could conceivably be relied upon as a useful tool for checking climate change-related harm. Moreover, an action in nuisance would not permit a plaintiff to evade the kind of scientific difficulties and problems of proof of causation with which he or she would be confronted in an action under negligence law.

Conclusion

It is not possible to conclude, at a level of abstract generality, that negligence-based litigation strategies seeking either to prevent or redress human-induced climate change are bound to fail. However, as Durrant observes, formidable legal and evidentiary problems are likely to be encountered in seeking to identify wrongdoers, recognise the wrong done, allocate blame and distribute losses.[196] Having reviewed the existing state of Australian tort law, with particular reference to a hypothetical scenario involving the long-term operation of a coal mine, coal fired electricity plant or other large-scale emitter of greenhouse gases in Australia, she concludes that '[a]s a regulatory tool, tortious actions for climate harm are expensive and unruly and the outcomes are indeterminate'.[197]

Under the law of New South Wales, particularly since the enactment of the *CLA*, there are multiple points at which a climate change-related negligence case could fail. That such a case would be so difficult to accommodate under the law as it stands is perhaps indicative of the fact that that law was not designed to deal with a problem of such magnitude. Where cause and consequence are so far removed from one another as they could be in a climate change-related case, and where the relationship between tortfeasor and victim may be so tenuous, existing legal principles are liable to be stretched to breaking point.

For this reason, we do not consider that the law of negligence, in its current form, necessarily has a significant role to play in facilitating litigation for the purpose of seeking to recover the human or economic cost of climate change-related harm in the longer term. The law is simply ill-adapted to dealing with problems on such a scale. If it is desirable from a public policy perspective to oblige emitters of GHGs to contribute in some manner to the cost of dealing with the consequences of their activities — a vexed political question — then the legislative imposition of an administrative or taxation-based solution, consistent with the international legal regime being developed to address climate change, may be far more sensible, efficient and just than leaving the field to isolated ad hoc litigation. So much has essentially been recognised in the United States, where 'public nuisance' claims against GHG emitters have (at least thus far) uniformly been dismissed as non-justiciable. There is considerable force to the comment of Boutrous and Lanza that:

> Whatever one thinks about global warming and its consequences, it cannot be denied that it is an issue of public and foreign policy fraught with scientific complexity, as well as profound political, social, and economic consequences. These exceedingly complex issues must be confronted at the national and international levels by [government]. They cannot rationally be addressed through piecemeal and ad hoc tort litigation seeking injunctive relief — or, even worse, billions of dollars in retroactive and future money damages — against targeted industries for engaging in lawful and comprehensively-regulated conduct.[198]

Be this as it may, in the shorter term, 'public interest' or 'ideological' organisations may yet potentially find benefit in initiating negligence-based litigation as part of a broader strategy seeking to draw attention to the severity of the potential consequences of climate change and to stimulate political debate. If one thing can be said with certainty, it is that if such an organisation was considering the use of negligence proceedings to attract political attention and/or shift corporate or governmental behaviour, it would be well-advised to pick its battleground very, very carefully.

Endnotes

1   P Huber, *Liability: The Legal Revolution and its Consequences* (1988); see also P Huber, *Galileo's Revenge: Junk Science in the Courtroom* (1991).

2   P Huber, *Liability: The Legal Revolution and its Consequences* (1988) 64.

3   (1868) LR 3 HL 330.

4   Ibid 66.

5   Ibid.

6   Ibid 67.

7   Ibid.

8   Ibid.

9   Ibid 68.

10 Ibid 70.

11 J J Spigelman, *Negligence: The Last Outpost of the Welfare State* (Summary of a speech delivered to the Judicial Conference of Australia Colloquium, Launceston, 27 April 2002).

12 N Durrant, Tortious Liability for Greenhouse Gas Emissions? Climate Change, Causation and Public Policy Considerations (2007) 7 *Queensland University of Technology Law and Justice Journal* 404, 403-4.

13 See e.g., *Greenpeace Australia Ltd v Redbank Australia Pty Ltd* [1994] NSWLEC 178; *Australian Conservation Foundation v Minister for Planning* [2004] VCAT 2029; *Wildlife Preservation Society of Queensland Proserpine/Whitsunday Branch Inc v Minister for the Environment and Heritage* [2006] FCA 736; *Gray v Minister for Planning* [2006] NSWLEC 720; *Re Xstrata Coal Queensland Pty Ltd* [2007] QLRT 33; *Queensland Conservation Council Inc v Xstrata Coal Queensland Pty Ltd* [2007] QCA 338; *Drake-Brockman v Minister for Planning* [2007] NSWLEC 490; *Anvil Hill Project Watch Association Inc v Minister for Environment and Water Resources* [2007] FCA 1480; *Anvil Hill Project Watch Association Inc v Minister for Environment and Water Resources* [2008] FCAFC 3; *Walker v Minister for Planning* [2007] NSWLEC 741; *Minister for Planning v Walker* [2008] NSWCA 224. See also P Biscoe, 'Climate Change Litigation' (Paper presented at the New South Wales Young Lawyers CLE Seminar, 26 May 2009).

14 Cf *Roads and Traffic Authority v Royal* [2008] HCA 19 [114] (Kirby J).

15 Cf eg *Civil Liability Act 2002* (NSW) s 21.

16 Shi-Ling Hsu, 'A Realistic Evaluation of Climate Change Litigation through the Lens of a Hypothetical Lawsuit' (2008) 79 *University of Colorado Law Review* 701, 716–17.

17 Ibid 717.

18 IPCC, *Climate Change 2007: Synthesis Report — Contribution of Working Groups I, II and III to the Fourth Assessment Report of the Intergovernmental Panel on Climate Change* (2007) ('IPCC Synthesis Report'); S Solomon et al (eds), *Climate Change 2007: The physical science basis — Contribution of Working Group I to the Fourth Assessment Report of the Intergovernmental Panel on Climate Change* (2007) ('IPCC Working Group I Report'); M Parry et al (eds), *Climate Change 2007: Impacts, adaptation and vulnerability — Contribution of Working Group II to the Fourth Assessment Report of the Intergovernmental Panel on Climate Change* (2007) ('IPCC Working Group II Report'); B Metz et al (eds), *Climate Change 2007: Mitigation — Contribution of Working Group III to the Fourth Assessment Report of the Intergovernmental Panel on Climate Change* (2007) ('IPCC Working Group III Report').

19 Ross Garnaut, *The Garnaut Climate Change Review — Final report* (2008) (*Garnaut Report*).

20 *IPCC Working Group I Report* (2007) 94, see also 667; *Garnaut Report* (2008) 27.

21 *IPCC Working Group I Report* (2007) 97.

22 Ibid 135. The atmospheric concentrations of ozone (O_3) and water vapour (H_2O) are also relevant, and human activities influence both: ibid.

23 See Garnaut, above n 19, 31–2 (Table 2.1).

24 *IPCC Working Group I Report* (2007) 135 (Figure 1).

25 Ibid 115–16.

26 *IPCC Working Group I Report (2007)* (Summary for Policymakers) 12–13.

27 See Garnaut, above n 19, 30 (Figure 2.4).

28 *IPCC Working Group II Report (2007)* (Summary for Policymakers) 19.

29 See ibid.

30 K Hennessy et al, 'Australia and New Zealand' in *IPCC Working Group II Report* (2007) 507, 529 (citation omitted).

31 Ibid 518.

32 Ibid 507, 524; see also Garnaut, above n 19, 130.

33 Hennessy et al, above n 30, 507, 525 (Table 11.7).

34 Garnaut, above n 19, 125; see also 129–31.

35 Ibid 133–5.

36 *IPCC Working Group II Report (2007)* (Summary for Policymakers) 13 [11.4].

37 Hennessy et al, above n 30, 507, 523, 527. The IPCC notes that '[t]ourism associated with the [Great Barrier Reef] generated over USD4.48 billion in the 12-month period 2004/5 and provided employment for about 63 000 full-time equivalent persons': 527. See also Garnaut, above n 19, 143–4 (Box 6.4).

38 Garnaut, above n 19, 134–5 (Box 6.2).

39 *IPCC Working Group I Report* (2007) 111.

40 Cf *IPCC Working Group II Report* (2007) (Summary for Policymakers) 12.

41 Hennessy et al, above n 30, 507, 520–1.

42 Ibid 520.

43 Ibid 507, 524, 525 (Table 11.7); see also Garnaut, above n 19, 139.

44 The IPCC defines 'adaptation' as '[a]djustment in natural or human systems in response to actual or expected climatic stimuli or their effects, which moderates harm or exploits beneficial opportunities' and 'adaptation costs' as the '[c]osts of planning, preparing for, facilitating, and implementing adaptation measures, including transition costs': see eg *IPCC Working Group II Report* (2007) 869.

45 K Richardson et al, *Climate Change — Global Risks, Challenges and Decisions: Synthesis* (2009) 14-15.

46 *IPCC Working Group II Report* (2007) 875. The IPCC definition continues: 'Definitions of "rare" vary, but an extreme weather event would normally be as rare as or rarer than the 10th or 90th percentile. By definition, the characteristics of what is called "extreme weather" may vary from place to place'.

47 Garnaut, above n 19, 40. Note that the *Garnaut Report* uses the term 'severe weather event', but it is similarly defined. See also Hennessy et al, above n 30, 507, 511 (Box 11.1) (noting some examples of extreme weather events in Australia and New Zealand).

48 *IPCC Working Group I Report* (2007) 97.

49 Garnaut, above n 19, 106 (emphasis added).

50 Hennessy et al, above n 30, 507, 509; see also 514–16.

51 See ibid 515.

52 See eg, ibid 522.

53 Ibid 507, 522, 524. For example, the IPCC estimates that by 2050 there could be 3200 to 5200 additional heat-related deaths per annum: ibid 525 (Table 11.7).

54 *IPCC Working Group II Report* (2007) (Summary for Policymakers) 12.

55 See ibid 18 (Table SPM.1).

56 Hennessy et al, above n 30, 507, 511 (citation omitted).

57 Cf Durrant, above n 12, 421.

58 Garnaut, above n 19, 106.

59 Ibid 108.

60 M Allen and R Lord, 'The Blame Game: Who Will Pay for the Damaging Consequences of Climate Change?' (2004) 432 *Nature* 551, 551.

61 [1932] AC 562.

62 [1932] AC 562, 580 ('You must take reasonable care to avoid acts and omissions which you can reasonably foresee would be likely to injure your neighbour. Who then, in law, is my neighbour? The answer seems to be persons who are so closely and directly affected by my act that I ought reasonably to have them in contemplation when I am directing my mind to the acts or omissions which are called in question'); cf *Heaven v Pender* (1883) 11 QBD 503, 509 (Brett MR).

63 Cf [1932] AC 562, 583–4.

64 See eg, *Anns v Merton LBC* [1978] AC 728; *Jaensch v Coffey* (1984) 155 CLR 549, 578–87 (Deane J).

65 *Caparo Industries plc v Dickman* [1990] 2 AC 605 per Lord Bridge of Harwich ('I think the law has now moved in the direction of attaching greater significance to the more traditional categorisation of distinct and recognisable situations as guides to the existence, the scope and the limits of the varied duties of care which the law imposes').

66 See eg, *Sutherland Shire Council v Heyman* (1984) 157 CLR 424, 481 (Brennan J) ('It is preferable, in my view, that the law should develop novel categories of negligence incrementally and by analogy with established categories, rather than by a massive extension of a prima facie duty of care restrained only by indefinable 'considerations which ought to negative, or to reduce or limit the scope of the duty or the class of person to whom it is owed'); *Crimmins v Stevedoring Industry Finance Committee* (1999) 200 CLR 1, 29 [61] (McHugh J).

67 Donoghue [1932] AC 562, 580.

68 *Berrigan Shire Council v Ballerini* (2005) 13 VR 111, 115 [8].

69 See eg, Heyman (1984) 157 CLR 424, 481 (Brennan J); *Sullivan v Moody* (2001) 207 CLR 562, 576 [42], 583 [64] (Gleeson CJ, Gaudron, McHugh, Hayne and Callinan JJ).

70 *Sullivan v Moody* (2001) 207 CLR 562, 576 [42] (Gleeson CJ, Gaudron, McHugh, Hayne and Callinan JJ).

71 See eg *Graham Barclay Oysters Pty Ltd v Ryan* (2002) 211 CLR 540, 555 [9] (Gleeson CJ).

72 (2001) 207 CLR 562.

73 Cf e.g., Heyman (1985) 157 CLR 424, 495 (Deane J).

74 *Sullivan* (2001) 207 CLR 562, 578–9 [48] (Gleeson CJ, Gaudron, McHugh, Hayne and Callinan JJ); cf *Pyrenees Shire Council v Day* (1998) 192 CLR 330, 360-1 [74]-[76] (Kirby J).

75 Cf *Caparo Industries Plc v Dickman* [1990] 2 AC 605 per Lord Bridge of Harwich.

76 *Sullivan* (2001) 207 CLR 562, 579 [49] (Gleeson CJ, Gaudron, McHugh, Hayne and Callinan JJ).

77 (2001) 207 CLR 562, 580 [53] (Gleeson CJ, Gaudron, McHugh, Hayne and Callinan JJ).

78 (2001) 207 CLR 562, 580 [53] (Gleeson CJ, Gaudron, McHugh, Hayne and Callinan JJ).

79 (2001) 207 CLR 562, 579–80 [50] (Gleeson CJ, Gaudron, McHugh, Hayne and Callinan JJ) (footnotes omitted).

80 *Sullivan* (2001) 207 CLR 562, 582 [62] (Gleeson CJ, Gaudron, McHugh, Hayne and Callinan JJ).

81 *Sullivan* (2001) 207 CLR 562, 582 [61] (Gleeson CJ, Gaudron, McHugh, Hayne and Callinan JJ).

82 *Berrigan Shire Council v Ballerini* (2005) 13 VR 111, 115 [8] (Callaway JA).

83 *Swain v Waverley Municipal Council* (2005) 220 CLR 517, 548 [79] (McHugh J).

84 Trindade et al, *The Law of Torts in Australia* (4th ed, 2008) 469.

85 Donoghue [1932] AC 562, 580.

86 Cf Durrant, above n 12, 406. See *Agar v Hyde* (2000) 201 CLR 552, [66]–[67] (Gaudron, Gummow and Hayne JJ).

87 (1999) 198 CLR 180 ('*Perre*').

88 See R Mulheron, 'The March of Pure Economic Loss ... but to Different Drums' (2003) 7 *Canberra Law Review* 87, 89–95.

89 Cf (1999) 198 CLR 180 [232] (Kirby J).

90 See in particular (1999) 198 CLR 180 [93] (McHugh J), [201] (Gummow J), [333] (Hayne J).

91 Mulheron, above n 88, 95–106.

92 Considerations which have been identified as relevant to the existence of a duty of care outside the context of purely economic loss include eg whether the duty contended for can be identified with sufficient particularity (see *Berrigan Shire Council v Ballerini* (2005) 13 VR 111, 124 [30] (Nettle JA)) and the degree of control capable of being exercised over the risk of harm by the defendant (cf *Agar v Hyde* (2000) 201 CLR 552, 564 [21] (Gleeson CJ)). Note that if action against a public authority was being contemplated, s 42 of the *Civil Liability Act 2002* (NSW) would be relevant in connection with the issue of duty of care.

93 Cf *Agar v Hyde* (2000) 201 CLR 552, 563 [19] (Gleeson CJ); *Sullivan* (2001) 207 CLR 562, 582 [61] (Gleeson CJ, Gaudron, McHugh, Hayne and Callinan JJ).

94 Cf *Waverley Council v Ferreira* [2005] NSWCA 418 [27] (Ipp JA); Council of the *City of Liverpool v Turano* [2008] NSWCA 270 [362] (McColl JA); *Bostik Australia Pty Ltd v Liddiard* [2009] NSWCA 167 [94] (Beazley JA), [158] (Basten JA).

95 (1980) 146 CLR 40 ('*Shirt*').

96 D Ipp et al, Review of the Law of Negligence (2002) ('*Ipp Report*') 105 [7.14].

97 *Wyong Shire Council v Shirt* (1980) 146 CLR 40, 47–8.

98 Durrant, above n 12, 411.

99 Ibid 410–11.

100 See eg B McDonald, 'Legislative Intervention in the Law of Negligence: The Common Law, Statutory Interpretation and Tort Reform in Australia' (2005) 27 *Sydney Law Review* 443, 465–6.

101 Cf *Ipp Report*, above n 96, 105 [7.15].

102 Council of the *City of Liverpool v Turano* [2008] NSWCA 270 [362] (McColl JA) citing *New South Wales v Fahy* [2007] HCA 20 [57] (Gummow and Hayne JJ).

103 McDonald, above n 100, 466.

104 Durrant, above n 12, 413.

105 Ibid.

106 Ibid (emphasis in original).

107 J Smith and D Shearman, *Climate Change Litigation: Analysis of the Law, Scientific Evidence and Impacts on the Environment, Health and Property* (2006) 107.

108 Cf Durrant, above n 12, 415–17.

109 Cf eg *Seltsam Pty Ltd v McGuiness* (2000) 49 NSWLR 262.

110 B Mank, 'Civil Remedies' in M Gerrard (ed), Global Climate Change and US Law (2008) 183, 201.

111 Cf eg M Allen and R Lord, 'The Blame Game: Who Will Pay for the Damaging Consequences of Climate Change?' (2004) 432 *Nature* 551, 552 (arguing that with improved monitoring and better mathematic analysis, 'we could one day see Californian farmers suing member states of the European Union for authorizing emissions that threatened the security of their water supplies').

112 Similar provisions have been enacted in other Australian jurisdictions: see *Wrongs Act 1958* (Vic) s 51; *Civil Liability Act 2002* (WA) s 5C; *Civil Liability Act 2003* (Qld) s 11; *Civil Liability Act 2002* (Tas) s 13; *Civil Liability Act 1936* (SA) s 34; *Civil Law (Wrongs) Act 2002* (ACT) s 45.

113 The onus of proving 'any fact relevant to the issue of causation' is explicitly placed on the plaintiff: *Civil Liability Act 2002* (NSW) s 5E; cf *Wrongs Act 1958* (Vic) s 52; *Civil Liability Act 2002* (WA) s 5D; *Civil Liability Act 2003* (Qld) s 12; *Civil Liability Act 2002* (Tas) s 14; *Civil Liability Act 1936* (SA) s 35; *Civil Law (Wrongs) Act 2002* (ACT) s 46.

114 In some jurisdictions, the relevant legislation provides that the court may make such a determination in an 'appropriate' case: *Wrongs Act 1958* (Vic) s 51(2); *Civil Liability Act 2002* (WA) s 5C(2).

115 In Western Australia, the corresponding legislation also specifically obliges the court to consider 'whether and why the harm should be left to lie where it fell': *Civil Liability Act 2002* (WA) s 5C(2).

116 The corresponding provision in South Australia — s 34(2) of the *Civil Liability Act 1936* (SA) — is quite different in form. It provides:

Where … a person (the 'plaintiff') has been negligently exposed to a similar risk of harm by a number of different persons (the 'defendants') and it is not possible to assign responsibility for causing the harm to any one or more of them —

(a) the court may continue to apply the principle under which responsibility may be assigned to the defendants for causing the harm[1] ; but

(b) the court should consider the position of each defendant individually and state the reasons for bringing the defendant within the scope of liability.

…

Note:

1 See *Fairchild v Glenhaven Funeral Services Ltd* [2002] 3 WLR 89.

The corresponding provision in the ACT is similar to the South Australian provision, although it omits specific reference to Fairchild: *Civil Law (Wrongs) Act 2002* (ACT) s 45(2).

117 Cf Melchior and Ors v *Sydney Adventist Hospital Ltd* [2008] NSWSC 1282 [129] (Hoeben J) (noting that '[n]either side referred to … sections [5D and 5E] nor were any submissions made in relation to them. Both sides seemed to assume that the principles embodied in s 5D … were in accord with the common law'); see also *O'Gorman v Sydney South West Area Health Service* [2008] NSWSC 1127 [147] (Hoeben J); *Nguyen v Cosmopolitan Homes* [2008] NSWCA 246 [56]–[57] (McDougall J); *Ginelle Finance Pty Ltd v Diakakis* [2007] NSWSC 60 [41] (Hoeben J).

118 (1991) 171 CLR 506 ('*March*').

119 (1954) 91 CLR 268, 277.

120 (1954) 91 CLR 268, 278.

121 (1991) 171 CLR 506, 515. See also *Bennett v Minister of Community Welfare* (1992) 176 CLR 408, 412–13 (Mason, Deane and Toohey JA) (noting that '[i]n the realm of negligence, causation is essentially a question of fact, to be resolved as a matter of common sense. In resolving that question, the 'but for' test, applied as a negative criterion of causation, has an important role to play but it is not a comprehensive and exclusive test of causation; value judgments and policy considerations necessarily intrude' (footnotes omitted)).

122 *Nguyen v Cosmopolitan Homes* [2008] NSWCA 246 [70] (McDougall J, McColl and Bell JJA agreeing); see also *Coastwide Fabrication & Erection Pty Ltd v Honeysett* [2009] NSWCA 134 [59] (McDougal J, Ipp and Young JJA agreeing); *Carpenter v Hinkley* [2008] WADC 161 [110]–[111] (Schoombee DCJ); *McDonald v Sydney South West Area Health Service* [2005] NSWSC 924 [48]–[50] (Harrison AJ) (observing that s 5D(1) was reminiscent of a passage from the reasons of *Hayne J in Pledge v RTA* [2004] HCA 13 [10]); *Finch v Rogers* [2004] NSWSC 39 [146] (Kirby J); cf *Cox v State of New South Wales* [2007] NSWSC 471 [116] (Simpson J) (observing that '[i]t has been said that the principles ... embodied [in s 5D] are in accord with the common law' (emphasis added) and citing Ruddock). In *O'Gorman v Sydney South West Area Health Service* [2008] NSWSC 1127 [147], Hoeben J noted that '[b]oth sides seemed to assume that the principles embodied in s 5D ... were in accord with the common law. That is certainly how I understand these sections. If there were any doubt on this issue, it was resolved in *Ruddock*'; cf Melchior and *Ors v Sydney Adventist Hospital Ltd* [2008] NSWSC 1282 [129] (Hoeben J) (simply noting that '[b]oth sides seemed to assume that the principles embodied in s 5D of the Act were in accord with the common law. That was the approach adopted by the Court of Appeal in *Ruddock*'). In other jurisdictions, corresponding provisions seem rarely to be mentioned at appellate level.

123 (2003) 58 *NSWLR* 269.

124 (2003) 58 *NSWLR* 269 [85]–[87] citing J Stapleton, 'Cause-in-Fact and the Scope of Liability for Consequences' (2003) 119 *Law Quarterly Review* 388.

125 (2003) 58 *NSWLR* 269 [89] (emphasis added).

126 *Ipp Report*, above n 96.

127 See ibid 108-9 [7.25]–[7.26].

128 See eg, S Bartie, 'Ambition Versus Judicial Reality: Causation and Remoteness Under Civil Liability Legislation' (2007) 33 *University of Western Australia Law Review* 415, 420-2 (arguing that the 'two-stage' approach was derived in substantial part from the work of Professor Jane Stapleton rather than the relevant common law authorities); McDonald, above n 100, 474 (arguing that in positing the 'necessary condition' test as the sole determinant of 'factual' causation, the *Ipp Report*, and by implication s 5D, are out of step with *March*, in which it was made clear that 'the common sense test was [itself] a test of factual causation'. As McDonald points out, a close reading of *March* belies the Panel's claim that appeals to 'common sense' may be neatly subsumed within the rubric of what it called 'scope of liability') cf e.g., *Roads and Traffic Authority v Royal* (2008) 245 ALR 653, 674 [81] (Kirby J) (pointing out that 'in the context of the law of negligence, causation is essentially a question of fact').

129 (2005) 224 CLR 627 ('*Tambree*').

130 (2005) 224 CLR 627, 643 [46]–[48] (Gummow and Hayne JJ); see also 653–4 [79]–[82] (Callinan J).

131 See *Ipp Report*, above n 96, 117 (Recommendation 29(b)(ii)); see also McDonald, above n 100, 472, 474–5; D Mendelson, *The New Law of Torts* (2006) 375 (suggesting that the provisions enacted do not cover the remoteness enquiry); Bartie, above n 128, 418–19.

132 This may explain why in other jurisdictions, the word 'appropriate' is substituted for 'exceptional': see above n 114.

133 Cf M Allen and R Lord, 'The Blame Game: Who Will Pay for the Damaging Consequences of Climate Change?' (2004) 432 *Nature* 551, 552 (suggesting that '[p]reliminary studies suggest that a substantial fraction of our current elevated level of carbon dioxide might be traced to products produced, sold or used by only a few dozen major companies').

134 *Cox v State of New South Wales* [2007] NSWSC 471 [154] (Simpson J); cf *Ipp Report*, above n 96, 109 [7.26].

135 See eg *Pledge v Roads and Traffic Authority* (2004) 205 ALR 56, 59 [9] (Hayne J); cf March (1991) 171 CLR 506, 512 (Mason CJ).

136 See eg D Mendelson, 'Australian Tort Law Reform: Statutory principles of causation and the common law' (2004) 11 *Journal of Law and Medicine* 492, 503–4.

137 New South Wales, Parliamentary Debates, Legislative Assembly, 23 October 2002, 5764.

138 [1956] AC 613.

139 *Ipp Report*, above n 96, 109 [7.28].

140 [2002] WLR 3 ('*Fairchild*').

141 *Ipp Report*, above n 96, 110 [7.29].

142 Ibid 110-11 [7.31]–[7.33] (footnote omitted).

143 Ibid 110 [7.30]; cf *Roads and Traffic Authority v Royal* (2008) 245 ALR 653, 677 [94] (Kirby J); McDonald, above n 100, 476–7.

144 *Ipp Report*, above n 96, 109-10 [7.28]; see also 110 [7.29].

145 Mendelson, above n 136, 500.

146 (2001) 206 CLR 459 [106] (footnote omitted); see also, e.g., *Roads and Traffic Authority v Royal* (2008) 245 ALR 653, 675 [85] (Kirby J), 689 [143] (Kiefel J).

147 [2006] 2 AC 572.

148 [2003] 1 AC 32, 40 [2].

149 [2003] 1 AC 32, 40 [2] referring to *Fairchild v Glenhaven Funeral Services Ltd* [2001] 1 All ER (D) 125; [2002] 1 WLR 1052.

150 [2003] 1 AC 32, 40 [2].

151 See eg J Stapleton, 'Lords A'Leaping Evidentiary Gaps' (2002) 10 *Torts Law Journal* 1, 11.

152 See ibid 17 citing *Fairchild* [2003] 1 AC 32, 40 [2], 66–7 [33].

153 Cf *Fairchild* [2003] 1 AC 32, 68 [36] (Lord Nicholls of Birkenhead).

154 [2003] 1 AC 32, 72–3 [55]–[56].

155 (1991) 171 CLR 506, 509.

156 *Ipp Report*, above n 96, 115 [7.45].

157 Ibid 117 (Recommendation 29).

158 McDonald, above n 100, 474.

159 *Ipp Report*, above n 96, 116–17 [7.48].

160 Cf Bartie, above n 128, 420–1.

161 Trindade et al, above n 84, 564.

162 [1961] AC 388.

163 However, the general rule must be regarded as qualified by the principle that a tortfeasor must take their victim as found: see ibid 566. See also *March v E & MH Stramare* (1990) 171 CLR 506, 534 (McHugh J).

164 See eg, *Hughes v Lord Advocate* [1963] 2 QB 405; *Doughty v Turner Manufacturing Co Ltd* [1964] 1 QB 518.

165 Trindade et al, above n 84, 569.

166 Durrant, above n 12, 421.

167 *Civil Liability Act 2002 (NSW)*, s 35(1)(a).

168 *Sindell v Abbott Laboratories*, 607 P 2d 924 (Supreme Court of California, 1980).

169 75 1 P 2d 470, 486 (1988); see also *Hymowitz v Eli Lilly* & Co, 539 NE 2d 1069 (1989) cited Barker [2006] AC 572, 593 [45].

170 Smith and Shearman, above n 107, 110-11.

171 E Penalver, 'Acts of God or Toxic Torts? Applying tort principles to the problem of climate change' (1998) 38 *Natural Resources Journal* 563, 592.

172 Smith and Shearman, above n 107, 111.

173 *Beecham Group Ltd v Bristol Laboratories Pty Ltd* (1968) 118 CLR 618; *Australian Broadcasting Corporation v O'Neill* (2006) 227 CLR 57.

174 Australian Law Reform Commission, Costs Shifting — Who pays for litigation, Report No 75 (1995) [13.3]–[13.4]. See eg, *Oshlack v Richmond River Council* (1998) 193 CLR 72; *Ruddock v Vadarlis* [2001] FCA 1865; *Save the Ridge Inc v Commonwealth* [2006] FCAFC 51.

175 (1998) 193 CLR 72 [134].

176 Cf *Oshlack v Richmond River Council* (1998) 193 CLR 72 [75] (McHugh J); *Buddhist Society of Australia (Inc) v Shire of Serpentine-Jarrahdale* [1999] WASCA 55 [11] (Kennedy, Wallwork and

Murray JJ); *QAAH of 2004 v Minister for Immigration and Multicultural and Indigenous Affairs* [2004] FCA 1644 [5] (Dowsett J).

177 See eg, *Oshlack v Richmond River Council* (1998) 193 CLR 72 [71] (McHugh J); *Ruddock v Vadarlis* [2001] FCA 1865 [19] (Black CJ and French J); *Northern Territory v Doepel* (No 2) [2004] FCA 46 [11] (Mansfield J).

178 *Oshlack v Richmond River Council* (1998) 193 CLR 72 [90] (McHugh J).

179 Cf Victorian Law Reform Commission, Civil Justice Review, Report No 14 (2008) 670. It should be noted that courts have not necessarily regarded the question of costs as necessarily requiring an 'all or nothing' decision. For example, in *Mees v Kemp* (No 2) [2004] FCA 549, Weinberg J ordered an unsuccessful applicant for judicial review to pay half of the Minister's costs, noting that the proceedings had raised difficult and important questions of construction, and had been brought 'selflessly' and conducted 'in a manner that was wholly commendable': see [20]–[21]. See also *North Australian Aboriginal Legal Aid Service Inc v Bradley* (No 2) [2002] FCA 564.

180 Cf e.g., *British Columbia (Minister of Forests) v Okanagan Indian Band* [2003] 3 SCR 371; *R (Corner House Research) v The Secretary of State for Trade and Industry* [2005] EWCA Civ 192. See also Liberty, Litigating the Public Interest: Report of the Working Group on Facilitating Public Interest Litigation (2006).

181 See Australian Law Reform Commission, *Costs Shifting — Who pays for litigation*, Report No 75 (1995) ch 13.

182 See Victorian Law Reform Commission, *Civil Justice Review*, Report No 14 (2008) 667–76.

183 See e.g., *Supreme Court Act 1986* (Vic) Pt 4A; *Uniform Civil Procedure Rules 2005* (NSW) rr 7.4–7.5. At federal level, class actions are enabled by pt IVA of the *Federal Court of Australia Act 1976* (Cth). However, proceedings may only be commenced if they involve a cause of action within the original jurisdiction of the Federal Court: *Federal Court of Australia Act 1976* (Cth) s 33G. If this requirement is satisfied, common law or other claims may be concurrently pursued in the proceedings.

184 See generally P Cashman, *Class Action Law and Practice* (2007).

185 The phrase 'without reasonable prospects of success' has been interpreted to mean 'so lacking in merit or substance as to be not fairly arguable': *Degiorgio v Dunn* (No 2) [2005] NSWSC 3 [27] (Barrett J); *Lemoto v Able Technical Pty Ltd* [2005] NSWCA 153 [132] (McColl JA).

186 See *Legal Profession Act 2004* (NSW) s 345

187 See *Batistatos v Roads and Traffic Authority* (NSW) (2006) 226 CLR 256 [217]; see also [213]–[216].

188 *Eurobodalla Shire Council v Wells* [2006] NSWCA 5 [28] (Ipp JA).

189 [2005] NSWCA 153 [27].

190 See eg *St Helen's Smelting Co v Tipping* (1865) 11 ER 1483; *Manchester Corp v Farnworth* [1930] AC 171; *Halsey v Esso Petroleum Co Ltd* [1961] 1 WLR 683.

191 Trindade et al, above n 84, 167 (emphasis added) (footnote omitted).

192 J Conaghan and W Mansell, *The Wrongs of Tort* (2nd ed, 1999) 132.

193 Of course, nuisance law could potentially be invoked against a GHG emitter if the relevant emissions have some direct effect on neighbouring land as well as making a general contribution to climate change. As Conaghan and Mansell point out, 'a successful suit in negligence often has the incidental effect of improving the environment': ibid 148.

194 Cf ibid 124.

195 See,eg, Trindade et al, above n 84, 169–73.

196 Durrant, above n 12, 406.

197 Ibid 423.

198 TJ Boutrous Jr and D Lanza, 'Global Warming Tort Litigation: The Real "Public Nuisance"' (2008) 35 *Ecology Law Currents* 80, 81.

www.ingramcontent.com/pod-product-compliance
Lightning Source LLC
Chambersburg PA
CBHW061142220326
41599CB00025B/4330